Antimicrobial Photodynamic Therapy

Antimicrobial Photodynamic Therapy: Concepts and Applications explores the novel antimicrobial therapeutic technique. As the world searches for new, efficient modalities for fighting microorganisms, this book offers a complete understanding of the concept, and knowledge about the emerging technique 'antimicrobial photodynamic therapy' (aPDT) for the scientific communities and budding researchers. The book aligns concepts, significance, and applications of the technique systematically. Chapters in the book cover microorganisms, pathogenesis, conventional treatment methods, and significance of new treatment approaches to the concept of antimicrobial photodynamic therapy. The authors describe the mechanism behind it, with applications and examples from research studies. The book discusses photosensitisers in detail, with one chapter emphasising natural photosensitisers. Use of nanostructures in the antimicrobial photodynamic therapy is elaborated on, and we conclude with a well-explored application of the therapeutic technique in dentistry.

Features:

- Efficiently covers the topic in detail with scientifically proven examples.
- Applications of the therapeutic approach are well discussed, and readers can learn about the research gaps, challenges, and the future of the technique.
- Starts from basics, enabling readers to understand why the approach is relevant and important for study.
- Simplistic elucidated concepts and applications make it accessible at all levels.

Antimicrobial Photodynamic Therapy

Concepts and Applications

Edited by

Siddhardha Busi
Ram Prasad

CRC Press
Taylor & Francis Group
Boca Raton London New York

CRC Press is an imprint of the
Taylor & Francis Group, an **informa** business

First edition published 2024
by CRC Press
6000 Broken Sound Parkway NW, Suite 300, Boca Raton, FL 33487-2742

and by CRC Press
4 Park Square, Milton Park, Abingdon, Oxon, OX14 4RN

CRC Press is an imprint of Taylor & Francis Group, LLC

ISBN: 978-1-032-38481-8 (hbk)
ISBN: 978-1-032-38485-6 (pbk)
ISBN: 978-1-003-34529-9 (ebk)

DOI: 10.1201/9781003345299

Typeset in Times
by SPi Technologies India Pvt Ltd (Straive)

Contents

Editors

Siddhardha Busi, Ph.D., is Assistant Professor, Department of Microbiology, School of Life Sciences, Pondicherry University, Pondicherry, India. His research interests are microbiology, drug discovery, quorum sensing, microbial chemical ecology, antimicrobial photodynamic therapy, and nanobiotechnology. His academic background and research experience align with applied and medical microbiology. He has in-depth knowledge about the control of bacterial biofilms and antibiotic-resistant pathogens through natural biomolecules and biogenic/conjugated nanoparticles. His numerous publications are testament to his expertise in food microbiology, infectious disease control, and novel antimicrobial therapies. He has published numerous papers on antimicrobial photodynamic therapy and its application on planktonic cells and biofilms. His studies include synthesis of dye/drug conjugated nanoparticles, photodynamic therapy, and understanding the mechanism of action. He has published in more than 80 reputed journals, and has contributed several chapters in books published internationally. He has been serving as editorial board member for various journals in the field of microbiology, and published three books related to microbial pathogenesis, nanotechnology, model organisms, and biofilms. Dr Siddhardha has 11 years of teaching experience, and has mentored Ph.D scholars and postgraduate students, mainly in the research areas of bioactive secondary metabolites, the anti-infective potential of bioactive metabolites, attenuation of quorum sensing regulated virulence factors, and molecular mechanism of anti-quorum sensing and antibiofilm activity.

Ram Prasad, Ph.D., is Associate Professor, Department of Botany, Mahatma Gandhi Central University, Motihari, Bihar, India. His research interest includes applied and environmental microbiology, plant-microbe-interactions, sustainable agriculture, chemistry of biology, and nanobiotechnology. Dr Prasad has more than 250 publications (Total citations 12805 with an h-index 56, i10-index 177) to his credit, including research papers, review articles and book chapters, and six patents issued or pending, as well as having edited or authored several books. Dr Prasad has 13 years of teaching experience and has been awarded the Young Scientist Award and Prof J.S. Datta Munshi Gold Medal by the International Society for Ecological Communications; Fellow of Biotechnology Research Society of India; Fellow of Agriculture Technology Development Society, India; Fellow of the Society for Applied Biotechnology; the American Cancer Society UICC International Fellowship for Beginning Investigators, USA; Outstanding Scientist Award in the field of Microbiology by Venus International Foundation; BRICPL Science Investigator Award and Research Excellence Award etc. He has been serving as editorial board member for: Nanotechnology Reviews; Green Processing and Synthesis; BMC Microbiology; BMC Biotechnology; Current Microbiology; Archives of Microbiology; Annals of Microbiology; IET Nanobiotechnology; Nanotechnology for Environmental Engineering; Journal of Renewable Material; Journal of Agriculture and Food Research; SN Applied Sciences; Agriculture amongst others, including Series editor of Nanotechnology in the Life Sciences, Springer Nature, USA. Previously, Dr Prasad served as Assistant Professor at Amity University, Uttar Pradesh, India; Visiting Assistant Professor, Whiting School of Engineering, Department of Mechanical Engineering at Johns Hopkins University, Baltimore, United States and Research Associate Professor at School of Environmental Science and Engineering, Sun Yat-sen University, Guangzhou, China.

Editors

Contributors

V. T. Anju
Pondicherry University
Pondicherry, India

Sayak Bhattacharya
Bijoy Krishna Girls' College, Howrah
West Bengal, India

Siddhardha Busi
Pondicherry University
Puducherry, India

Indranil Chattopadhyay
Central University of Tamil Nadu
Thiruvarur, India

Manjari Datta
Central University of Tamil Nadu
Thiruvarur, India

Madhu Dyavaiah
Pondicherry University
Pondicherry, India

Pitchaipillai Sankar Ganesh
Saveetha Institute of Medical and Technical Sciences (SIMATS)
Chennai, India

G. Ganga
Sree Ayyappa College
Alappuzha, India

Nikhilesh Ghantala
GITAM (Deemed to be) University
Visakhapatnam, India

Madangchanok Imchen
Pondicherry University
Puducherry, India

Archana Priyadarshini Jena
Pondicherry University
Puducherry, India

Ranjith Kumavath
Pondicherry University
Puducherry, India

Suseela Lanka
Krishna University
Machilipatnam, India

Monika Mishra
Sambalpur University
Sambalpur, India

Mahima S. Mohan
Pondicherry University
Puducherry, India

Jamseel Moopantakath
Central University of Kerela
Kasaragod, India

Pradeep Kumar Naik
Sambalpur University
Sambalpur, India

Santhi Latha Pandrangi
GITAM (Deemed to be) University
Visakhapatnam, India

Paramanantham Parasuraman
Pondicherry University
Puducherry, India

Subhaswaraj Pattnaik
Sambalpur University
Sambalpur, India

Simi Asma Salim
Pondicherry University
Puducherry, India

Sathish Sankar
Saveetha Institute of Medical and Technical Sciences (SIMATS)
Chennai, India

Esaki Muthu Shankar
Central University of Tamil Nadu
Thiruvarur, India

Shaik Hamad Sharif
GITAM (Deemed to be) University
Visakhapatnam, India

Indu Sharma
Assam University
Silchar, India

Hanna Yumnam
Assam University
Silchar, India

Monal Yuwanati
Saveetha Institute of Medical and Technical Sciences (SIMATS)
Chennai, India

1 Introduction to Microorganisms and Pathogenesis

Indu Sharma and Hanna Yumnam
Assam University, Silchar, India

CONTENTS

1.1 INTRODUCTION: MICROORGANISMS

Robert Hooke and Antoni Van Leeuwenhoek made the discovery of microscopic life in 1665 and 1685 respectively (Gest, 2004). A varied microbial population known as the microbiota has colonised the surface of the human body, and is susceptible to its surroundings (Maraz and Khan, 2021). Microorganisms are primitive organisms that, according to the size, shape, and morphology of the

DOI: 10.1201/9781003345299-1

cell surface, can be divided into two groups: prokaryotes (such as bacteria and archaea (Wajid et al., 2017), which are cells without a nucleus or membrane-bound organelles) and eukaryotes (such as fungi, viruses, and some types of parasites like protozoa which are cells with a nucleus or membrane-bound organelles (Shi et al., 2018)). Many of these organisms are believed to outnumber human cells (Maraz and Khan, 2021). The microbes have a variety of effects on the individual. They can be found everywhere in our environment, including the soil, air, water, animals, plants, food products, dead wood, clothing, and skin (Dubey and Maheshwari, 1999).

1.1.1 Bacteria

Among all the microorganisms, bacteria have the simplest structures, yet the most diverse range of metabolic options. According to research into the conserved 16S rRNA sequence, there are more than 50 different bacterial phyla. There are bacteria in almost every ecosystem on Earth, even inside and on the surface of humans. The majority of bacteria are beneficial or harmless; however, some are pathogens that cause disease to both people and animals. They are classified as prokaryotic organisms because the genetic material DNA in bacteria does not reside inside a veracious nucleus. Most bacteria consist of a cell wall surrounded by a peptidoglycan layer (Schloss and Handelsman, 2004). Bacteria are often classified based on their general form, which includes rod-shaped (bacillus), spherical-shaped (coccus) or curved-shaped or spirally coiled (spirochete, or vibrio).

Basically, bacteria have a fundamental composite, and are typically 0.2 µm in diameter and 2–8 µm in length. They are distinguished primarily through many replications, a high surface-to-volume ratio, and inherited adaptability. The relative simplicity of the bacterial cell enables it to react and adapt quickly to shifting environmental conditions (Schloss and Handelsman, 2004; Timberly et al., 2009).

1.1.2 Archaea

Archaea are microorganisms that resemble bacteria in terms of size and structure, but differ significantly in terms of their genetic makeup and biochemically. They seem to be a more basic type of life, and perhaps inhabit the planet's earliest formation of microbes. Archaeans are classified based on their niche, and are methanogenic organisms, halotolerant organisms which don't require salt but can grow under saline condition, extremophiles organisms that thrive at relatively high temperatures, and psychrophilic bacteria, thriving at low temperature. The energy sources used by archaeans include gaseous hydrogen, carbon dioxide, and sulphur (Maraz and Khan, 2021). The word 'extremophile' came from the idea that archaea only lived in extreme conditions, although more recent research has proven that they can either live in normal or non-extreme conditions. Archaea are abundant in cold marine habitats as well (Giovannoni and Stingl, 2005). Various habitats, such as soil, ocean, or even sewage, have yielded non-extreme archaea. Archaea that are non-pathogenic have not yet been isolated (Cavivvhioli et al., 2003).

1.1.3 Fungi

Fungi are highly diverse groups of eukaryotic microorganisms. Although some multicellular fungi have characteristics of plants, they are essentially quite different. Since fungi lack the ability to photosynthesise, their cell walls are made up of chitin, rather than cellulose. A long row of cells forms a middle strand which several spherical cell clusters are joined to, and therefore a nucleus can be found inside each cell, which is about 5 µm in size. Even though bacteria may be the most numerous microorganisms, fungus have the biggest biomass among eukaryotic microorganisms, and are physically a larger group. The number of known fungal species, which is estimated at 1.5 million, is only 7%. Fungi have been recognised based on their appearance, structure of spore, and fatty acid makeup of their membranes. However, the 18S rRNA gene which is an equivalent eukaryotic gene has been used for fungal identification, similar to how the bacterial identification and categorisation are done using the 16S rRNA gene (Crous et al., 2006). Moulds are multicellular fungus, whereas yeasts are unicellular fungi, and they generate filamentous (Pelczar and Pelczar, 2022).

Mould cells have a cylindrical shape and are joined end to end to create threadlike filaments (hyphae), which may contain spores. Hyphae are minute microscopic structures. A mycelium, on the other hand, is a fuzzy mass that forms when several hyphae gather in vast numbers. Unicellular yeasts come in a variety of shapes, including spherical, egg-shaped, and filamentous. In goods such as wine and bread, yeasts are known for their capacity to ferment carbohydrates into alcohol and carbon dioxide (Pelczar and Pelczar, 2022). Some fungi, such as *Aspergillus*, *Penicillium*, and *Rhizopus*, are considered to be filamentous fungus. A single fungus cell with filaments, known as a hypha (hyphae in plural form) that develops in masses to create clusters of hyphae or mycelia (Crous et al., 2006). In humans, fungi cause a skin disease which includes athlete's foot, ringworm, and thrush. In crops, fungal diseases can cause a significant monetary loss for the farmer.

1.1.4 Protozoa

Protozoa are microscopic single-celled eukaryotes with internal complexity and intricate metabolic processes. All protozoans require water to survive, thus, freshwater and marine environments are where they are most frequently found. However, some can also be discovered on land with soggy soils, while others can only be discovered in the digestive tracts of animals. A complex lifecycle of some protozoa includes the formation of cysts or oocysts. Cysts are able to prolong the life of the organism, much like spores in bacteria and fungi. The kingdom Protista is considered to include the protozoa as a sub-kingdom. Protozoa have been classified into more than 50,000 species, most of which are free-living species and are present in every natural environment. The taxonomy of eukaryotic microorganisms based on the 18S rRNA gene has shown that protozoa exhibit fundamental genetic variations. A polyphyletic group of eukaryotic microorganisms known as single-celled protozoa has recently come to light. The majority of human parasitic protozoa are less than 50 μm in size. Almost all people have protozoa in or on them at some point in their lives, and most shall contract one or more forms of infection as a result. The protozoa were divided into six phyla according to a taxonomy scheme published in 1985 by the Society of Protozoologists. The most significant organisms causing human disease are found in two of these phyla, specifically *Sarcomastigophora* and *Apicomplexa*. Using scanning, electron and light microscopy, morphology was used to develop this plan; for example: the family Entamoebidae included *Dientamoeba fragilis*, since it was formerly believed to be an amoeba. Electron microscopy images of the interior components, however, revealed that it belongs in the flagellate protozoa order Trichomonadida (Timberly et al., 2009; Martinez-Giron et al., 2008).

1.1.5 Viruses

The term "virus" refers to a type of microbe that lacks cells. Genetic material and inert proteins that exist independently of their host organism, such as DNA or RNA, which are the fundamental building blocks of viruses. Virus is a class of living things made up of nucleic acids that vary in size and morphology and are enclosed in protein capsids. It is necessary to have knowledge of viruses and their interactions on a molecular level in order to comprehend bacteria, archaea, and eukaryotic microorganisms, along with their genetics and ecological systems. This is due to the significant impact that viruses of microorganisms have microbiology of the population and natural environments on a worldwide scale. Single or double stranded DNA or RNA can be found in viral nucleic acids. Even though certain viruses do include some enzymes, they are obligatory parasites that lack the ability to conduct their own metabolism and rely on hosts to build viral components that self-assemble. The virological cycle is illustrated by the following stages: attachment or absorption, penetration or entry into the cell, replication, maturity, and release. Bacteriophage and its prokaryotic hosts can interact in a variety of ways. Lytic phages are prokaryote predators. Lysogenic and persistent infections, on the other hand, are essentially parasitic interactions that might be referred to as mutualism (Timberly et al., 2009; Rohwer and Thurbe, 2009; Suttle, 2007).

1.2 PATHOGENICITY OF MICROORGANISMS

The term "pathogen" refers to any microorganism that can infect its host and cause disease. The majority of living things on Earth are microbes, which use a range of strategies to support their life on both inanimate and living things. For an infection to develop, which then results in an infectious disease and ultimately the affected patient's mortality, the host cell must be in contact with microorganisms. According to the WHO, microbial infections are the second leading cause of death globally (Meena et al., 2020). There are two types of microbial infection: acute and chronic. Acute infections can be treated quickly; however, the host system may suffer long-term repercussions from persistent infections (Thakur et al., 2019). Episodes of the previously existing illness over time and the impetus on mainstream researchers to uncover the basic mechanism of microbial pathogenesis have increased because of antimicrobial resistance to treatments currently available. Understanding the critical course of microbial alliance, incursion, and injury to have framework requires a thorough understanding of microbial pathogenesis. Interpreting the relationship between bacteria's harmful components and the variables that influence them with having protective framework has shed light on numerous aspects of treating illnesses and repairing damage caused by prior prescription interactions (Solanki et al., 2018).

1.3 INTERACTION BETWEEN HOST AND MICROBE

The outermost layer of the host defence system, the skin, exposes people to germs early, and aids the immune system in preventing infection at the first line of defence. According to the pathogen and host cell's cooperative arrangement for survival in the host system, there are three different types of symbiosis that might take place: mutualism, commensalism, and parasitism (Meena et al., 2020).

1.3.1 MUTUALISM

Mutualism describes a relationship between two organisms in which the host provides sustenance and a habitat for the resilient of the bacterium, while the microbe provides nutrients and energy for the growth and development of the host. There are specific interactions between the host body system and the gut flora. The surroundings and metabolic schedule of the host have an effect on the microbiota capacity for survival; a small alteration in either can result in drastic changes in microbial behaviour and various types of sickness (Visconti et al., 2019; Wu and Wang, 2019). Due to abundance of nutrients in the environment, the gut microbiota contains a large number of bacterial species that are categorised into dominant groups (e.g., *Actinobacteria, Bacteroides*, and *Firmicutes*) and facultative or tolerant species (e.g., *Streptococci, Enterobacteriaceae, Enterococci*, and Lactobacilli). An anaerobic, a Gram-positive bacterium, *Bifidobacteria*, has a variety of advantageous characteristics that include reduced lactose intolerance, immunomodulation, gut immunostimulant, and serum cholesterol level (Candela et al., 2008). The intrinsic bacterium, *Bifidobacterium longum*, dominates the gastrointestinal tract (GIT) microbiome. The glycosyl hydrolases and phosphotransferases released by *B. longum* are linked to the utilisation of carbohydrates which provides a higher metabolic rate for ingesting food and maintain a healthy balance of nutrients throughout the dietary cycle (O'Callaghan and Van Sinderen, 2016).

1.3.2 Commensalism

This describes the relationship of a microbe with its host, in which a parasite that is not harmful to the host lives inside it. Commensal microbes, such as *Streptococcus pyogenes* and Lactobacilli, can be found on the skin's epidermis and in the mucus layers of the respiratory and gastrointestinal tracts. *Streptococcus pyogenes* is a bacterium that is frequently found in the skin flora and digestive system. It maintains an acidic pH level and prevents the formation of other harmful strains. Under stressful conditions, *S. pyogenes* migrates from the usual flora to the area around the epiglottis, where it produces swelling and inflammation that lead to sore throats (Eloe-Fadrosh and Rasko, 2013). Species of commensal Lactobacili (*L. jensenii, L. crispatus*, and *L. iners*) that live in the vaginal system maintain low pH condition and emit toxic chemicals to protect women from urinary tract infections usually caused by *Escherichia coli* or through sexually transmitted infections (STIs) and bacterial vaginosis (Antikainen et al., 2007; Eloe-Fadrosh and Rasko, 2013).

1.3.3 Parasitism

Humans provide favourable habitats for healthy growth and development for microbes, including protein- and nutrient-rich conditions. Parasitism is the coevolution of a bacterium and a host in which the microbe depends entirely on the host for survival. *Leishmania, Trypanosoma, Toxoplasma*, and *Schistosoma* are a few examples of the uncommon pathogens that live inside the human body and cause disease by consuming the energy from the host's metabolic pathways (Tedla et al., 2019). Since the host system is essential to the parasites' survival, they do not cause harm to it. *Schistosoma mansoni* and *Leishmania major* are the two main protozoan species that can cause cutaneous leishmaniasis and schistosomiasis, respectively. Bacteria can enter the dermis layer of the skin and occur an infection by secreting an enzyme called protein disulfide isomerase (PDI), which dissolves the extracellular matrix (ECM) of the skin. Leishmania promastigotes are bound to host cells and are encouraged to proliferate by the release of matrycryptin endostatin by PDI and ECM interaction (Miele et al., 2019; Philippsen and Marco, 2019).

1.4 TRACES OF INFECTION AND ITS DEVELOPMENT

Depending on how the microbial infection manifests the disease in the host, it can spread via a variety of channels. According to the research that is currently accessible, some of the well known methods include microbial adherence to the outer surface, invasion, colonisation, and toxin release (Meena et al., 2020).

1.4.1 Adhesion

The most common way for bacteria to infect a host is by adhering to its cell membrane. Microbial surface protein assists in chronic infection, promotes microbial invasion, and comprehends numerous signalling pathways involved in pathogenesis, and also detects the protein on the cell surface that starts a microbial infection (Boyle and Finlay, 2003). Projectors for lectin which selectively attach to glycan molecules on the surface of the host are responsible for mediating bacterial adhesion. Bacterial lectins can change in response to altered glycome dynamics and external factors. Lectins are produced by several microorganisms, each of which performs a different function. For example, *P. aeruginosa* lectins A and B aid in the formation of biofilms, whereas *Clostridium botulinum* lectin B serves as a bacteriocins (Moonens and Remaut, 2017).

Campylobacter flaA, *ciaB*, *cadF*, and *pldA* are few examples of surface anchoring molecules that have been reported for their adherence to host tissues. Additionally, it has been discovered that these substances are crucial for bacterial invasion and adherence to epithelial cells. Flagellin, which is produced by the *flaA* gene and is involved in cell adhesion, and the surface protein linked with fibronectin are produced by the *cadF* gene. Cell invasion by a different outer membrane, phospholipase synthase, has been linked to *Campylobacter* (Ganan et al., 2010). The seven cell adhesion genes (*ompA*, *inv*, *sip*, *aut*, *hly*, *fliC*, and *cpa*) were used to discriminate between *Cronobacter sakaakii* and *Cronobacter malonaticus* species due to their significance in adhesion and invasion processes. Enteropathogenic *E. coli* (EPEC), one of the causes of diarrhoea, expressed the *pilS* gene during adhesion to intestinal epithelium and microvilli effacement, which caused a histopathological lesion (A/E) to attach (Garcia et al., 2019). Extracellular adherence protein (Eap) from *Staphylococcus aureus*, is another example of a cell adhesion protein (Palankar et al., 2018).

1.4.2 Internalisation

Bacterial pathogenicity is brought on by bacterial invasion or internalisation, which enables microbial access into host cells and permits multiplication. Gram-positive and Gram-negative bacteria have various structural arrangements on the cell surface, based on their genetic and physiological characteristics. The membrane channel that facilitates the transport of cellular molecules to the host cell controls adhesion and internalisation in Gram-negative bacteria. There are two distinct types of bacterial invasion systems based on the pathogen's usage: trigger and zipper mechanisms. Bacterial cells can enter host cells utilising the type III secretion system (T3SS), which uses a trigger mechanism. The procedure involves a bacterial cell depositing its effectors in its cytoplasm, and effector protein release into the host cell, by altering the host cell membrane. The invader is then prevented from being defended by the outer membrane (e.g., type III secretory system of *P. aeruginosa)* (Bohn et al., 2019; Kochut and Dersch, 2013). The innate immune system and intestinal homeostasis are controlled by the human-defensin protein. Then, it channels Shigella's entrance so that it can adhere and internalise, increasing its pathogenicity (Xu et al., 2018). Attachment and invasion locus (Ail), Yersinia adhesion A (YadA), and invasion (Inv) are cell surface appendages that *Yersinia pestis* and *Y. psuedotuberculosis* use to mediate effector translocation. The combination of adhesion and invasion proteins directs leukocytes towards myeloid cells, resulting in host cell interaction. In contrast, the invasion molecules produced by Gram-positive bacteria contains the LPXTG motif which attached to the C-terminus in order to interact with the host cell receptor reservoir. Another bacterial surface molecule from *Streptococcus pneumonia* that has a C-terminus attached PfbA (plasmin and fibronectin binding protein A) recognises host cell receptors like fibronectin, plasminogen, and serum albumin changes the disease progression (Beulin et al., 2014).

There are three stages involved in the zipper mechanism: the first stage involves the interaction of a ligand attached to the surface of the bacterial cell with a specific cell receptor on the host cell; the second stage involves the activation of signalling pathways; the third stage involves the presence of a vacuole with a membrane that helps the host cell to internalise the bacterial cell (Kochut and Dersch, 2013). Gram-positive bacterium such as *Listeria monocytogenes*, the causative agent of listeriosis, use a zipper mechanism to help bacterial cells invade epithelial cells or the blood-cerebrospinal fluid barrier and enable proliferation in the host cytoplasm by inducing phagolysis. This zipper mechanism involves two surfaces of invasion factors: In1A and In1B and the pore-forming enzyme listeriolysin O (Phelps et al., 2018; Grundler et al., 2013).

1.4.3 Colonisation

Organism colonisation is the development of bacteria on host tissue, where it ultimately overcomes the host's defences, then reproduces and spreads. Bacterial colonisation can be divided into three stages: (i) attachment to the host cell surface; (ii) invasion of the hosts cell; and (iii) immune system suppression in the host (Ribet and Cossart, 2015). During the adhesion and invasion processes, the bacteria's metabolic rate may be negatively impacted if patterns on the surface of the host cell and cellular signalling pathways are not recognised, which in turn prevents the synthesis of virulence components and hinders colonisation (Stones and Krachler, 2016). *Prophyromonas gingivalis*, a bacterium which causes plaque on teeth, controls the succession of microorganisms in dental health by allowing the growth of other bacterial species which then produce biofilm on the tooth enamel (Periasamy and Kolenbrander, 2009). *Bifidobacterium* is the predominant bacterial species that crosses from the placenta of the mother to the newborn baby. *Bifidobacterium* is typically found in the epithelial layers of the skin, urinary tract, and gastrointestinal tract. *Bifidobacterium longum* can colonise due to a number of conditions, including the availability of glycosyl hydrolases and phosphotransferases for the consumption of carbohydrates, adaptability to the concentration of bile salts, a lot of adhesins and pili on the mucus layer to entrap the bacteria (Rodriguez et al., 2013; Zhang et al., 2019; Grimm et al., 2014). *E. coli*, a Gram-negative bacterium, colonises the respiratory system and the epithelial layer of the epidermis which causes bacteraemia and meningitis. (Alfaro-Viquez et al., 2019). The human gastrointestinal system is occupied by *Enterococcus faecalis* and *E. faecium*, which can transform from a benign microbe to a pathogenic strain depending on the environmental factors (Banla et al., 2019). Infection and damage to the cornea's epithelial layer are brought on by *P. aeruginosa*, and microbial keratitis is the next condition that develops. The biofilm colonisation on the epithelial cells is another criticism of microbial keratitis (Wu et al., 2019). *Staphylococcus aureus*, a well-known coloniser of Gram-positive bacteria, attaches itself to the epithelium layer of the nasal cavity by a zipper mechanism. A severe infection is produced by the interaction of loricrin, cytokeratin 10 (K10), involucrin, filaggrin, and small proline-rich proteins with bacterial cell surface adhesins, iron regulated surface determinant A, and clumping factor B (C1Fb) (Mulcahy and McLoughin, 2016).

1.4.4 Mechanism for Microbial Secretion

Exotoxin A, exoenzymes, cell-lytic enzymes (protease, elastase, chitinase), and redox-active phenazine chemicals (Pyocyanin and Pyoverdine) are the key virulence agents released during the critical infection phase of *P. aeruginosa* infections. To transfer effector proteins to the host cell through a pipeline resembling a needle-shaped, *P. aeruginosa* uses the type III secretory system (T3SS), also known as the needle complex. The first step is the development of the basal compartment on the surface of the bacteria, the formation of effector proteins within the cytoplasm and the process of creating the needle-shaped bridge by causing host cell surface configuration modifications that result in the formation of a pore. The effector proteins eventually move energy-driven to the cytoplasm of the host cell from the bacterial cell (Lombardi et al., 2019). The expression of the exsCEBA operon is regulated by the transcription factor ExsA, which controls the regulation of the T3SS system and aids in translocation. As a result of the cellular medium's calcium content, ExsC serves as an antiactivator, while ExsD serves as an antiactivator. Among the effector proteins, ExoU is known for its cytotoxic properties and phospholipase activity that inhibit the immune system and trigger inflammatory cascades. Adenylate cyclase activity is present in ExoY, which also keeps the cell's level of cyclic adenosine monophosphate (Camp) constant. ExoS and ExoT, two effectors proteins with dual functions, cause cells to round and impair the healing of wounds (Ahator and Zhang, 2019; Sawa, 2014). The virulence factor of the pathogenic bacteria *Vibrio cholera* produces cholera toxin which is spread by the consumption of contaminated water and known to have a significant effect on pathogenicity. Cholera toxin is released by *V. cholera* at host body temperatures as a consequence of bacterial cells adhering to the small intestine's mucus layer's microvilli by the action of pili. The mucolytic characteristic of cholera toxin causes disruptions in the integrity of the mucous layer and regulates bacterial penetration into the host cell. The complex structure of cholera toxin is made up of subunit A and five B subunits, and both have played a significant role. Subunit A acts as a binding factor, and subunit B has adenylate cyclase activity. In order to start phagocytosis, binding subunits bind to the GM1ganglioside mucous layer receptor on the surface of the host. Through enhanced chloride channel permeability, the cAMP concentration exceeds a threshold value and facilitates the release of Na+, H+, and K+. A build-up of Na+, H_2O, and K+ in the cytoplasm causes severe diarrhoea and worsens dehydration levels in an infected host (Tirumale and Thomas, 2018).

1.5 MICROBE-RELATED INFECTIONS

1.5.1 Bacterial Infections

Staphyloccocus aureus is a prevalent opportunistic bacterium that causes soft tissue and skin infections all over the world. They caused a variety of illnesses, occurring from minor skin infections to severe infections, such as pneumonia, bacteraemia, and endocarditis (Moran et al., 2006). Numerous bacteria can infect the skin. When bacteria enter the central nervous system (CNS), a life-threatening disorder with the highest fatality rate occurs. Furthermore, it can leave survivors with long-term neurological abnormalities (Geyer et al., 2019). More precisely, methicillin-resistant *S. aureus* (MRSA), together with *Acinetobacter baumannii*, is a notorious source of community-related and healthcare-related infections, with an estimated frequency of roughly 30% (David and Daum, 2010).

Streptococcus pneumoniae and *Neisseria meningitides* are both responsible for the deadly condition known as bacterial meningitis. *S. pneumoniae*, the second most prevalent infection, is more deadly, and causes more morbidity than *N. meningitides* (Ramakrishna et al., 2009; Gessner et al., 2010; Mihret et al., 2016). Both bacteria and fungi can cause keratitis, a prevalent type of corneal blindness that affects people all over the world. Patients with cystic fibrosis (CF) are susceptible to chronic lung infections brought on by the human pathogen *Pseudomonas aeruginosa* (Winstanley and Fothergill, 2009). They also affect patients with burn wounds, those with human immunodeficiency virus (HIV), those receiving chemotherapy for cancer, and immune-weakened people (Lau et al., 2004).

In recent years, epidemics of necrotising enterocolitis have been linked to a number of bacterial species, particularly Clostridia such as *Clostridium butyricum*, *C. perfringes*, and *C. neoatale* (Hosny et al. 2017; Rozé et al., 2017). Plague is a lethal illness that is caused by *Yersinia pestis*. The Black Death, a pandemic known for its significant death rate and enduring social and economic impacts, was one of the epidemics responsible for spreading it in Europe (Bramanti et al., 2019). *C. perfringens* type A is the causative agent of gas gangrene, also known as clostridial myonecrosis, which results in fast tissue necrosis spread and infiltration of leukocytes was absent at the infection area (Rood, 1998).

1.5.2 Fungal Infections

Fungi are just as similarly efficient as bacteria at causing a wide range of human infections. The most typical pathogens isolated from CF patients which are filamentous fungus such as *Aspergillus fumigatus* and *Scedosporium apiospermum* species complex (15–50% and 10–12%, respectively) (Noni et al., 2017; Reece et al., 2017). Fusarium spp. are responsible for the localised illnesses that affect healthy hosts, including keratitis and onychomycosis. The necrotic skin lesions, sinuses and visceral organs were all impacted by invasive fusariosis, and 60% of people with immune-compromised diseases developed positive blood cultures (Nucci et al., 2003; Nelson et al., 1994; Lionakis and Kontoyiannis, 2004). Not only healthy people, but occasionally immune-compromised people can also be impacted. The most common fungus in nature, *Aspergillus*, affects only the lungs of an immune-compromised host (Camargo and Husain, 2014; Steinbach et al., 2012; Husain et al., 2017).

1.5.3 Virus Infections

Among all other microbes, viruses are the most important microorganism causing diseases in humans. Recently, there was an outbreak of pandemic Corona virus disease in 2019 which has spread all over and affected many lives. Patients with COVID symptoms can range from minor to severe, and a substantial percentage of the population being asymptomatic carriers. Shortness of breath, coughing, and fever are the most often reported symptoms (Wang et al., 2020). Human cytomegalovirus (HCMV) is another lethal virus, which infects people chronically and has a lifelong latent period. This HCMV is an opportunistic disease that infects immunocompromised people and causes birth abnormalities (Mocarski et al., 2007). Uncontrolled inflammation to the central nervous system causes Japanese encephalitis (JE). Japanese encephalitis (JE) is caused by the neurotropic flavivirus known as JE virus (Misra and Kalita, 2010). The mosquito vector is the means by which the JE virus is spread. Pigs act to amplify, and birds act as reservoir hosts in the zoonotic cycle that transmits the JE virus (Solomon, 2004). Another mosquito-borne illness, dengue, is a major public health issue, by which 390 million of people are affected each year (Bhatt et al., 2013). Dengue virus (DENV) infection occurs ranging from asymptomatic cases to plasma leakage leads to hypovolemic shock which is life-threatening.

1.6 FACTORS RELATED TO VIRULENCE WITH MICROBIAL INFECTIONS

Pathogenic microbes create virulence factors to induce illnesses. These factors include proteins, carbohydrates, or lipids composition. The immune systems of host cells that are secretory, cell-mediated, or cytoplasmic in origin can be affected by the virulence components. The cytosolic factors enable the bacteria make fast physiological, metabolic, and morphological changes in order to adapt and survive inside host cells. The membrane-related virulence factors help in bacterial adhesion and host cell evasion. The bacterium is helped by secretory substances to avoid innate and adaptive immunological responses of the host. The secretory virulence components of the extracellular pathogen combine to damage the host cells. Sometimes they poison food by releasing toxins into it, and humans that consume this become infected. A disease is brought on by an organism's ability to infect the host cell (Bhattacharya and Mukherjee, 2020).

1.6.1 Extracellular Virulence Factors

Extracellular proteases and enzymes that break down cell walls are poisons secreted by a significant number of pathogenic bacteria, and they play an important role in virulence. Both plasmid and chromosomes can encode them. Furthermore, due to diverse host conditions modifying the virulence characteristics of some enteropathogenic pathogen strains, circumstances like pH, bile, alkalinity and high osmolarity can change the aspects of virulence factors in the gastrointestinal tract (Chiang, 2013; Hofmann, 1999; Begley et al., 2005). The gene A associated with cytotoxin and the cytotoxin A vacuolisation are the two key virulence components found in Gram-negative *Helicobacter pylori*. These virulence factors result in an elevated risk of gastric cancer. As a result, they adjust their virulence property to the infection niche that is most favourable.

Shiga toxin is produced by *E. coli* (STEC), which is the leading cause of food-borne illness globally. Enterohemorrhagic *E. coli* (EHEC), also known as pathogenic STEC, is the source of diarrhoea that is bloody or watery. These conditions may eventually progress to fatal illnesses like thrombotic thrombocytopenic purpura or haemolytic uremic syndrome (HUS). For the majority of infections, adhesion is another virulence factor for pathogens. The rapid eradication of adhesion-deficient *S. aureus* indicated that adherence played a vital role in continuing the colonisation process

(Mulcahy et al., 2012; Weidenmaier et al., 2004). Alpha-hemolysin (Hla), a toxin that creates pores, is secreted by the majority of *S. aureus* isolates (Li et al., 2009). Hla invades by disrupting cell connections and adopting a disintegrin and metalloprotease 10 (ADAM10) host molecule (Kennedy et al., 2010; Inoshima et al., 2012; Malachowa et al., 2013).

Toxin is another released virulence aspect which allows pathogenic bacteria to infiltrate the immune system. Anthrax is a disease caused by *Bacillus anthracis*, an rod-shaped Gram-positive bacterium. *Bacillus anthracis* possesses the pXO1 and pXO2 extrachromosomal plasmids. An oedema and fatal factor are protective antigen and anthrax toxin activator A, which are encoded by the plasmid pXo1 (AtxA) (Hammerstrom et al., 2011). The above secreted toxins cause a systemic infection in a short duration. They all function as a key regulator for the synthesis of toxins. Additionally, the plasmid pXo2 encodes the proteins that synthesise capsules. Anthrolysin O, another important virulence factor, compromises the membrane integrity of the host cell (Shannon et al., 2003). Streptococcal toxic shock syndrome (TSS), caused by a new pyrogenic toxin with a molecular weight of 12 kDa produced by Group B *Streptococcus* strains (Schlievert et al., 1993). Food-borne illnesses are one of the diseases caused by the enterotoxin that *S. aureus* produces. The majority of food-borne illnesses are caused by staphylococci (Bean et al., 1990; Archer and Young 1988; Bunning et al., 1997).

The toxins described above are considered to be exotoxins, whereas a gram-negative bacterium secretes a type of toxin called endotoxin, as they are not secreted out of the cells. Gram-negative microbes release lipopolysaccharide (LPS), which has endotoxic properties. Lipid A, O antigen, and core polysaccharide are the three components that make up LPS. The hazardous sign of severe Gram-negative infections and burning sensation are known to be caused by LPS, a significant contributor (Alving, 1993). Exotoxins are typically heat-labile proteins (polypedptide) released by gram-negative bacteria that permeate the medium in the area. Endotoxin and exotoxin have many differences, including exotoxin levels of toxicity being higher than that of endotoxins. Additionally, exotoxin has strong and highly antigenic characteristics, in contrast to those of endotoxins. Toxins produced by *Streptococcus pyogenes*, *Bacillus anthracis* (alpha-toxin/Hla), and *Staphylococcus aureus* are examples of exotoxins. Toxins produced by *Escherichia coli*, *Salmonella typhi*, and *Vibrio cholera* are examples of endotoxin.

1.6.2 Virulence Factors Specific to Cells

The capsule is the best illustration of a virulence factor unique to cells. *Bacillus anthracis* capsules contain poly-gamma-d-glutamic acid. It is poorly immunogenic in nature, thus it prevents *B anthracis* from being phagocytosed during infection (Makino et al., 2002). Various mycobacterial cell surface ligands, such as HSP70, 19kDa lipoarabinomannan (LAM), phosphatidylinositol mannoside (PIM) and lipoprotein. These are recognised by specialised pattern recognition receptors (PRRs) and host cell phagocytise receptors (Dorhoi et al., 2011). The pathogen's ability to survive, however, is aided by some receptors' absorption. *Streptococcus*, another encapsulated bacterium, is responsible for certain extremely harmful invasive diseases, such as septicaemia and meningitis as they contain capsular substance, which is a virulence agent and contains sialic acid (Jacques et al., 1990).

1.7 GENERAL CHARACTERISTICS OF ACQUIRED/ADAPTIVE AND INNATE IMMUNITY

The environment in which humans and other mammals live is highly populated by both pathogenic and non-pathogenic bacteria, and contains a wide range of poisonous or allergic compounds that pose a threat to normal homeostasis. The host must be able to tolerate and keep in check the community of microbes in order to ensure healthy tissue and organ function. This community contains both obligatory pathogens and helpful commensal organisms. Pathogenic bacteria have a wide range of ways by which they reproduce, disseminate, and endanger the stability of host functions. The immune system must prevent reactions that cause excessive damage to self-tissues or those that could eliminate helpful, commensal microorganisms, while simultaneously destroying pathogenic microbes and poisonous or allergenic proteins. A wide variety of poisonous compounds and pathogenic bacteria can challenge the host in our environment. These pathogens can employ a wide range of pathogenic methods to attack the host. It follows that the immune system employs a wide range of intricate defence systems to manage and typically get rid of these pathogens and poisons. It is a characteristic of the immune system as a whole that these processes rely on identifying structural characteristics of the pathogen or toxin that distinguish them from host cells. Such host-pathogen or host-toxin discrimination is necessary to allow the host to get rid of the threat without causing harm to its own tissues (Chaplin, 2010).

The mechanisms allowing the recognition of microbial, toxic, or allergenic structures can be divided into two main categories: (i) hard-wired responses encoded by genes in the host's germ line and that recognise molecular patterns shared by numerous microbes and toxins that are not present in the mammalian host; and (ii) reactions provided by gene elements that rearrange somatically to form antigen-binding molecules with excellent specificity for the target antigen. Innate immune response is made up of the initial set of reactions. The innate system acts as the first line of defence against an invading virus or toxin because it has widespread expression on a variety of cells that express the recognition molecules it uses. As a result, this system is prepared to react quickly. The adaptive immune response is made up of the second set of reactions. The responding cells must multiply after coming into contact with the antigen in order to reach sufficient numbers to mount an effective response against the microbe or the toxin because the adaptive system is made up of tiny numbers of cells with specificity for any one pathogen, toxin, or allergen. As a result, the innate reaction to host defence typically occurs before the adaptive response. The ability of long-lived cells to survive in an apparent inactive state while yet being able to quickly express effector activities in response to a subsequent encounter with their particular antigen is a crucial characteristic of the adaptive response. This gives the adaptive response the capacity to develop immunological memory, enabling it to significantly contribute to a more potent host response when those infections or toxins are met again, even decades after the initial sensitising encounter.

All parts of the host's immune defence mechanisms that are encoded in their fully developed functional forms by the host's germ-line genes are collectively referred to as the innate immune system. Physical barriers such as those found in the respiratory, gastrointestinal, and genitourinary tracts that express tight cell-cell contacts, the mucus layer secreted over the epithelium there, and the epithelial cilia that sweep away this mucus layer to allow it to be continuously refreshed after becoming contaminated with inhaled or ingested particles are a few examples of these. As part of the innate response, soluble proteins and bioactive small molecules are also present. These substances are either naturally present in biological fluids or are released by activated cells such as cytokines that control the activity of other cells, chemokines that draw in leukocytes that cause inflammation, lipid mediators of inflammation, reactive free radical species, bioactive amines, and enzymes are additional factors in tissue inflammation. The innate immune system also contains cytoplasmic proteins and membrane-bound receptors that bind molecular patterns expressed on the surfaces of invasive microorganisms. Aspects of the innate host defences are either constitutively active or become active when host cells or host proteins come into contact with chemical compounds that are present in invading microorganisms, but missing in host cells. The adaptive immune system exhibits remarkable specificity for its target antigens, unlike the innate host defence mechanisms. Antigen-specific receptors that are expressed on the surfaces of T- and B-lymphocytes serve as the main foundation for adaptive responses. Contrary to the germ-line-encoded recognition molecules of the innate immune response, intact T cell receptor (TCR) and immunoglobulin (B cell antigen receptor; Ig) genes are formed by somatic rearrangement of germ-line gene elements to code for the antigen-specific receptors of the adaptive immune response. Millions of distinct antigen receptors with potentially distinct selectivity for various antigens can be formed by assembling antigen receptors from a collection of a few hundred germ-line-encoded gene components. The innate immune system serves as the first line of defence for the host, and the adaptive immune system only becomes noticeable after a few days, when antigen-specific T and B cells have gone through clonal expansion. Despite the fact that the innate and adaptive immune systems are frequently described as opposing, distinct arms of the host response, they typically work together. The innate system's elements help the antigen-specific cells get activated. In order to completely control invasive microorganisms, the antigen-specific cells also strengthen their responses by enlisting innate effector pathways. As a result, even though the innate and adaptive immune responses have fundamentally different modes of action, cooperation between them is crucial for a complete, efficient immune response (Chaplin, 2010).

1.7.1 Immune Response against Pathogens

Inflammation, which is characterised by its cardinal indications of redness, swelling, heat, pain, and impaired function, is a protective reaction to extreme challenges to homeostasis, such as infection, tissue stress, and damage (Kotas and Medzhitov, 2015; Nathan, 2002). The elements of a typical inflammatory reaction are as follows. Inflammatory inducers: depending on the type of infection (bacterial, viral, fungal, or parasitic organisms), inflammatory inducers are detected by sensors such as TLRs, NLRs, and RLRs of the innate immune system. Inflammatory mediators are then induced by the sensors, such as cytokines, chemokines, and the complement system, which then affect the target tissues. Each component has several forms, and when combined, they work in different inflammatory pathways (Medzhitov, 2008, 2010; Yano and Kurata, 2011; Lavelle et al., 2010).

The inflammatory response is divided into four phases: (i) a silent phase, where cells in the injured tissue release the first inflammatory mediators; (ii) a vascular phase, where vasodilation and increased vascular permeability take place; (iii) a cellular phase, which is characterised by the infiltration of leukocytes to the site of injury; and (iv) a resolution phase, which is the process of bringing the tissues back to homeostasis (Vergnolle, 2003; Gilroy et al., 2004; Headland and Norling, 2015).

1.8 CONCLUSION

Microorganisms are primitive organisms with varied populations of microbes that have colonised the surface of every human body, as well as that of every animal. The majority of human microorganisms co-exist peacefully with human cells, but illness and infection can develop when this balance is disrupted, or when the body's immune system is weak. Insights into the aetiology and course of disease provided by microbial pathogenesis are the two main components of pathogenesis crucial for the prevention, management, and treatment of a variety of diseases.

REFERENCES

Ahator, Stephen Dela, and Lian Hui Zhang. 2019. "Small is mighty-chemical communication systems in *Pseudomonas aeruginosa*". *Annual Review of Microbiology* 73: 559.

Alfaro-Viquez, Emilia, Daniel Esquivel-Alvarado, Sergio Madrigal-Carballo, Christina G Kruger, and Jess D Reed. 2019. "Proanthocyanidin-chitosan composite nanoparticles prevent bacterial invasion and colonization of gut epithelial cells by extra-intestinal pathogenic *Escherichi coli*". *International Journal of Biological Macromolecules* 135: 630–636.

Alving, C R 1993. "Lipopolysaccharide, lipid A, and liposomes containing lipid A as immunologic adjuvants". *Immunobiology* 187: 430–446.

Antikainen, Jenni, Veera Kuparinea, Kaarina Lahteenmaki, and Timo K Korhonen. 2007. "Enolases from gram-positive bacterial pathogens and commentary Lactobacilli share functional similarity in virulence-associated traits". *FEMS Immunology and Medical Microbiology* 51: 526–534.

Archer, D L, and F E Young 1988. "Contemporary issues: Diseases with a food vector". *Clinical Microbiology Reviews* 1: 377–398.

Banla, Leou Ismael, Nita H Salzman, and Christopher J Kristich. 2019. "Colonization of the mammalian intestinal tract by Enterococci". *Current Opinion in Microbiology* 47: 26–31.

Bean, Nancy H, Patricia M Griffin, Joy S Goulding, and Cecile B Ivey. 1990. "Foodborne disease outbreaks, 5-year summary, 1983-1987". *Journal of Food Protection* 39: 15–57.

Begley, Maire, Cormac G M Gahan, and Colin Hill. 2005. "The interaction between bacteria and bile". *FEMS Microbiology Reviews* 29: 625–651.

Beulin, D S Jemina, Masana Yanguchi, Shigetada Kawabata, and Karthe Ponnuraj. 2014. "Crystal structure of PfbA, a surface adhesin of *Streptococcus pneumoniae*, provides hints into its interaction with fibronectin". *International Journal of Biological Macromolecules* 64: 168–173.

Bhatt, Samir, Peter W Gething, Oliver J Brady, Jane P Messina, Andrew W Farlow, Catherine L Moye, John M Drake, et al. 2013. "The global distribution and burden of dengue". *Nature* 496: 504–507.

Bhattacharya, Sayak, and Joydeep Mukherjee. 2020. *Microbial Infections and Virulence Factors*. Edited by Busi Siddhardha, Madhu Dyavaiah, and Asad Syed. Singapore: Springer.

Bohn, Erwin, Michael Sonnabend, Kristina Klein, and Ingo B Autenrieth. 2019. "Bacterial adhesion and host cell factors leading to effector protein injection by type III secretion system". *International Journal of Medical Microbiology* 309: 344–350.

Boyle, Erin C, and B Brett Finlay. 2003. "Bacterial pathogenesis: Exploiting cellular adherence". *Current Opinion in Cell Biology* 15: 344–350.

Bramanti, Barbara, Katharine R Dean, Lars Walløe, and Nils Chr Stenseth. 2019. "The third plague pandemic in Europe". *Proceddings Biological Sciences* 286: 20182429.

Bunning, V K, J A Lindsay, and D L Archer. 1997. "Chronic health effects of microbial foodborne disease". *World Health Statistics Quarterly* 50: 51–56.

O'Callaghan, Amy, and Douew Van Sinderen. 2016. "*Bifidobacteria* and their role as members of the human gut microbiota". *Frontiers in Microbiology* 7: 925.

Camargo, Jose F, and Sahid Husain. 2014. "Immune correlates of protection in human invasive aspergillosis". *Clinical Infectious Diseases* 59: 569–577.

Candela, Marco, Elena Beigi, Sivia Turroni, Beatrice Vitali, and Palriza Brigidi. 2008. "Mechanisms involved in the intestinal interaction between host and bifidobacterial". *Microbial Ecology in Health and Disease* 20: 189–192.

Cavicchioli, Ricardo, Paul M G Curmi, Neil Saunders, and Torsten Thomas. 2003. "Pathogenic archaea: Do they exist?" *BioEssays* 25: 1119–1128.

Chaplin, David D 2010. "Overview of the immune response". *The Journal of Allergy and Clinical Immunology* 125: S3–S23.

Chiang, John Y L. 2013. "Bile acid metabolism and signalling". *Comprehensive Physiology* 3: 1191–1212.

Crous, Pedro W, Isabella H Rong, Alan Wood, Seonju Lee, Hugh Glen, Wilhelm Botha, Bernard Slippers, Wilhelm Z de Beer, Michael J Wingfield, and David L Hawksworth. 2006. "How many species of fungi are there at the tip of Africa?" *Studies in Mycology* 55: 13–33.

David, Michael Z, and Robert S Daum. 2010. "Community-associated Methicillin-resistant *Staphylococcus aureus*" *Clinical Microbiology Reviews* 23: 616–687.

Dorhoi, Anca, Stephen T Reece, and Stefan H E Kaufmann 2011. "For better or for worse: The immune response against Mycobacterium tuberculosis balances pathology and protection". *Immunological Reviews* 240: 235–251.

Dubey, R C, and D K Maheshwari. 1999. *A textbook of Microbiology*. New Delhi: S Chand & Company.

Eloe-Fadrosh, Emiley A, and David A Rasko. 2013. "The human microbiome: From symbiosis to pathogenesis". *Annual Review of Medicine* 64: 145–163.

Ganan, M, G Campos, R Munoz, A V Carrascosa, S de Pascual-Teressa, and A J Martinez-Rodriguez. 2010. "Effect of growth phase on the adherence to and invasion of Caco-2-epithelial cells by *Campylobacter*". *International Journal of Food Microbiology* 140: 14–18.

Garcia, Bruna G, Felipe S Castro, Monica A M Vieria, Dennys M Girao, Lucas T Uenishi, Maria C Cergole-Novella, Luis F Dos Santos, Roxane M F Piazza, Rodrigo T Hernandes, and Tania A T Gomes. 2019. "Distribution of the pil S gene in *Escherichia coli* pathovars, its transfer ability and influence in the typical enteropathogenic E.coli adherence phenotype". *International Journal of Medical Microbiology* 309: 66–72.

Gessner, Bradford D, Judith E Mueller, and Seydou Yaro. 2010. "African meningitis belt pneumococcal disease epidemiology indicates a need for an effective serotype 1 containing vaccine, including for older children and adults". *BMC Infectious Diseases* 10: 22–31.

Gest, Howard. 2004. "The discovery of microorganisms by Robert Hooke and Antoni Van Leewenhoek, fellows of the Royal Society". *Notes and Records of the Royal Society of London* 58: 187–201.

Geyer, Sohair, Muazzan Jacobs, and Nai-Jin Hsu. 2019. "Immunity against bacterial infection of the central nervous system: An astrocyte perspective". *Front Molecular Neurosciences* 12: 57.

Gilroy, Derek W, Toby Lawrence, Mauro Perretti, and Adriano G Rossi. 2004. "Inflammatory resolution: New opportunities for drug discovery". *Nature Reviews Drug Discovery* 3: 401–416.

Giovannoni, Stephen J, and Uli Stingl. 2005. "Molecular diversity and ecology of microbial plankton". *Nature* 437: 343–348.

Grimm, Verema, Christina Westermann, and Christian U Riedel. 2014. "Bifidobacteria-host interactions: An update on colonization factors". *BioMed Research International* 2014: 960826.

Grundler, Tim, Natascha Quednau, Carolin Stump, Veronique Orian-Rousseau, Hiroshi Ishikawa, Hartwig Wolburg, Horst Schroten, Tobias Tenenbaum, and Christian Schwerk. 2013. "The surface proteins In1A and InIB are independently required for polar basolateral invasion by *Listeria monocytogenes* in a human model of the blood-cerebrospinal fluid barrier". *Microbes and Infection* 15: 291–301.

Hammerstrom, Troy G, Jung Hyeob Roh, Edward P Nikonowicz, and Theresa M Koehler. 2011. "Bacillus anthracis virulence regulator AtxA: Oligomeric state, function and CO(2) – signalling". *Molecular Microbiology* 82: 634–647.

Headland, Sarah E, and Lucy V Norling 2015. "The resolution of inflammation: Principles and challenges". *Seminars in Immunology* 27: 149–160.

Hofman, Allan F. 1999. "Bile acids: The good, the bad and the ugly". *News in Physiological Sciences* 14: 24–29.

Hosny, Michel NadimCssir, and Bernard La Scola. 2017. "Updating on gut microbiota and its relationship with the occurrence of necrotizing enterocolitis". *Human Microbiome Journal* 4: 14–19.

Husain, Shahid, Fernanda P Silveria, Nkechi E Azie, Billy Franks, and David L Horn. 2017. "Epidemiological features of invasive mild infections among solid organ transplant recipients: PATH Alliance registry analysis". *Medical Mycology* 55: 269–277.

Jacques M, M Gottschalk, B Foiry, and R Higgins. 1990. "Ultrastructural study of surface components of *Streptococcus suis*". *Journal of Bacteriology* 172: 2833–2838.

Kennedy, Adam D, Juliane Bubeck Wardenburg, Donald J Gardner, Daniel Long, Adeline R Whitney, Kevin R Braughton, Olaf Scheewind, and Frank R DeLeo. 2010. "Targeting of alpha-hemolysin by active or passive immunization decreases severity of U S A 300 skin infection in a mouse model". *The Journal of Infectious Diseases* 202: 1050–1058.

Kochut, Annika, and Petra Dersch. 2013. "Bacterial invasion factors: Tools for crossing biological barriers and drug delivery?" *European Journal of Pharmaceutics and Biopharmaceutics* 84: 242–250.

Kotas, Maya E, and Ruslan Medzhitov. 2015. "Homeostasis, inflammation, and disease susceptibility". *Cell* 160: 816–827.

Lavelle, E C, C Murphy, L A J O'Neill, E M Creagh. 2010. "The role of TLRs, NLRs, and RLRs in mucosal innate immunity and homeostasis". *Mucosal Immunology* 3: 17–28.

Lau, Gee W, Daniel J Hassett, Huimin Ran, and Fansheng Kong. 2004. "The role of pyocyanin in *Pseudomonas aeruginosa* infection". *Trends in Molecular Medicine* 10: 599–606.

Li, Min, Bin An Diep, Amer E Villaruz, Kevin R Braughton, Xiaofei Jiang, Frank R DeLeo, Henry F Chambers, Yuan Lu, and Michael Otto. 2009. "Evolution of virulence in epidemic community-associated methicillin-resistant *Staphylococcus aureus*". *Proceedings of the National Academy of Sciences U S A* 106: 5883–5888.

Lionakis, Michail, and Dimitrios P Kontoyiannis. 2004. "Fusarium Infections in Critically Ill Patients". *Seminars in Respiratory and Critical Care Medicine*, 159–169.

Lombardi, Charlotte, James Tolchard, Stephanie BouillotmLuca Signor, Caroline Cebus David Liebl, Daphna Fenel, Jean-Marie Teulon, Julliane Brock Bergit Habenetein, Jean-Luc Pelleguer Eric Faudry, I Attree Andrea Antoine Loquet, and Viviana Job. 2019. "Structural and functional characterization of the type three secretion system (T3SS) needle of *Pseudomonas aeruginosa*". *Frontiers of Microbiology* 10: 573.

Makino, Sou-Ichi, Masahisa Watarai, Hyeng-Il Cheun, Toshikazu Shirahata, and Ikuo Uchida. 2002. "Effect of the lower molecular capsule released from the cell surface of Bacillus anthracis on the pathogenesis of anthrax". *The Journal of Infectious Diseases* 186: 227–233.

Malachowa, Natalia, Scott D Kobayashi, Kevin R Braughton, and Frank R DeLeo. 2013. "Mouse model of Staphylococcus aureus skin infection". *Methods in Molecular Biology* 1031: 109–116.

Maraz, Kazi Madina, and Ruhul Amin Khan. 2021. "An overview on impact and application of microorganisms on human health, medicine and environment". *GSC Biological and Pharmaceutical Sciences* 16: 089–104.

Martinez-Giron, R, A Ribas, L Dogani, and J G Esteban. 2008. "Protozoa in respiratory pathology: A review". *European Respiratory Journal* 32: 1354–1370.

Medzhitov, Ruslan. 2008. "Origin of physiological roles of inflammation". *Nature* 454: 428–435.

Medzhitov, Ruslan. 2010. "Inflammation 2010- new adventure of an old flame". *Cell* 140: 771–776.

Meena, Himani, Asad Syed, and Busi Siddhardha. 2020. *Model Organisms of Microbial Pathogenesis, Biofilm formation and Antimicrobial Drug Discovery*. Edited by Busi Siddhardha, Madhu Dyavaiah and Asad Syed. Singapore: Springer.

Miele, Adriana E, Sofiane Badaoui, Lorenzo Margliani, Romain Salza, Giovanna Boumis, Selvia Chichiarelli, Berfrand Duclos, and Sylvie Ricard-Blum. 2019. "A comparative analysis of secreted protein disulfide isomerases from the tropical co-endemic parasites *Schistosoma mansoni* and *Leishmania major*". *Scientific Reports* 9: 9568.

Mihret, Wude, Tsehaynesh Lema, Yared Merid, Afework Kassu, Workeabeba Abebe, Beyene Moges, Admasu Tenna, et al. 2016. "Surveillance of bacterial meningitis, Ethiopia, 2012–2013". *Emerging Infectious Diseases* 22: 75–78.

Misra, Usha Kant, and Jayantee Kalita. 2010. "Overview: Japanese encephalitis". *Progress in Neurobiology* 91: 108–120.

Mocarski, Edaward S, Thomas Shenk, and R F Pass. 2007. *Cytomegaloviruses*. Philadelphia: Lippincott Williams & Wilkins.

Moonens, Kristof, and Han Remaut. 2017. "Evolution and structural dynamics of bacterial glycan binding adhesins". *Current Opinion in Structural Biology* 44: 48–58.

Moran, Gregory J, Anusha Krishnadasan, Rchel J Gorwitz, Gregory E Fosheim, Linda K Mc Dougal, Roberta B Carey, and David A Talan. 2006. "Methicillin-resistant *S. aureus* infections among patients in the emergency department". *The New England Journal of Medicine* 355: 666–674.

Mulcahy, Michelle E, Joan A Geoghegan, Ian R Monk, Kate M O'Keeffe, Evelyn J Walsh, Timothy J Foster, and Rachel M McLoughlin. 2012. "Nasal colonisation by *Staphylococcus aureus* depends upon clumping factor B binding to the squamous epithelial cell envelope protein loricrin". *PLoS Pathogens* 8: e1003092.

Mulcahy, Michelle E, and Rachel M McLoughin. 2016. "Host-bacterial crosstalk determines *Staphylococccus aureus* nasal colonization". *Trends in Microbiology* 24: 872–886.

Naoko Inoshima, Yang Wang, and Juliane Bubeck Wardenburg. 2012. "Genetic requirement for ADAM10 in severe Staphylococcus aureus skin infection". *The Journal of Investigative Dermatology* 132: 1513–1516.

Nathan, Carl. 2002. "Points of control in inflammation". *Nature* 420: 846–852.

Nelson, P E, M C Dignani, and E J Anaissie. 1994. "Taxonomy, biology and clinical aspect of Fusarium species". *Clinical Microbiology Reviews* 7: 479–504.

Noni, Maria, Anna Katelari, Aikaterini Kapi, Angeliki Stathi, George Dimopoulos, and Stavros-Eleftherios Doudounakis. 2017. "*Scedosporium apiospermum* complex in cystic fibrosis; should we treat?" *Mycoses* 60: 594–599.

Nucci, Marcio, Elias J Anaissie, Flavio Querioz-Telles, Carlos A Martin, Plinio Trabasso, Cris Ha-Na Solza, Claudia Mangini, et al. 2003. "Outcome predictor of patients with hematologic malignancies and Fusarium infection". *Cancer* 98: 315–319.

Palankar, Raghavendra, Ulrike Binsker, Bianca Haracska, Jan Wesche, Andreas Greinacher, and Suen Hammerschmidt. 2018. "Interaction between the *Staphylococcus aureus* extracellular adherence protein Eap and its subdomains with platelets". *International Journal of Medical Microbiology* 308: 683–691.

Pelczar, Michael Joseph, and Rita M Pelczar. 2022. "Microbiology". *Encyclopedia Britannica*.

Periasamy, Saravanan, and Paul E Kolenbrander. 2009. "Mutualistic biofilm communities develop with *Porphyromonas gingivalis* and initial, early and late colonizers of enamel". *Journal of Bacteriology* 191: 6804–6811.

Phelps, Christopher C, Stephen Vadia, Eusandia Arnett, Yubo Tan, Xiaoli Zhang, Sarika Pathak-Sharma, Mikhail A. Gavrillin, and Stephanie Seveau. 2018. "Relative roles of listeriolysin O, In1A and InlB in *Listeria monocytogenes* uptake by host cells". *Infection and Immunity* 86: e00555–e00518.

Philippsen, Gisele S, and Ricardo De Marco. 2019. "Impact of transposable elements in the architecture of genes of the human parasite *Schistosoma mansoni*". *Molecular and Biochemical Parasitology* 228: 27–31.

Ramakrishnan, Meenakshi, Aaron J Ulland, Laura C Steinhardt, Jennifer C Moïsi, Fred Were, and Orin S Levine. 2009. "Sequelae due to bacterial meningitis among African children: A systematic literature review". *BMC Medicine* 7: 47.

Reece, Emma, Ricardo Segura, Abaigeal Jackson, Siobhan McClean, and Julie Renwick Peter Greally. 2017. "Co-colonisation with *Aspergillus fumigatus* and *Pseudomonas aeruginosa* is associated with poorer health in cystic fibrosis patients: An Irish registry analysis". *BMC Pulmonary Medicine* 17: 1–8.

Ribet, David, and Pascale Cossart. 2015. "How bacterial pathogens colonize their hosts and invade deeper tissues". *Microbes and Infection* 17: 173–183.

Rodriguez, Irene Gonzalez, Lorena Ruiz, Miguel Gueimonde, Abelardo Margolles, and Borja Sanchez. 2013. "Factors involved in the colonization and survival of bifidobacteria in the gastrointestinal tract". *FEMS Microbiology Letters* 340: 1–10.

Rohwer, Forest, and Rebecca Vega Thurbe. 2009. "Viruses manipulate the marine environment". *Nature* 459: 207–212.

Rood, Julian I. 1998. "Virulence genes of *Clostridium perfringens*". *Annual Review of Microbiology* 52: 333–360.

Rozé, Jean-Christophe, Pierre-Yves Ancel, Patricia Lepage, Laetitia Martin-Marchand, Ziad Al Nabhani, Johanne Delannoy, Jean-Charles Picaud, et al. 2017. "Nutritional strategies and gut microbiota composition as risk factors for necrotizing enterocolitis in very-preterm infants". *The American Journal of Clinical Nutrition* 36: 821–830.

Sawa, Teiji, Masaru Shimizu, Kiyoshi Moriyama, and Jeanine P Weiner-Kronish. 2014. "Association between *Pseudomonas aeruginosa* type III secretion, antibiotic resistance and clinical outcome: A review". *Critical Care* 18: 668.

Schlievert, P M, J E Gocke, and J R Deringer. 1993. "Group B streptococcal toxic shock-like syndrome: Report of a case and purification of an associated pyrogenic toxin". *Clinical Infectious Diseases* 17: 26–31.

Schloss, Patrick D, and Jo Handelsman. 2004. "Status of the microbial census". *Microbiology and Molecular Biology Reviews* 68: 686–691.

Shannon, Jeffrey G, Cana L Ross, Theresa M Koehler, and Richard F Rest. 2003. "Characterization of anthrolysin O, the Bacillus anthracis cholesterol-dependent cytolysin". Infection and Immunity 71: 3183–3189.

Shi, Zhijun, Xudian Shi, Ullah Muhammed Wajid, Sixiang Li, Victor V Revin, and Guang Yang. 2018. "Fabrication of nanocomposites and hybrid materials using microbial biotemplates". *Advanced Composites and Hybrid Materials* 1: 79–93.

Solanki, Vandana, Monalisa Tiwari, and Vishvanath Tiwari. 2018. "Host-bacteria interaction and adhesin study for development of therapeutics". *International Journal of Biological Macromolecules* 112: 54–64.

Solomon, Tom. 2004. "Flavivirus encephalitis". *The New England Journal of Medicine* 351: 370–378.

Steinbach, William J, Kieren A Marr, Elias J Anaissie, Nkechi Azie, Shra Ping Quan, Herwing Meier-Kriesche, Senee Apeworkin, and David L Horn. 2012. "Clinical epidemiology of 960 patients with invasive aspergillosis from the PATH alliance registry". *Journal of Infection* 65: 453–464.

Stones, Daniel Henry, and Annie Marie Krachler. 2016. "Against the tide: The role of bacterial adhesion in host colonization". *Biochemical Society Transactions* 44: 1571–1580.

Suttle, Curtis A. 2007. "Marine viruses—major players in the global ecosystem". *Nature Reviews Microbiology* 5: 801–812.

Tedla, Mebrahtu G, Alison L Every, and Jean-Piere Y Scheerlinck. 2019. "Investigating immune responses to parasites using transgenesis". *Parasites & Vectors* 12: 303.

Thakur, Aneesh, Heidi Mikkelsen, and Gregers Jungersen. 2019. "Intracellular pathogens: Host immunity and microbial persistence strategies". *Journal of Immunology* 2019: 1356540.

Timberly, Roane M, Raina M Maier, Kelly A Reynolds, and Ian L Pepper. 2009. "Microorganisms". *Environmental Microbiology*, 9–36.

Tirumale, Sharmila, and Tessy Anu Thomas. 2018. "Pathogenesis of cholera: Recent prospective in rapid detection and prevention of cholera". *Bacterial Pathogenesis and Antibacterial Control* 172: 2833–2838.

Vergnolle, Nathalie. 2003. "The inflammatory response". *Drug Development Research* 59: 375–381.

Visconti, Alessia, Caroline I Le Roy, Fabio Rosa, Nicolo Rossi, Robert P Mohnrey Tiphana C Martin, Weizhong Li, Emanuele de Rinaldis, et al. 2019. "Interplay between the human gut microbiome and host metabolism". *Nature Communications* 10: 4505.

Wajid, Ullah Muhammed, Zhijun Shi, Xudian Shi, Di Zeng, Sixiang Li, and Guang Yang. 2017. "Microbes as structural templates in biofabrication: Study of surface chemistry and applications". *ACS Sustainable Chemistry & Engineering* 5: 11163–11175.

Wang, Dawei, Bothe Hu, Chang Hu, Fangfang Zhu, Xing Hiu, Sing Zhang, Binbin Wang, et al. 2020. "Clinical characteristics of 138 hospitalized patients with 2019 novel coronavirus-infected pneumonia in Wuhan China". *JAMA* 323: 1061–1069.

Weidenmaier, Christopher, John F Kokai-Kun, Sascha A Kristian, Tanya Chanturiya, Hubert Kalbacher, Matthias Gross, Graeme Nicholson, Birgid Neumeister, James J Mond, and Andreas Peschel. 2004. "Role of teichoic acids in Staphylococcus aureus nasal colonization, a major risk factor in nosocomial infections". *Nature Medicine* 10: 243–245.

Winstanley, Craig, and Joanne L Fothergill. 2009. "The role of quorum sensing in chronic cystic fibrosis *Pseudomonas aeruginosa* infections". *FEMS Microbiology Letters* 290: 1–9.

Wu, Yvonne T, Tan N Truong, Connie Tam, Myra N Mandoza, Lucia Zha, David J Evans, and Suzanne M J Fleiseig. 2019. "Impact of topical corticosteroid pre-treatment on susceptibility of the injured murine cornea to *Pseudomonas aeruginosa* colonization and infection". *Experimental Eye Research* 179: 1–7.

Wu, Ze Ang, and Hai Xia Wang. 2019. "A systemic review of the interaction between gut microbiota and host health from a symbiotic perspective". *SN Comprehensive Clinical Medicine* 1: 224–235.

Xu, Dan, Chongbing Liao, Bing Zhang, W David Tolbert, Wangxiao He, Zhijun Dai, Wei Zhang, et al. 2018. "Human enteric alpha-defensin 5 promotes Shigella infection by enhancing bacterial adhesion and invasion". *Immunity* 48: 1233–1244.

Yano, Tamaki, and Shoichiro Kurata. 2011. "Intracellular recognition of pathogens and autophagy as an innate immune host defence". *The Journal of Biochemistry* 150: 143–149.

Zhang, Hushan, Luman Wang, and Yiwei Chu. 2019. "Reactive oxygen species: The signal regulator of B cell". *Free Radical Biology and Medicine* 142: 16–22.

2 Conventional Treatments Modalities for Bacterial Infection

Manjari Datta and Indranil Chattopadhyay
Central University of Tamil Nadu, Thiruvarur, India

CONTENTS

2.1 INTRODUCTION

In 1928, after the discovery of penicillin by Alexander Fleming, antibiotics were the most common type of antibacterial agents. They had the ability to treat almost all kinds of bacterial infections. From 1950 to 1960 was the 'golden era' of antibiotics. Many antibiotics were discovered during this period. But now, due to the widespread and frequent use of broad-spectrum antibiotics, they are gradually losing their effectiveness, and there is a continuous rising of multidrug-resistant (MDR) bacteria. This is a serious health issue worldwide. 'Superbugs' such as *Enterococcus faecium, Klebsiella pneumoniae*, *Pseudomonas aeruginosa*, and *Staphylococcus aureus*, are resistant to most conventional antibiotics (Kortright et al., 2019). This multidrug resistance property is more prominent in Gram-negative bacteria (Liu et al., 2019a). Therefore, now is a crucial time to develop alternative therapeutic options over traditional antibiotics. Some of these novel therapeutic strategies include the use of nanomaterials in combination with antibiotics. Nanomaterials show antimicrobial activity by directly interacting with the cell wall, producing reactive oxygen species (ROS), or by interfering with other cellular functions. They also can inhibit bacterial biofilm formation (Galbadage et al., 2019; Wang et al., 2016). Nanomaterials protect antibiotics from the attack of

DOI: 10.1201/9781003345299-2

hydrolytic enzymes which are secreted by the microorganisms. Positively charged nanoparticles can interact with the negatively charged bacterial cell membrane and therefore alter the permeability so the antibiotic can easily enter into the cell (Wang et al., 2020). In addition to nanoparticles antibiotic adjuvants, monoclonal antibody, probiotics, antimicrobial peptides (AMPs), vaccines, bacteriophage therapy are also promising alternatives to antibiotics. Adjuvants increase the lifespan of antibiotics and help in their accumulation within the cell. They also block the development of resistance property (Liu et al., 2019). The use of anti-virulence drugs is also a new concept (Maura et al., 2016). These drugs can inhibit the total process of infection. Inhibition of the quorum sensing process is an important example of anti-virulence approach (Kalia, 2013). They are very much specific to their target so don't cause any harm to our normal microflora (Neville and Jia, 2019). Various essential oils such as citrus can prevent the development of biofilm formation (Lahiri et al., 2022). Recent studies have shown that biosurfactants from microbial and plant origin can be used to prevent several foetal diseases. These compounds are produced as secondary metabolites and they can alter the permeability of the cell membrane. (Srivastava et al., 2022). Targeting the ADP ribosylation process is also an important mechanism to treat microbial infections (Catara et al., 2019). Some natural compounds like terpenoids, lectins isolated from different medicinal plants can kill or prevent the growth of pathogenic microorganisms more efficiently than conventional antibiotics (Enioutina et al., 2017). Human AMP LL-37 has both antimicrobial and antibiofilm activity. It is effective against both Gram-positive and Gram-negative bacteria (Ridyard and Overhage, 2021). So, currently, numerous alternative therapeutic strategies are under investigation (Streicher, 2021) and in this book chapter we want to illuminate these novel therapeutic options over conventional antibiotics, which may open new doors of hope in the post-antibiotic era.

2.2 MODE OF ANTIBIOTIC RESISTANCE DEVELOPMENT IN BACTERIA

Bacteria have several mechanisms to develop resistance against antibiotics and other antimicrobial agents. Resistance can be achieved by the mutation in genomic DNA that may occur either spontaneously or due to the selective pressure of antibiotics (Wang et al., 2020). Horizontal transfer of resistance genes like R (resistance) plasmid is responsible for the spreading of resistance property between closely related bacterial species (Blair et al., 2015; Lerminiaux and Cameron, 2019; Munita and Arias 2016). Some other mechanisms include the inactivation of antibiotics by secreting different enzymes such as beta-lactamase enzymes, causing the hydrolysis of beta-lactam antibiotics, such as penicillin and cephalosporin (Thomson and Bonomo, 2005; Livermore and Woodford, 2006). Acyltransferases, phosphotransferases these types of enzymes cause covalent modification of the antibiotic structures (Wright, 2005). In addition, preventing the entry of antibiotics by altering cell permeability, alteration of the target site, efflux of antibiotics from the cell, biofilm formation these are some other important mechanisms. Some bacteria become resistant to tigecycline by over expression of the *ramA*gene, which is a positive regulator of AcrAB efflux pump (Osei Sekyere et al., 2016; Wang et al., 2018; Zhang et al., 2018; Akiyama et al., 2013; Taniguchi et al., 2017). Bacterial biofilms are 1,000 times more antibiotic resistant than planktonic cells (Wang et al., 2020). Many Gram-negative bacteria use major facilitator superfamily (MFS) and resistance-nodulation cell division (RND) systems to remove antibiotics from the cell. In Gram-negative bacteria, the lipopolysaccharide (LPS) layer is a vital component of the outer membrane. They modify the structure of this LPS layer to become resistant against antibiotics like colistin (Dijkmans et al., 2015; Moffatt et al., 2019).

2.3 PHAGE THERAPY

The concept of phage therapy is not new. The first human phage therapy test was conducted in 1921 by Dr Jose da Costa Cruz (Almeida and Sundberg, 2020). After that because of the discovery of antibiotics to treat bacterial infection this phage therapy concept became unpopular. But now as the development of antibiotic resistance among bacteria is a great public health concern so the development of these alternative therapeutic approaches is urgently required (Royer et al., 2021).

Use of phage therapy is more advantageous than traditional antibiotics. It is target-specific, so does not harm the normal microflora (Azeredo and Sutherland, 2008; Pires et al., 2017) and bacteriophage does not harm eukaryotic cells (Domingo-Calap and Delgado-Martínez, 2018). Virulent phage exhibit the lytic lifecycle and causes lysis of the host bacterial cell (Kortright et al., 2019) and this is useful for phage therapy. Some bacteriophages produce proteins such as amurins, which inhibit bacterial cell wall synthesis. They can also use holin protein to degrade the host cell (Cisek et al., 2017). Phage therapy also has the ability to eradicate bacterial biofilm (Royer et al., 2021). It also has the synergistic effect with antibiotics. For example, phage SBW25φ2 can be combined with kanamycin to treat *P. fluorescens* SBW25 infection (Valério et al., 2017). It is useful for the treatment of gastrointestinal infection, chronic lung infections, different other Systemic and local infections (Kortright et al., 2019). Lyticbacteriophages not only cause lysis of the bacterial cell, but they also inhibit other cellular processes like DNA, RNA, and protein synthesis (Loc-Carrillo and Abedon, 2011). In 2013 European Commission funded to an interesting project work called Phagoburn which is based on the application of phage cocktails for the treatment of burn victims infected with *E. coli* and *Pseudomonas aeruginosa* (Reardon, 2014).

Some phage-derived enzymes like lysine also have bactericidal activity. They can inhibit bacterial cell wall synthesis and cause cell lysis. They also help the phage to penetrate the bacterial cell. These types of enzymes work mostly against Gram-positive bacteria, but in the case of Gram-negative bacteria, due the presence of outer membrane, they can't penetrate the cell (São-José, 2018). Lysins are target specific, so the chance of developing bacterial resistance is rare (Maciejewska et al., 2018). Holin is another example of a phage-derived enzyme (São-José, 2018). Depolymerase enzymes help to degrade bacterial biofilm (Pires et al., 2017). Hydrolases can attack both Gram-positive and Gram-negative bacteria (Criscuolo et al., 2017). So, enzymes derived from bacteriophages are an attractive therapeutic option against antibiotic-resistant bacteria. In addition to natural bacteriophages, phage can be modified by using genetic engineering techniques to increase their efficiency and specificity (Olsen, 2016). However, the use of phage therapy and phage-derived enzymes are useful treatment options against bacterial infections over antibiotics.

2.4 NANOMEDICINE

The combination of nanoparticles with antibiotics is one of the most useful treatment options to cope with MDR bacteria (Huang et al., 2016; Su et al., 2019). Nanomaterials have both antimicrobial and anticancer properties. They have different mechanisms of action to treat the bacterial infection. They can directly interfere with the cell wall formation and inhibit the function of proteins, kill microbes through the production of reactive oxygen species (ROS), and also help in the eradication of bacterial biofilm (Galbadage et al., 2019; Wang et al., 2016). When we mix nanomaterials with antibiotics they increase the interaction capability of antibiotics with pathogenic bacteria, nanomaterials protect antibiotics from the attack of hydrolytic enzymes and also prevent their efflux from the bacterial cells. Nanoparticles are positively charged like silver nanoparticles (Ag NPs), so they can easily interact with the negatively charged bacterial cell membranes and therefore alter the permeability so that antibiotics can easily enter the cell. Polymeric nanoparticles are used for the encapsulation and targeted delivery of antibiotics, they enhance the solubility and internalisation of antibiotics and also help to reduce the toxicity of the drug (Chen et al., 2015; Chen et al., 2020). For example, polymyxin B, which is effective against Gram-negative bacteria, has nephrotoxic and neurotoxic activity when administrated alone, but if it is mixed with polymeric nanoparticles, its toxicity is reduced (Insua et al., 2017a). In addition to antibiotics, antimicrobial peptides also can be coated with nanoparticles to prevent them from *in vivo* enzymatic degradation (Borro and Malmsten, 2019). Recently mesoporous silica nanoparticles (MSN) are used for the delivery of drugs to the infected tissue (Wang et al., 2015; Llopis-Lorente et al., 2019). Silver nanoparticles and gold nanoparticles are also used for antimicrobial applications. Iron oxide nanoparticle (SPION) is used with the antibiotic methicillin to treat infections associated with medical devices (Geilich et al., 2017). CeO2 nanoparticles increase the permeability of the outer membrane of *Escherichia coli* so the antibiotic can easily penetrate (Bellio et al., 2018). Magnetic-iron oxide nanoparticle (MIONP) is used for the eradication of bacterial biofilm (Quan et al., 2019). So there are many applications of different nanoparticles. Different disadvantages are also associated with the use of nanoparticles like many studies showed that high toxicity is a result of using inorganic nanoparticles. Their instability and cost of production are also associated with this. However, several nano-antibiotics are now in the clinical trial stage, and long-term research work is required for the complete application of nano-antibiotic systems to treat MDR bacterial infection.

2.5 ANTIBIOTIC ADJUVANT

Antibiotic adjuvant doesn't have any bactericidal property itself, but is used to increase the *in vivo* lifespan of antibiotics. They help in their intracellular accumulation and prevent the development of the resistant property. They also boost the host immune response against bacterial infection (Liu et al., 2019). Antibiotic adjuvants are used mainly to treat MDR Gram-negative bacteria, because this drug-resistant property is more common in Gram-negative bacteria than Gram-positive bacteria. According to recent studies, stigmasterol is a common type of antibiotic adjuvant that is used with beta-lactam antibiotics to treat beta-lactamase-positive *E. coli* and *P. aeruginosa* (Yenn et al., 2017). Adjuvants are also used to inhibit the efflux pump in bacteria and therefore help to accumulate antibiotics within the cell. For example, adjuvant A22 is used with the antibiotic novobiocin to inhibit the MreB efflux pump in *E. coli*. For the treatment of Gram-negative bacterial infections adjuvants increase the permeability of the outer membrane so antibiotics can easily enter into the cell. Antibiotic adjuvants are useful to eradicate bacterial biofilm, the combination of triclosan with the antibiotic tobramycin can act against *P. aeruginosa* biofilm (Maiden et al., 2018). The use of antibiotic adjuvant is therefore a new and promising concept to treat MDR Gram-negative bacteria (Liu et al., 2019).

2.6 SEMISYNTHETIC AND FULLY SYNTHETIC ANTIBIOTICS

Natural antibiotics are modified to increase their efficiency and stability. Modifications of antibiotics are also helpful to reduce their toxic side effects. Semisynthesis of antibiotics first started in 1946 with the hydrogenation of streptomycin and its result was the formation of dihydrostreptomycin, which was chemically more stable than natural streptomycin. Chloramphenicol was the first fully synthetic antibiotic. Metronidazole is another example of a fully synthetic antibiotic; having antibacterial as well as antiprotozoal activity, this antibiotic has great clinical application (Alauzet et al., 2019). Carbapenem is also a class of beta-lactam antibiotic which is a slightly modified form of penicillin. It is more resistant to the action of beta-lactamase enzymes. So these types of modifications of natural antibiotics help to overcome the phenomenon of antibiotic resistance and also increase the potentiality of antibiotics (Ribeiro da Cunha et al., 2019).

2.7 ANTIMICROBIAL PEPTIDE

Antimicrobial peptide (AMP), also known as host defence peptides, is a part of the innate immune system. AMPs have a wide range of antibacterial activity against both Gram-positive and Gram-negative bacteria (Yazici et al., 2018). They are also helpful in the destruction of bacterial biofilm (Yasir et al., 2018). They interfere with the quorum sensing procedure and therefore inhibit biofilm formation (Chung and Khanum, 2017). They also inhibit bacterial cell signalling systems within the biofilm and cause the degradation of the exopolysaccharide layer. Most AMPs are cationic in nature so they can easily interact with the negatively charged bacterial cell membrane. Although the main target site of AMPs is the cell membrane, they can also disrupt other cellular functions also (Bechinger and Gorr, 2017). One of the major advantages of using AMP is that it works with high specificity, and does not harm normal tissue, because eukaryotic cell membranes are neutral in charge due to the presence of zwitterionic (uncharged) lipids so cationic AMPs can't interact with them. By using AMPs there is a low chance of bacterial resistance. They are effective against multidrug-resistant (MDR) bacterial species (Chung and Khanum, 2017). They also show synergistic effects with antibiotics. For example, AMP-1018 in combination with ceftazidime, ciprofloxacin, imipenem, or tobramycin can inhibit 50% of the biofilm produced by *A. baumannii*, *E. coli*, *K. pneumoniae*, *P. aeruginosa*, *S. enterica*, and methicillin-resistant *S. aureus* (Reffuveille et al., 2014). AMP in combination with antibiotics can degrade biofilm matrix and disperse biofilms embedded cells in a more efficient way than AMPs can do this alone (Chung and Khanum, 2017). Bacteriocin is a type of ribosomally synthesised antimicrobial peptide produced by several bacteria like lactic acid bacteria as a secondary metabolite that inhibits the growth of closely related bacterial species (Nishie et al., 2012). Lantibiotic is a class of bacteriocin that is extensively modified, and it is highly effective against Gram-positive bacteria (Nowakowski et al., 2018). Some AMPs, such as Defensin and Lactoferrin, have anti-tuberculosis properties. Although AMPs have several promising features to be used as potential therapeutic agents, some drawbacks are also associated. Their antimicrobial activity is significantly decreased in the presence of biological fluids such as saliva. Natural AMPs are also susceptible to proteolytic degradation. Complex procedures are required for their isolation, extraction, and purification, and that's why they are quite expensive. Their large-scale production is also difficult (Chung and Khanum, 2017). However, the potential use of AMPs in medical science may be new hope for the post-antibiotic era.

2.8 HUMAN ANTIMICROBIAL PEPTIDE LL-37

Human AMP LL-37 has broad-spectrum antimicrobial as well as anti-biofilm activity. It can act as a peptide antibiotic. It belongs to the cathelicidin family, and it is synthesised and secreted by the epithelial cells. It has antibacterial activity against both Gram-positive and Gram-negative bacteria like *Pseudomonas, Escherichia, Staphylococcus* etc (Neshani et al., 2019). It mainly causes the disruption of cell membranes like other AMPs. It also has an immunomodulatory activity which is helpful for its bactericidal property. Biofilm formation can be inhibited in different bacteria like *S. aureus, P. aeruginosa* by using LL-37 (Bandurska et al., 2015; Dean et al., 2011; Overhage et al., 2008). Sometimes it shows the synergistic effects with other antimicrobial agents such as antibiotics. It increases the permeability of bacterial cell membrane so antibiotics can easily penetrate the cell (Leszczyńska et al., 2010; Zharkova et al., 2019; Nuding et al., 2014; Abou Alaiwa et al., 2014). LL-37 also can be used in combination with nanoparticles and bacteriocins. For example, gold nanoparticles and LL-37 combination is helpful to treat *P. aeruginosa* and *S. aureus* (Niemirowicz et al., 2016). Different techniques are used to enhance the potentiality of LL-37, for example, immobilisation technique can increase their ability to inhibit bacterial biofilm formation (Song et al., 2016; Mishra and Wang, 2017a, 2017b). LL-37 derivatives can be formed by removing a certain portion of the peptide and that's why derivatives of LL-37 are shorter in length than the original peptide. This process not only reduces the production cost, but also improves their antibacterial activity, because shorter peptides can more easily penetrate the bacterial cytoplasmic membrane, and also the bacterial biofilm (Saporito et al., 2018). For example, LL-32 is more potent than LL-37 (Bandurska et al., 2015). Total removal of the N-terminus of the peptide can reduce its cytotoxic activity (Kang et al., 2019) and haemolytic property (Saporito et al., 2018) but this process does not compromise the stability of the peptide (Saporito et al., 2018). Some limitations are there of using LL-37 as a therapeutic agent, the first disadvantage is their production cost, synthesis, and recovery process. In addition, they are highly sensitive to protease activity. They have toxicity and haemolytic activity against eukaryotic cells (Nagant et al., 2012). But the development of resistance property against LL-37 and its derivatives is a rare event (Ridyard and Overhage, 2021) (see Tables 2.1 and 2.2).

TABLE 2.1
Antibiotics That Show Synergistic Activity with Ll-37 and Their Respective Susceptible Pathogens

Name of Bacteria	Name of Antibiotic Which Can Be Combined with LL-37	Reference
Pseudomonasaeruginosa	Colistin	Dosler and Karaaslan (2014); Geitani et al. (2019)
	Ciprofloxacin	Dosler and Karaaslan (2014)
	Tobramycin	Payne et al. (2017)
Enterococcusfaecium	Ampicillin	Smith et al. (2015); Sakoulas et al. (2012)
	Ertapenem	Smith et al. (2015)
Staphylococcusaureus	Vancomycin	Shurko et al. (2018)
	Nafcillin	Le et al. (2016)
Clostridioidesdifficile	Tigecycline	Nuding et al. (2014)
Acinetobacterbaumannii	Tazobactam	Sakoulas et al. (2017)
	Azithromycin	Lin et al. (2015)
Stenotrophomonasmaltophilia	Colistin	Kumaraswamy et al. (2016)
Klebsiellapneumonaie	Azithromycin	Lin et al. (2015)
	Avibactam	Ulloa et al. (2019)

TABLE 2.2
LL-37 Derivatives That Show Superior Antimicrobial Activity in Comparison to the Original LL-37 Peptide

Derivative of LL-37	Sequence	Reference
LL-32	LLGDFFRKSKEKIGKEFKRIVQRIKDFLRNLV	Dannehl et al. (2013)
IG-19	LLGDFFRKSKEKIGKEFKR	Kanthawong et al. (2010)
RK-25	RKSKEKIGKEFKRIVQRIKDFLRNL	Nagant et al. (2012)
KR-12-a2	KRIVQRIKKWLR	Jacob et al. (2013)
L31-P113	LGDFFRKSKEKIGKEFKRIVQRIKDFLRNLAK	Wanmakok et al. (2018)
P38	TSVRQRWRWRQRVRTS	Aghazadeh et al. (2019)
SAAP-276	LKRVWKAVFKLLKRYWRQLKKPVR	de Breij et al. (2018)

2.9 COMBINATION THERAPY

Mitotane which is a non-antibiotic drug has a synergistic effect with polymyxin B to treat MDR Gram-negative bacteria such as *K. pneumonia* and *P. aeruginosa* (Tran et al., 2018). SPR741 increases the spectrum of activity of several antibiotics; it enhances the potentiality of rifampicin (Zurawski et al., 2017). Phenylalanyl arginyl β-naphthylamide (PAβN) is used in combination with antibiotics to treat fluoroquinolones resistant *P. aeruginosa* (Barrero et al., 2014; Lomovskaya et al., 2001). Another common example is Curcumin, which is isolated from *Curcuma longa* can act an inhibitor of efflux pump. It is used with antibiotics to treat MDR *P. aeruginosa* (Negi et al., 2014). Bacterial two-component systems are useful for their resistance property, biofilm formation, and virulence (Liu et al., 2019; Hoch, 2000). They also protect bacteria from the action of the host immune system (Liu et al., 2019; Pardo-Esté et al., 2018; Zschiedrich et al., 2016). In *Salmonella enterica* the two-component system PhoP/PhoQ helps them to establish the infection (Navarre et al., 2005). Some compounds, such as quinazoline, down-regulate the transcription of PhoP/PhoQ activated genes when used with antibiotics.

2.10 MICROBIOME MANIPULATION

The most successful way to manipulate the microbiome is the faecal microbiota transplant (FMT). Microflora is collected from a healthy donor and transferred to the recipient patient. A human stool sample is used for FMT, and this is useful for the treatment of *C. difficile* infections (CDI) (Chai and Lee, 2018). The success rate of this FMT is very high and the newly incorporated microbiota remain stable in the host system for several months (Cammarota et al., 2019; Shogbesan et al., 2018; Wang et al., 2019). In severe cases, infusion of multiple FMT protocol is more efficient than a single protocol (Ianiro et al., 2018). Probiotics are used to alter the microbiome. Some studies show that *Lactobacillus sp* and *Saccharomyces boulardii* are helpful for the treatment of CDI (Sartor, 2005). Some antagonistic microorganisms also compete with pathogens for the attachment site (Van Hoogmoed et al., 2008). In case of the oral cavity the attachment of *Porphyromonas gingivalis* ATCC 33277 is inhibited by different antagonistic strains like *Actinomyces naeslundii*, *Haemophilus parainfluenzae*, *Streptococcus mitis*, and *Streptococcus sanguinis* (Van Hoogmoed et al., 2008). This FMT treatment acts like broad-spectrum antibiotics which eradicate pathogenic microorganisms by competing for food and niche (Wang et al., 2020).

2.11 ANTIVIRULENCE DRUGS

Targeting the virulence factors of pathogens is an alternative treatment approach. Virulence is the ability of microorganisms to cause disease. Antivirulence drugs prevent the infection process but unlike antibiotics, they do not kill or inhibit the growth of bacteria; these drugs cause the inactivation of bacterial toxins like tetanus toxin and diphtheria toxin (Keller and Stiehm, 2000). They also can inhibit the quorum sensing process (Kalia, 2013), bacterial colonisation and adhesion with the host cell (Cusumano and Hultgren, 2009), synthesis of pili and fimbriae. Use of these types of drugs does not create any selective pressure on bacteria, so the chance of development of resistance is also less (Totsika, 2017). They are very much target-specific, so don't harm normal host microbiota. These antivirulence drugs may therefore be a useful alternative to traditional antibiotics, which directly kill or prevent the growth of bacteria, thus creating a strong selective pressure, and this selective pressure is responsible for the development of resistance properties in bacteria. But in the case of these antivirulence drugs, the chance of the development of resistance is very less (Neville and Jia, 2019).

2.12 ESSENTIAL OILS AGAINST BACTERIAL BIOFILM

Several natural compounds like essential oils have antimicrobial as well as antibiofilm activity. Bacterial biofilm is an aggregation of different bacterial species. It is surrounded by an exo-polysaccharide (EPS) layer (Dula et al., 2021), which protects the biofilm from the action of antibacterial agents and from the host immune system. Bacteria within the biofilm are extremely resistant to antibiotics. Essential oils such as citrus play a key role in the eradication of microbial biofilm. Essential oils are the mixture of various aromatic compounds which are from plant origin. They can interfere with the quorum sensing process and therefore inhibit biofilm formation (Alves et al., 2016). Rutaceae family plants are known as citrus plants, they are a rich source of various essential oils (Mahato et al., 2019). Sesquiterpene and monoterpenes are some common examples of citrus essential oils (Jing et al., 2014). Essential oils are hydrophobic in nature so they can easily degrade the cytoplasmic membrane of bacteria and therefore alter the permeability. They can directly attack Gram-positive bacteria, because they lack the outer membrane covering. They also cause the inactivation of various bacterial enzymes (Selim et al., 2014). Citrus essential oils which are collected from the lemon peel have the ability to destroy the biofilm of *Streptococcus mutans* (Sun et al., 2018). Citrus essential oil from grapes can inhibit the quorum sensing process of *Pseudomonas aeruginosa* (Luciardi et al., 2020). The essential oil collected from *Dodartia orientalis* can reduce the adherence capability of bacterial cells which is required for the biofilm formation. Essential

TABLE 2.3
Effect of Several Citrus Essential Oils (CEOs) on Different Types of Bacterial Biofilm

Name of CEO	Collected From	Name of Biofilm-Forming Microorganism	Reference
Geraniol	*Cymbopogon nardus*	*S. aureus*	Oliveira et al. (2014)
Cinnamaldehyde	*Cymbopogon nardus*	*S. aureus*	Budri et al. (2015)
Limonene	*Citrus lemon*	*E. coli*	Kerekes et al. (2013)
α-pinene	*Origanum majorana*	*B. cereus*	Kerekes et al. (2013)
Carvacrol	*Origanum vulgare*	*S. aureus*	Lahiri et al. (2022)
Eugenol	*Pimenta officinalis*	*E. coli*	Kim et al. (2015)
Linalool	*Salvia sclarea*	*E. coli*	Kačániová et al. (2020)
Eugenol	*Syzygium aromaticum*	*S. aureus*	Lahiri et al. (2022)

oil from *Origanum vulgare* has the capability to alter the motility of *P. fluorescens*. Essential oils alter cell membrane permeability by directly interacting with the unsaturated fatty acids (Burt and Reinders, 2003). They also inhibit enzymes associated with fatty acid biosynthesis. Some proteins required for the cell division process can be inactivated by various essential oils. For example, cinnamaldehyde, which is an essential oil inhibits the cell division process in *B. cereus*. It blocks the activity of FtsZ protein which regulates the cell division process in prokaryotes (Domadia et al., 2007). Essential oils target different virulence proteins in bacteria and inhibit their disease-causing ability; i.e., pathogenesis (Carneiro et al., 2011). The amount of different organic acids is reduced in the presence of essential oils (Carneiro et al., 2011). Nano-emulsified essential oils are more potent than pure essential oil to eradicate bacterial biofilm. Several drawbacks associated with the use of essential oils can be avoided easily by using nano-emulsification and microencapsulation techniques (Lahiri et al., 2022) (see Table 2.3).

2.13 VACCINE

Vaccination is also a valuable treatment option to treat MDR bacterial infections. Here, the occurrence of resistance is a rare event (Mishra et al., 2012), because antibiotics are used after the occurrence of infection or after the development of symptoms, whereas vaccines are used as prevention, not treatment (Buchy et al., 2020). Vaccines do not have any significant impact on the normal microflora in our body, but broad-spectrum antibiotics destroy the microbiome, leading to the development of antibiotic-resistant properties in bacteria (Bloom et al., 2018). Vaccination also reduces the need of using antibiotics, so the risk of developing antibiotic resistance is also less (Jansen et al., 2018). Viral vaccines are more common than bacterial vaccines, because by using viral vaccines we can get total protection, but in the case of bacterial vaccines their success rate is lower. Recently recombinant DNA technology, whole genome sequencing, phage display techniques, and bioinformatics tools, are approaches used for the enhancement of vaccinology and to overcome the drawbacks associated with traditional vaccines (Rappuoli, 2001; Bragg et al., 2018). Bacteriophages can also be used for the delivery of vaccines. Vaccines not only protect people from the disease, but also effects protection to the unvaccinated and immunocompromised, through the process of herd immunity. Every year, vaccination prevents millions of deaths (Klatt and Hopp, 2012). Vaccines prevent the transmission of disease within the population, and therefore inhibit the development of antibiotic-resistant property in bacteria. *Haemophilus influenzae* type b vaccine (Hib) was the first effective vaccine for bacteria. After the discovery of the Hib vaccine, the number of beta-lactamase-containing bacterial strains became significantly reduced (Heilmann et al., 2005). In addition to Hib vaccine, pneumococcal polysaccharide conjugate vaccine is also important to control antibiotic resistance. Not only bacterial vaccines, but viral vaccines such as influenza vaccine can also reduce the need for antibiotics. Currently, several bacterial vaccines like *Clostridiumdifficile* vaccine, *Staphylococcusaureus* vaccines, Group B *Streptococcus* vaccine, *Mycobacterium tuberculosis* vaccines are under development. Hopefully, these types of vaccines will be helpful to overcome bacterial antibiotic resistance (Jansen et al., 2018). This is a new concept of using the combination of antibiotics and vaccines to treat MDR bacteria (Tekle et al., 2012).

2.14 MONOCLONAL ANTIBODY

The first clinically approved monoclonal antibody was Orthoclone, an anti-CD3 monoclonal antibody that is used to prevent rejection in case of kidney transplantation (McConnell, 2019). Monoclonal antibodies are mostly useful for the treatment of cancers and rheumatic diseases. Only a few monoclonal antibodies are now approved for the treatment of bacterial diseases; for example, Obiltoxaximab and Raxibacumab, which are useful for the treatment of inhalational anthrax (Hou and Morrill, 2017; Tsai and Morris, 2015). In the case of bacterial infections, the main target of monoclonal antibodies is the surface antigen or the toxins which are secreted by microorganisms. There is no chance of cross-resistance between antibiotics and monoclonal antibodies because they have a completely distinct mode of action (Wang-Lin and Balthasar, 2018; Bebbington and Yarranton, 2008). They are target specific, and so do not harm the normal microflora of the host. Therapeutic bactericidal antibodies can directly cause lysis of the bacterial cells, killing the bacteria (Wang-Lin and Balthasar, 2018). By using monoclonal antibodies we can easily reduce the need of using conventional antibiotics, so the chance of developing resistance is also less (Gagliotti et al., 2015; Goossens et al., 2005; van de Sande-Bruinsma et al., 2008). In the case of Gram-positive bacterial infections, monoclonal antibodies cause the neutralisation of exotoxin which is secreted by bacteria. They prevent the binding of bacterial toxins to the particular receptor present on the host cell surface (Kufel et al., 2017). In Gram-negative bacteria they target specific antigens present on the bacterial cell surface. Several monoclonal antibodies are in the developmental stage which can be applied for the treatment of MDR Gram-negative bacterial infections, for example, infection caused by MDR *A. baumannii*, *K. pneumonia*, and *P. aeruginosa*. The use of monoclonal antibody therapy to treat MDR Gram-positive and Gram-negative bacteria represents an alternative therapeutic approach over traditional antibiotics. But the resistance against monoclonal antibodies is not a rare event; some bacteria, such as *S. aureus, Streptococcus pyogenes, Streptococcus pneumonia, Haemophilus influenza, P. aeruginosa*, secrete some proteases that cleave the antibody by separating the Fc and Fab regions (Wang-Lin and Balthasar, 2018). Some other bacterial species have proteins that bind with the Fc region of the antibody and prevent the process of opsonisation and complement activation. Examples of these proteins include protein A in *S. aureus*, protein G in *Streptococcus*. In addition, masking the epitope and removing the binding site of the antibody are also common resistance mechanisms. Another potential drawback of using monoclonal antibodies is that they are expensive (Streicher, 2021). However, the concept of using monoclonal antibodies for the treatment of bacterial infection is mostly in the preclinical stage. More trials and research work are necessary (McConnell, 2019) (see Table 2.4).

2.15 CRISPR-CAS9 SYSTEM

CRISPR-Cas9 system, published by Jennifer Doudna and Emmanuelle Charpentier in 2012,was one of the most significant discoveries in modern biological science. This is an adaptive immune system in bacteria that protects them from the attack of invading viral DNA and foreign DNA. This

TABLE 2.4
Application of Monoclonal Antibodies (mAbs) Which Are Either Approved or Are in the Preclinical Trial Stage

Name of the mAb	Mode of Action	Current Stage	Reference
Obiltoxaximab	Can be used with antibiotics to neutralise anthrax toxin	Approved by FDA (2017)	Hou and Morrill (2017)
Bezlotoxumab	Neutralise enterotoxin B	Approved by FDA (2017)	Wilcox et al. (2017)
MEDI3902	Binds with exopolysaccharide Psl and PcrV antigen, a component of the type III secretion system	Phase II clinical trial (2019)	DiGiandomenico et al. (2014); Ali et al. (2019)
MEDI4893	Binds with the α-toxin and prevents it to adopt the conformation of lytic transmembrane	Phase II clinical trial (2019)	Oganesyan et al. (2014)
514G3	Enhances the process of opsonisation by binding with the protein A	Phase II clinical trial (2018)	Varshney et al. (2018)
DSTA4637S monoclonal antibody– antibiotic conjugate	Monoclonal antibody causes the opsonisation of *S. aureus* and after phagocytosis, antibiotic is released which kills the bacteria	Phase I clinical trial (2019)	Zhou et al. (2019)

gene-editing tool has opened new avenues in biomedical research. This system is helpful for the development of novel antimicrobials so this is a great promising strategy for the treatment of infectious diseases caused by multidrug-resistant microorganisms. Bacterial resistant plasmid DNA can be cleaved by CRISPR-Cas9 system (Garneau et al., 2010). This system can kill bacteria based on their sequence specificity (Bikard et al., 2012). CRISPR can be used to develop 'smart' antibiotics that can combat MDR bacteria and can differentiate between pathogenic and beneficial microorganisms (Gomaa et al., 2014). CRISPR can specifically target the virulence factors which are associated with pathogenesis and the genes that are responsible for drug resistance. The major drawback of using CRISPR as a therapeutic agent is its *in vivo* delivery system. Immunogenicity is associated with the use of viral vectors (for example adeno-associated viral vector), so currently we are searching for the non-viral based methods for gene delivery. Nanoparticle is one of the important examples of this non-viral based gene delivery system. The combination of CRISPR-Cas9 system with nanotechnology is completely a new concept to overcome MDR bacterial infections. Nanoparticles are also widely used for the targeted delivery of antibacterial agents (Santos et al., 2018; Naskar and Kim, 2019). Nanoparticles disrupt cell membrane potential and create oxidative stress by the production of reactive oxygen species (ROS; Wang et al., 2017; Tang and Zheng, 2018; Liao et al., 2019). Inorganic nanoparticles such as silver (AgNPs) and gold (AuNPs) are potent as antimicrobial agents. Copper nanoparticles are effective against MDR *Staphylococcus aureus*. Different types of nanoparticles such as polymeric nanoparticles and lipid nanoparticles are used as carriers for the delivery of genes into target cells. The nanoparticle protects the gene of interest from degradation. CRISPR-Cas9 gene-editing system also can be encapsulated within nanoparticles. Lipid nanoparticles are most potent for the *in vivo* delivery of the CRISPR system (Patel et al., 2017). CRISPR-Cas9 is a novel and promising therapeutic strategy to treat antibiotic-resistant bacteria. Application of CRISPR for the treatment of infectious diseases may a new hope of light for the post-antibiotic era. But this concept is in the early stage of research, so more investigations are required before successful therapeutic applications (Wan et al., 2021) (see Table 2.5).

TABLE 2.5
Applications of the CRISPR-Cas9 System

Purpose	Application of CRISPR	Example	Reference
For the development of 'smart' antibiotics that can specifically kill MDR pathogenic bacteria without causing any harm to beneficial microbiota	Killing of pathogenic bacteria with sequence specificity by using phagemids	*Staphylococcus aureus*	Bikard et al. (2014)
Identification of some essential genes present in bacteria	Whole genome screening of the pathogens	*Bacillus subtilis*	Peters et al. (2016)
Functional study about the virulence factors of the pathogens	Genome editing of the pathogen	*Mycobacterium tuberculosis* *Toxoplasma gondii*	Choudhary et al. (2015) Sidik et al. (2014)

2.16 CONCLUSION

The continuous rising of antibiotic-resistant bacteria is an alarming signal that indicates a post-antibiotic era is near. After the discovery of penicillin in 1928 by Alexander Fleming, lots of antibiotics were identified from the1940 to the 1960s, and it was the golden age of the antibiotic. But now, most antibiotics are gradually losing effectiveness, and the discovery of new antibiotics is also diminishing. It is difficult to discover new antibiotics with the unique target sites. According to the current scenario, almost 23,000 deaths occur per year in North America (Ribeiro da Cunhaet al., 2019) and yearly 33,000 in Europe (Cassini et al., 2019) because of antibiotic-resistant bacterial infections. The total concept of replacing antibiotics by using non-antibiotic therapies is still in the research stage, and long-term research is required for their successful *in vivo* application. Antimicrobial peptides (AMPs), monoclonal antibody, probiotics, natural compounds, nanoparticles, and the CRISPR-Cas9 system are perhaps some reliable alternatives for the post-antibiotic era. These types of research are still in their infancy, so more attention should be paid to the development and potent clinical applications of these novel therapeutic strategies.

REFERENCES

Abou Alaiwa MH, Reznikov LR, Gansemer ND, Sheets KA, Horswill AR, Stoltz DA, Zabner J, Welsh MJ. 2014. pH modulates the activity and synergism of the airway surface liquid antimicrobials β-defensin-3 and LL-37. *Proc Natl Acad Sci U S A*. 111(52):18703–8.

Aghazadeh H, Memariani H, Ranjbar R, Pooshang Bagheri K. 2019. The activity and action mechanism of novel short selective LL–37-derived anticancer peptides against clinical isolates of *Escherichia coli. Chem Biol Drug Des*. 93(1):75–83.

Akiyama T, Presedo J, Khan AA. 2013. The tetA gene decreases tigecycline sensitivity of Salmonella enterica isolates. *Int J Antimicrob Agents*. 42(2):133–40.

Alauzet C, Lozniewski A, Marchandin H. 2019. Metronidazole resistance and nim genes in anaerobes: A review. *Anaerobe*. 55:40–53.

Ali SO, Yu XQ, Robbie GJ, Wu Y, Shoemaker K, Yu L, Digiandomenico A, Keller AE, Anude C, Hernandez-Illas M, Bellamy T, Falloon J, Dubovsky F, Jafri HS. 2019. Phase 1 study of MEDI3902, an investigational anti-Pseudomonas aeruginosa PcrV and Psl bispecific human monoclonal antibody, in healthy adults. *Clin Microbiol Infect*. 25(5):629.e1–629.e6.

Almeida GMF, Sundberg LR. 2020. The forgotten tale of Brazilian phage therapy. *Lancet Infect Dis*. 20(5):e90–e101.

Alves S, Duarte A, Sousa S, Domingues FC. 2016. Study of the major essential oil compounds of Coriandrum sativum against Acinetobacter baumannii and the effect of linalool on adhesion, biofilms and quorum sensing. *Biofouling*. 32(2):155–65.

Azeredo J, Sutherland IW. 2008. The use of phages for the removal of infectious biofilms. *Curr Pharm Biotechnol.* 9(4):261–6.

Bandurska K, Berdowska A, Barczyńska-Felusiak R, Krupa P. 2015. Unique features of human cathelicidin LL-37. *Biofactors.* 41(5):289–300.

Barrero MA, Pietralonga PA, Schwarz DG, Silva A Jr, Paula SO, Moreira MA. 2014. Effect of the inhibitors phenylalanine arginyl ß-naphthylamide (PAßN) and 1-(1-naphthylmethyl)-piperazine (NMP) on expression of genes in multidrug efflux systems of Escherichia coli isolates from bovine mastitis. *Res Vet Sci.* 97(2):176–81.

Bebbington C, Yarranton G. 2008. Antibodies for the treatment of bacterial infections: Current experience and future prospects. *Curr Opin Biotechnol.* 19(6):613–9.

Bechinger B, Gorr SU. 2017. Antimicrobial peptides: Mechanisms of action and resistance. *J Dent Res.* 96(3):254–60.

Bellio P, Luzi C, Mancini A, Cracchiolo S, Passacantando M, Di Pietro L, Perilli M, Amicosante G, Santucci S, Celenza G. 2018. Cerium oxide nanoparticles as potential antibiotic adjuvant. Effects of CeO_2 nanoparticles on bacterial outer membrane permeability. *Biochim Biophys Acta Biomembr.* 1860(11):2428–35.

Bikard D, Euler CW, Jiang W, Nussenzweig PM, Goldberg GW, Duportet X, Fischetti VA, Marraffini LA. 2014. Exploiting CRISPR-Cas nucleases to produce sequence-specific antimicrobials. *Nat Biotechnol.* 32(11):1146–50.

Bikard D, Hatoum-Aslan A, Mucida D, Marraffini LA. 2012. CRISPR interference can prevent natural transformation and virulence acquisition during *in vivo* bacterial infection. *Cell Host Microbe.* 12(2):177–86.

Blair JM, Webber MA, Baylay AJ, Ogbolu DO, Piddock LJ. 2015. Molecular mechanisms of antibiotic resistance. *Nat Rev Microbiol.* 13(1):42–51.

Bloom DE, Black S, Salisbury D, Rappuoli R. 2018. Antimicrobial resistance and the role of vaccines. *Proc Natl Acad Sci U S A.* 115(51):12868–71.

Borro BC, Malmsten M. 2019. Complexation between antimicrobial peptides and polyelectrolytes. *Adv Colloid Interface Sci.* 270:251–60.

Bragg RR, Meyburgh CM, Lee JY, Coetzee M. 2018. Potential treatment options in a post-antibiotic era. *Adv Exp Med Biol.* 1052:51–61.

Buchy P, Ascioglu S, Buisson Y, Datta S, Nissen M, Tambyah PA, Vong S. 2020. Impact of vaccines on antimicrobial resistance. *Int J Infect Dis.* 90:188–96.

Budri PE, Silva NC, Bonsaglia EC, Fernandes Júnior A, Araújo Júnior JP, Doyama JT, Gonçalves JL, Santos MV, Fitzgerald-Hughes D, Rall VL. 2015. Effect of essential oils of Syzygium aromaticum and Cinnamomum zeylanicum and their major components on biofilm production in Staphylococcus aureus strains isolated from milk of cows with mastitis. *J Dairy Sci.* 98(9):5899–904.

Burt SA, Reinders RD. 2003. Antibacterial activity of selected plant essential oils against Escherichia coli O157:H7. *Lett Appl Microbiol.* 36(3):162–7.

Cammarota G, Gallo A, Bibbò S. 2019. Fecal microbiota transplant for C. difficile infection: Just say yes. *Anaerobe.* 60:102109.

Carneiro S, Villas-Bôas SG, Ferreira EC, Rocha I. 2011. Metabolic footprint analysis of recombinant Escherichia coli strains during fed-batch fermentations. *Mol Biosyst.* 7(3):899–910.

Cassini A, Högberg LD, Plachouras D, Quattrocchi A, Hoxha A, Simonsen GS, Colomb-Cotinat M, Kretzschmar MÈ, Devleesschauwer B, Cecchini M, Ouakrim DA, Oliveira TC, Struelens MJ, Suetens C, Monnet DL, Burden of AMR Collaborative Group. 2019. Attributable deaths and disability-adjusted life-years caused by infections with antibiotic-resistant bacteria in the EU and the European Economic Area in 2015: A population-level modelling analysis. *Lancet Infect Dis.* 19(1):56–66.

Catara G, Corteggio A, Valente C, Grimaldi G, Palazzo L. 2019. Targeting ADP-ribosylation as an antimicrobial strategy. *Biochem Pharmacol.* 167:13–26.

Chai J, Lee CH. 2018. Management of primary and recurrent *Clostridium difficile* infection: An update. *Antibiotics (Basel).* 7(3):54.

Chen J, Ding J, Xiao C, Zhuang X, Chen X. 2015. Emerging antitumor applications of extracellularly reengineered polymeric nanocarriers. *Biomater Sci.* 3(7):988–1001.

Chen X, Jia F, Li Y, Deng Y, Huang Y, Liu W, Jin Q, Ji J. 2020. Nitric oxide-induced stromal depletion for improved nanoparticle penetration in pancreatic cancer treatment. *Biomaterials.* 246:119999.

Choudhary E, Thakur P, Pareek M, Agarwal N. 2015. Gene silencing by CRISPR interference in mycobacteria. *Nat Commun.* 6:6267.

Chung PY, Khanum R. 2017. Antimicrobial peptides as potential anti-biofilm agents against multidrug-resistant bacteria. *J Microbiol Immunol Infect.* 50(4):405–10.

Cisek AA, Dąbrowska I, Gregorczyk KP, Wyżewski Z. 2017. Phage therapy in bacterial infections treatment: One hundred years after the discovery of bacteriophages. *Curr Microbiol.* 74(2):277–83.

Criscuolo E, Spadini S, Lamanna J, Ferro M, Burioni R. 2017. Bacteriophages and their immunological applications against infectious threats. *J Immunol Res.* 2017:3780697.

Cusumano CK, Hultgren SJ. 2009. Bacterial adhesion – a source of alternate antibiotic targets. *IDrugs.* 12(11):699–705.

Dannehl C, Gutsmann T, Brezesinski G. 2013. Surface activity and structures of two fragments of the human antimicrobial LL-37. *Colloids Surf B Biointerfaces.* 109:129–35.

de Breij A, Riool M, CordfunkeRA, MalanovicN, de BoerL, Koning RI, Ravensbergen E, Franken M, van der Heijde T, Boekema BK, Kwakman PHS, Kamp N, El Ghalbzouri A, Lohner K, Zaat SAJ, Drijfhout JW, Nibbering PH. 2018. The antimicrobial peptide SAAP-148 combats drug-resistant bacteria and biofilms. *Sci Transl Med.* 10(423):eaan4044.

Dean SN, Bishop BM, van Hoek ML. 2011. Natural and synthetic cathelicidin peptides with anti-microbial and anti-biofilm activity against Staphylococcus aureus. *BMC Microbiol.* 11:1–13.

DiGiandomenico A, Keller AE, Gao C, Rainey GJ, Warrener P, Camara MM, Bonnell J, Fleming R, Bezabeh B, Dimasi N, Sellman BR, Hilliard J, Guenther CM, Datta V, Zhao W, Gao C, Yu XQ, Suzich JA, Stover CK. 2014. A multifunctional bispecific antibody protects against Pseudomonas aeruginosa. *Sci Transl Med.* 6(262):262ra155.

Dijkmans AC, Wilms EB, Kamerling IM, Birkhoff W, Ortiz-Zacarías NV, van Nieuwkoop C, Verbrugh HA, Touw DJ. 2015. Colistin: Revival of an old polymyxin antibiotic. *Ther Drug Monit.* 37(4):419–27.

Domadia P, Swarup S, Bhunia A, Sivaraman J, Dasgupta D. 2007. Inhibition of bacterial cell division protein FtsZ by cinnamaldehyde. *Biochem Pharmacol.* 74(6):831–40.

Domingo-Calap P, Delgado-Martínez J. 2018. Bacteriophages: Protagonists of a post-antibiotic era. *Antibiotics (Basel).* 7(3):66.

Dosler S, Karaaslan E. 2014. Inhibition and destruction of Pseudomonas aeruginosa biofilms by antibiotics and antimicrobial peptides. *Peptides.* 62:32–7.

Dula S, Ajayeoba TA, Ijabadeniyi OA. 2021. Bacterial biofilm formation on stainless steel in the food processing environment and its health implications. *Folia Microbiol (Praha).* 66(3):293–302.

Enioutina EY, Teng L, Fateeva TV, Brown JCS, Job KM, Bortnikova VV, Krepkova LV, Gubarev MI, Sherwin CMT. 2017. Phytotherapy as an alternative to conventional antimicrobials: Combating microbial resistance. *Expert Rev Clin Pharmacol.* 10(11):1203–14.

Gagliotti C, Buttazzi R, Di Mario S, Morsillo F, Moro ML. 2015. A regionwide intervention to promote appropriate antibiotic use in children reversed trends in erythromycin resistance to Streptococcus pyogenes. *Acta Paediatr.* 104(9):e422–4.

Galbadage T, Liu D, Alemany LB, Pal R, Tour JM, Gunasekera RS, Cirillo JD. 2019. Molecular nanomachines disrupt bacterial cell wall, increasing sensitivity of extensively drug-resistant *Klebsiella pneumoniae* to meropenem. *ACS Nano.* 13(12):14377–87.

Garneau JE, Dupuis MÈ, Villion M, Romero DA, Barrangou R, Boyaval P, Fremaux C, Horvath P, Magadán AH, Moineau S. 2010. The CRISPR/cas bacterial immune system cleaves bacteriophage and plasmid DNA. *Nature.* 468(7320):67–71.

Geilich BM, Gelfat I, Sridhar S, van de Ven AL, Webster TJ. 2017. Superparamagnetic iron oxide-encapsulating polymersome nanocarriers for biofilm eradication. *Biomaterials.* 119:78–85.

Geitani R, Ayoub Moubareck C, Touqui L, Karam Sarkis D. 2019. Cationic antimicrobial peptides: Alternatives and/or adjuvants to antibiotics active against methicillin-resistant Staphylococcus aureus and multidrug-resistant Pseudomonas aeruginosa. *BMC Microbiol.* 19(1):54.

Gomaa AA, Klumpe HE, Luo ML, Selle K, Barrangou R, Beisel CL. 2014. Programmable removal of bacterial strains by use of genome-targeting CRISPR-Cas systems. *mBio.* 5(1):e00928–13.

Goossens H, Ferech M, Vander Stichele R, Elseviers M, ESAC Project Group. 2005. Outpatient antibiotic use in Europe and association with resistance: A cross-national database study. *Lancet.* 365(9459):579–87.

Heilmann KP, Rice CL, Miller AL, Miller NJ, Beekmann SE, Pfaller MA, Richter SS, Doern GV. 2005. Decreasing prevalence of beta-lactamase production among respiratory tract isolates of Haemophilus influenzae in the United States. *Antimicrob Agents Chemother.* 49(6):2561–4.

Hoch JA. 2000. Two-component and phosphorelay signal transduction. *Curr Opin Microbiol.* 3(2):165–70.

Hou AW, Morrill AM. 2017. Obiltoxaximab: Adding to the treatment arsenal for bacillus anthracis infection. *Ann Pharmacother.* 51(10):908–13.

Huang Y, Ding X, Qi Y, Yu B, Xu FJ. 2016. Reduction-responsive multifunctional hyperbranched polyaminoglycosides with excellent antibacterial activity, biocompatibility and gene transfection capability. *Biomaterials.* 106:134–43.

Ianiro G, Maida M, Burisch J, Simonelli C, Hold G, Ventimiglia M, Gasbarrini A, Cammarota G. 2018. Efficacy of different faecal microbiota transplantation protocols for *Clostridium difficile* infection: A systematic review and meta-analysis. *United European Gastroenterol J.* 6(8):1232–44.

Insua I, Majok S, Peacock AF, Krachler AM, Fernandez-Trillo F. 2017a. Preparation and antimicrobial evaluation of polyion complex (PIC) nanoparticles loaded with polymyxin B. *Eur Polym J*. 87:478–86.

Insua I, Zizmare L, Peacock AFA, Krachler AM, Fernandez-Trillo F. 2017b. Polymyxin B containing polyion complex (PIC) nanoparticles: Improving the antimicrobial activity by tailoring the degree of polymerisation of the inert component. *Sci Rep*. 7(1):9396.

Jacob B, Park IS, Bang JK, Shin SY. 2013. Short KR-12 analogs designed from human cathelicidin LL-37 possessing both antimicrobial and antiendotoxic activities without mammalian cell toxicity. *J Pept Sci*. 19(11):700–7.

Jansen KU, Knirsch C, Anderson AS. 2018. The role of vaccines in preventing bacterial antimicrobial resistance. *Nat Med*. 24(1):10–19.

Jing L, Lei Z, Li L, Xie R, Xi W, Guan Y, Sumner LW, Zhou Z. 2014. Antifungal activity of citrus essential oils. *J Agric Food Chem*. 62(14):3011–33.

Kačániová M, Terentjeva M, Galovičová L, Ivanišová E, Štefániková J, Valková V, Borotová P, Kowalczewski PŁ, Kunová S, Felšöciová S, Tvrdá E, Žiarovská J, Benda Prokeinová R, Vukovic N. 2020. Biological activity and antibiofilm molecular profile of *Citrus aurantium* essential oil and its application in a food model. *Molecules*. 25(17):3956.

Kalia VC. 2013. Quorum sensing inhibitors: An overview. *Biotechnol Adv*. 31(2):224–45.

Kang J, Dietz MJ, Li B. 2019. Antimicrobial peptide LL-37 is bactericidal against Staphylococcus aureus biofilms. *PLoS One*. 14(6):e0216676.

Kanthawong S, Bolscher JG, Veerman EC, van Marle J, Nazmi K, Wongratanacheewin S, Taweechaisupapong S. 2010. Antimicrobial activities of LL-37 and its truncated variants against Burkholderia thailandensis. *Int J Antimicrob Agents*. 36(5):447–52.

Keller MA, Stiehm ER. 2000. Passive immunity in prevention and treatment of infectious diseases. *Clin Microbiol Rev*. 13(4):602–14.

Kerekes EB, Deák É, Takó M, Tserennadmid R, Petkovits T, Vágvölgyi C, Krisch J. 2013. Anti-biofilm forming and anti-quorum sensing activity of selected essential oils and their main components on food-related microorganisms. *J Appl Microbiol*. 115(4):933–42.

Kim YG, Lee JH, Kim SI, Baek KH, Lee J. 2015. Cinnamon bark oil and its components inhibit biofilm formation and toxin production. *Int J Food Microbiol*. 195:30–9.

Klatt TE, Hopp E. 2012. Effect of a best-practice alert on the rate of influenza vaccination of pregnant women. *Obstet Gynecol*. 119(2 Pt 1):301–5.

Kortright KE, Chan BK, Koff JL, Turner PE. 2019. Phage therapy: A renewed approach to combat antibiotic-resistant bacteria. *Cell Host Microbe*. 25(2):219–32.

Kufel WD, Devanathan AS, Marx AH, Weber DJ, Daniels LM. 2017. Bezlotoxumab: A novel agent for the prevention of recurrent clostridium difficile infection. *Pharmacotherapy*. 37(10):1298–308.

Kumaraswamy M, Lin L, Olson J, Sun CF, Nonejuie P, Corriden R, Döhrmann S, Ali SR, Amaro D, Rohde M, Pogliano J, Sakoulas G, Nizet V. 2016. Standard susceptibility testing overlooks potent azithromycin activity and cationic peptide synergy against MDR Stenotrophomonas maltophilia. *J Antimicrob Chemother*. 71(5):1264–9.

Lahiri D, Nag M, Dey A, Pandit S, Joshi S, Upadhye VJ, Ray RR. 2022. Citrus essential oils: A treasure trove of antibiofilm agent. *Appl Biochem Biotechnol*. 194(10):4625–38.

Le J, Dam Q, Schweizer M, Thienphrapa W, Nizet V, Sakoulas G. 2016. Effects of vancomycin versus nafcillin in enhancing killing of methicillin-susceptible Staphylococcus aureus causing bacteremia by human cathelicidin LL-37. *Eur J Clin Microbiol Infect Dis*. 35(9):1441–7.

Lerminiaux NA, Cameron ADS. 2019. Horizontal transfer of antibiotic resistance genes in clinical environments. *Can J Microbiol*. 65(1):34–44.

Leszczyńska K, Namiot A, Janmey PA, Bucki R. 2010. Modulation of exogenous antibiotic activity by host cathelicidin LL-37. *APMIS*. 118(11):830–6.

Liao S, Zhang Y, Pan X, Zhu F, Jiang C, Liu Q, Cheng Z, Dai G, Wu G, Wang L, Chen L. 2019. Antibacterial activity and mechanism of silver nanoparticles against multidrug-resistant *Pseudomonas aeruginosa*. *Int J Nanomedicine*. 14:1469–87.

Lin L, Nonejuie P, Munguia J, Hollands A, Olson J, Dam Q, Kumaraswamy M, Rivera H Jr, Corriden R, Rohde M, Hensler MÈ, Burkart MD, Pogliano J, Sakoulas G, Nizet V. 2015. Azithromycin synergizes with cationic antimicrobial peptides to exert bactericidal and therapeutic activity against highly multidrug-resistant gram-negative bacterial pathogens. *EBioMedicine*. 2(7):690–8.

Liu Y, Li R, Xiao X, Wang Z. 2019a. Antibiotic adjuvants: An alternative approach to overcome multi-drug resistant Gram-negative bacteria. *Crit Rev Microbiol*. 45(3):301–314.

Liu Y, Li S, Li W, Wang P, Ding P, Li L, Wang J, Yang P, Wang Q, Xu T, Xiong Y, Yang B. 2019b. RstA, a two-component response regulator, plays important roles in multiple virulence-associated processes in enterohemorrhagic *Escherichia coli* O157:H7. *Gut Pathog*. 11:53.

Livermore DM, Woodford N. 2006. The beta-lactamase threat in enterobacteriaceae, pseudomonas and acinetobacter. *Trends Microbiol*. 14(9):413–20.

Llopis-Lorente A, García-Fernández A, Murillo-Cremaes N, Hortelão AC, Patiño T, Villalonga R, Sancenón F, Martínez-Máñez R, Sánchez S. 2019. Enzyme-powered gated mesoporous silica nanomotors for on-command intracellular payload delivery. *ACS Nano*. 13(10):12171–183.

Loc-Carrillo C, Abedon ST. 2011. Pros and cons of phage therapy. *Bacteriophage*. 1(2):111–14.

Lomovskaya O, Warren MS, Lee A, Galazzo J, Fronko R, Lee M, Blais J, Cho D, Chamberland S, Renau T, Leger R, Hecker S, Watkins W, Hoshino K, Ishida H, Lee VJ. 2001. Identification and characterization of inhibitors of multidrug resistance efflux pumps in Pseudomonas aeruginosa: Novel agents for combination therapy. *Antimicrob Agents Chemother*. 45(1):105–16.

Luciardi MC, Blázquez MA, Alberto MR, Cartagena E, Arena MÈ. 2020. Grapefruit essential oils inhibit quorum sensing of Pseudomonas aeruginosa. *Food Sci Technol Int*. 26(3):231–41.

Maciejewska B, Olszak T, Drulis-Kawa Z. 2018. Applications of bacteriophages versus phage enzymes to combat and cure bacterial infections: An ambitious and also a realistic application? *Appl Microbiol Biotechnol*. 102(6):2563–581.

Mahato N, Sharma K, Koteswararao R, Sinha M, Baral E, Cho MH. 2019. Citrus essential oils: Extraction, authentication and application in food preservation. *Crit Rev Food Sci Nutr*. 59(4):611–25.

Maiden MM, Hunt AMA, Zachos MP, Gibson JA, Hurwitz MÈ, Mulks MH, Waters CM. 2018. Triclosan is an aminoglycoside adjuvant for eradication of pseudomonas aeruginosa biofilms. *Antimicrob Agents Chemother*. 62(6):e00146–18.

Maura D, Ballok AE, Rahme LG. 2016. Considerations and caveats in anti-virulence drug development. *Curr Opin Microbiol*. 33:41–46.

McConnell MJ. 2019. Where are we with monoclonal antibodies for multidrug-resistant infections? *Drug Discov Today*. 24(5):1132–38.

Mishra B, Wang G. 2017a. Titanium surfaces immobilized with the major antimicrobial fragment FK-16 of human cathelicidin LL-37 are potent against multiple antibiotic-resistant bacteria. *Biofouling*. 33(7):544–55.

Mishra B, Wang G. 2017b. Individual and combined effects of engineered peptides and antibiotics on pseudomonas aeruginosa biofilms. *Pharmaceuticals (Basel)*. 10(3):58.

Mishra RP, Oviedo-Orta E, Prachi P, Rappuoli R, Bagnoli F.2012. Vaccines and antibiotic resistance. *Curr Opin Microbiol*. 15(5):596–602.

Moffatt JH, Harper M, Boyce JD. 2019. Mechanisms of polymyxin resistance. *Adv Exp Med Biol*. 1145:55–71.

Munita JM, Arias CA. 2016. Mechanisms of antibiotic resistance. *Microbiol Spectr*. 4(2). 10.1128/microbiolspec.VMBF-0016-2015.

Nagant C, Pitts B, Nazmi K, Vandenbranden M, Bolscher JG, Stewart PS, Dehaye JP. 2012. Identification of peptides derived from the human antimicrobial peptide LL-37 active against biofilms formed by Pseudomonas aeruginosa using a library of truncated fragments. *Antimicrob Agents Chemother*. 56(11):5698–708.

Naskar A, Kim KS. 2019. Nanomaterials as delivery vehicles and components of new strategies to combat bacterial infections: Advantages and limitations. *Microorganisms*. 7(9):356.

Navarre WW, Halsey TA, Walthers D, Frye J, McClelland M, Potter JL, Kenney LJ, Gunn JS, Fang FC, Libby SJ. 2005. Co-regulation of Salmonella enterica genes required for virulence and resistance to antimicrobial peptides by SlyA and PhoP/PhoQ. *Mol Microbiol*. 56(2):492–508.

Negi N, Prakash P, Gupta ML, Mohapatra TM. 2014. Possible role of curcumin as an efflux pump inhibitor in multi-drug resistant clinical isolates of Pseudomonas aeruginosa. *J Clin Diagn Res*. 8(10):DC04-7.

Neshani Alireza, Zare Hosna, Eidgahi Mohammad Reza Akbari, Kakhki Reza Kamali, Safdari Hadi, Khaledi Azad, Ghazvini Kiarash.2019. LL-37: Review of antimicrobial profile against sensitive and antibiotic-resistant human bacterial pathogens. *Gene Reports*. 17:100519.

Neville N, Jia Z. 2019. Approaches to the structure-based design of antivirulence drugs: Therapeutics for the post-antibiotic era. *Molecules*. 24(3):378.

Niemirowicz K, Piktel E, Wilczewska AZ, Markiewicz KH, Durnaś B, Wątek M, Puszkarz I, Wróblewska M, Niklińska W, Savage PB, Bucki R. 2016. Core-shell magnetic nanoparticles display synergistic antibacterial effects against *Pseudomonas aeruginosa* and *Staphylococcus aureus* when combined with cathelicidin LL-37 or selected ceragenins. *Int J Nanomedicine*. 11:5443–455.

Nishie M, Nagao J, Sonomoto K. 2012. Antibacterial peptides "bacteriocins": An overview of their diverse characteristics and applications. *Biocontrol Sci.* 17(1):1–16.

Nowakowski M, Jaremko Ł, Wladyka B, Dubin G, Ejchart A, Mak P. 2018. Spatial attributes of the four-helix bundle group of bacteriocins—The high-resolution structure of BacSp222 in solution. *Int J Biol Macromol.* 107(Pt B):2715–724.

Nuding S, Frasch T, Schaller M, Stange EF, Zabel LT. 2014. Synergistic effects of antimicrobial peptides and antibiotics against Clostridium difficile. *Antimicrob Agents Chemother.* 58(10):5719–25.

Oganesyan V, Peng L, Damschroder MM, Cheng L, Sadowska A, Tkaczyk C, Sellman BR, Wu H, Dall'Acqua WF. 2014. Mechanisms of neutralization of a human anti-α-toxin antibody. *J Biol Chem.* 289(43):29874–80.

Oliveira SA, Zambrana JR, Iorio FB, Pereira CA, Jorge AO. 2014. The antimicrobial effects of Citrus limonum and Citrus aurantium essential oils on multi-species biofilms. *Braz Oral Res.* 28:22–7.

Olsen I. 2016. Modification of phage for increased antibacterial effect towards dental biofilm. *J Oral Microbiol.* 8:33089.

Osei Sekyere J, Govinden U, Bester LA, Essack SY. 2016. Colistin and tigecycline resistance in carbapenemase-producing Gram-negative bacteria: Emerging resistance mechanisms and detection methods. *J Appl Microbiol.* 121(3):601–17.

Overhage J, Campisano A, Bains M, Torfs EC, Rehm BH, Hancock RE. 2008. Human host defense peptide LL-37 prevents bacterial biofilm formation. *Infect Immun.* 76(9):4176–82.

Pardo-Esté C, Hidalgo AA, Aguirre C, Briones AC, Cabezas CE, Castro-Severyn J, Fuentes JA, Opazo CM, Riedel CA, Otero C, Pacheco R, Valvano MA, Saavedra CP. 2018. The ArcAB two-component regulatory system promotes resistance to reactive oxygen species and systemic infection by Salmonella Typhimurium. *PLoS One.* 13(9):e0203497.

Patel S, Ashwanikumar N, Robinson E, DuRoss A, Sun C, Murphy-Benenato KE, Mihai C, Almarsson Ö, Sahay G. 2017. Boosting intracellular delivery of lipid nanoparticle-encapsulated mRNA. *Nano Lett.* 17(9):5711–718.

Payne JE, Dubois AV, Ingram RJ, Weldon S, Taggart CC, Elborn JS, Tunney MM. 2017. Activity of innate antimicrobial peptides and ivacaftor against clinical cystic fibrosis respiratory pathogens. *Int J Antimicrob Agents.* 50(3):427–35.

Peters JM, Colavin A, Shi H, Czarny TL, Larson MH, Wong S, Hawkins JS, Lu CHS, Koo BM, Marta E, Shiver AL, Whitehead EH, Weissman JS, Brown ED, Qi LS, Huang KC, Gross CA. 2016. A comprehensive, CRISPR-based functional analysis of essential genes in bacteria. *Cell.* 165(6):1493–506.

Pires DP, Melo L, Vilas Boas D, Sillankorva S, Azeredo J. 2017. Phage therapy as an alternative or complementary strategy to prevent and control biofilm-related infections. *Curr Opin Microbiol.* 39:48–56.

Quan K, Zhang Z, Chen H, Ren X, Ren Y, Peterson BW, van der Mei HC, Busscher HJ. 2019. Artificial channels in an infectious biofilm created by magnetic nanoparticles enhanced bacterial killing by antibiotics. *Small.* 15(39):e1902313.

Rappuoli R.2001. Reverse vaccinology, a genome-based approach to vaccine development. *Vaccine.* 19(17–19):2688–91.

Reardon S. 2014. Phage therapy gets revitalized. *Nature.* 510(7503):15–6.

Reffuveille F, de la Fuente-Núñez C, Mansour S, Hancock RE. 2014. A broad-spectrum antibiofilm peptide enhances antibiotic action against bacterial biofilms. *Antimicrob Agents Chemother.* 58(9):5363–71.

Ribeiro da Cunha B, Fonseca LP, Calado CRC. 2019. Antibiotic discovery: Where have we come from, where do we go? *Antibiotics (Basel).* 8(2):45.

Ridyard KE, Overhage J. 2021. The potential of human peptide LL-37 as an antimicrobial and anti-biofilm agent. *Antibiotics (Basel).* 10(6):650.

Royer S, Morais AP, da Fonseca Batistão DW. 2021. Phage therapy as strategy to face post-antibiotic era: A guide to beginners and experts. *Arch Microbiol.* 203(4):1271–79.

Sakoulas G, Bayer AS, Pogliano J, Tsuji BT, Yang SJ, Mishra NN, Nizet V, Yeaman MR, Moise PA. 2012. Ampicillin enhances daptomycin- and cationic host defense peptide-mediated killing of ampicillin- and vancomycin-resistant Enterococcus faecium. *Antimicrob Agents Chemother.* 56(2):838–44.

Sakoulas G, Rose W, Berti A, Olson J, Munguia J, Nonejuie P, Sakoulas E, Rybak MJ, Pogliano J, Nizet V. 2017. Classical β-Lactamase Inhibitors Potentiate the Activity of Daptomycin against Methicillin-Resistant Staphylococcus aureus and Colistin against Acinetobacter baumannii. *Antimicrob Agents Chemother.* 61(2):e01745–16.

Santos RS, Figueiredo C, Azevedo NF, Braeckmans K, De Smedt SC. 2018. Nanomaterials and molecular transporters to overcome the bacterial envelope barrier: Towards advanced delivery of antibiotics. *Adv Drug Deliv Rev.* 136–137:28–48.

São-José C. 2018. Engineering of phage-derived lytic enzymes: Improving their potential as antimicrobials. *Antibiotics*. 7:29.

Saporito P, Vang Mouritzen M, Løbner-Olesen A, Jenssen H. 2018. LL-37 fragments have antimicrobial activity against Staphylococcus epidermidis biofilms and wound healing potential in HaCaT cell line. *J Pept Sci*. 24(7):e3080.

Sartor RB. 2005. Probiotic therapy of intestinal inflammation and infections. *Curr Opin Gastroenterol*. 21(1):44–50.

Selim SA, Adam MÈ, Hassan SM, Albalawi AR. 2014. Chemical composition, antimicrobial and antibiofilm activity of the essential oil and methanol extract of the Mediterranean cypress (Cupressus sempervirens L.). *BMC Complement Altern Med*. 14:179.

Shogbesan O, Poudel DR, Victor S, Jehangir A, Fadahunsi O, Shogbesan G, Donato A. 2018. A systematic review of the efficacy and safety of fecal microbiota transplant for *Clostridium difficile* infection in immunocompromised patients. *Can J Gastroenterol Hepatol*. 2018:1394379.

Shurko JF, Galega RS, Li C, Lee GC. 2018. Evaluation of LL-37 antimicrobial peptide derivatives alone and in combination with vancomycin against S. aureus. *J Antibiot (Tokyo)*. 71(11):971–74.

Sidik SM, Hackett CG, Tran F, Westwood NJ, Lourido S. 2014. Efficient genome engineering of Toxoplasma gondii using CRISPR/Cas9. *PLoS One*. 9(6):e100450.

Smith JR, Barber KE, Raut A, Rybak MJ. 2015. β-Lactams enhance daptomycin activity against vancomycin-resistant Enterococcus faecalis and Enterococcus faecium in in vitro pharmacokinetic/pharmacodynamic models. *Antimicrob Agents Chemother*. 59(5):2842–8.

Song DW, Kim SH, Kim HH, Lee KH, Ki CS, Park YH. 2016. Multi-biofunction of antimicrobial peptide-immobilized silk fibroin nanofiber membrane: Implications for wound healing. *Acta Biomater*. 39:146–55.

Srivastava RK, Bothra N, Singh R, Sai MC, Nedungadi SV, Sarangi PK. 2022. Microbial originated surfactants with multiple applications: A comprehensive review. *Arch Microbiol*. 204(8):452.

Streicher LM. 2021. Exploring the future of infectious disease treatment in a post-antibiotic era: A comparative review of alternative therapeutics. *J Glob Antimicrob Resist*. 24:285–95.

Su L, Li Y, Liu Y, An Y, Shi L. 2019. Recent advances and future prospects on adaptive biomaterials for antimicrobial applications. *Macromol Biosci*. 19(12):e1900289.

Sun Y, Chen S, Zhang C, Liu Y, Ma L, Zhang X. 2018. Effects of sub-minimum inhibitory concentrations of lemon essential oil on the acid tolerance and biofilm formation of Streptococcus mutans. *Arch Oral Biol*. 87:235–41.

Tang S, Zheng J. 2018. Antibacterial activity of silver nanoparticles: Structural effects. *Adv Healthc Mater*. 7(13):e1701503.

Taniguchi Y, Maeyama Y, Ohsaki Y, Hayashi W, Osaka S, Koide S, Tamai K, Nagano Y, Arakawa Y, Nagano N. 2017. Co-resistance to colistin and tigecycline by disrupting mgrB and ramR with IS insertions in a canine Klebsiella pneumoniae ST37 isolate producing SHV-12, DHA-1 and FosA3. *Int J Antimicrob Agents*. 50(5):697–98.

Tekle YI, Nielsen KM, Liu J, Pettigrew MM, Meyers LA, Galvani AP, Townsend JP. 2012. Controlling antimicrobial resistance through targeted, vaccine-induced replacement of strains. *PLoS One*. 7(12):e50688.

Thomson JM, Bonomo RA. 2005. The threat of antibiotic resistance in Gram-negative pathogenic bacteria: Beta-lactams in peril! *Curr Opin Microbiol*. 8(5):518–24.

Totsika M. 2017. Disarming pathogens: Benefits and challenges of antimicrobials that target bacterial virulence instead of growth and viability. *Future Med Chem*. 9(3):267–69.

Tran TB, Wang J, Doi Y, Velkov T, Bergen PJ, Li J. 2018. Novel polymyxin combination with antineoplastic mitotane improved the bacterial killing against polymyxin-resistant multidrug-resistant gram-negative pathogens. *Front Microbiol*. 9:721.

Tsai CW, Morris S. 2015. Approval of Raxibacumab for the Treatment of Inhalation Anthrax Under the US Food and Drug Administration "Animal Rule". *Front Microbiol*. 6:1320.

Ulloa ER, Dillon N, Tsunemoto H, Pogliano J, Sakoulas G, Nizet V. 2019. Avibactam sensitizes carbapenem-resistant NDM-1-producing klebsiella pneumoniae to innate immune clearance. *J Infect Dis*. 220(3):484–93.

Valério N, Oliveira C, Jesus V, Branco T, Pereira C, Moreirinha C, Almeida A. 2017. Effects of single and combined use of bacteriophages and antibiotics to inactivate Escherichia coli. *Virus Res*. 240:8–17.

van de Sande-Bruinsma N, Grundmann H, Verloo D, Tiemersma E, Monen J, Goossens H, Ferech M. 2008. Antimicrobial drug use and resistance in Europe. *Emerg Infect Dis*. 14(11):1722–30.

Van Hoogmoed CG, Geertsema-Doornbusch GI, Teughels W, Quirynen M, Busscher HJ, Van der Mei HC. 2008. Reduction of periodontal pathogens adhesion by antagonistic strains. *Oral Microbiol Immunol*. 23(1):43–8.

Varshney AK, Kuzmicheva GA, Lin J, Sunley KM, Bowling RA Jr, Kwan TY, Mays HR, Rambhadran A, Zhang Y, Martin RL, Cavalier MC, Simard J, Shivaswamy S. 2018. A natural human monoclonal antibody targeting Staphylococcus Protein A protects against Staphylococcus aureus bacteremia. *PLoS One*. 13(1):e0190537.

Wan F, Draz MS, Gu M, Yu W, Ruan Z, Luo Q. 2021. Novel strategy to combat antibiotic resistance: A sight into the combination of CRISPR/Cas9 and nanoparticles. *Pharmaceutics*. 13(3):352.

Wang CH, Hsieh YH, Powers ZM, Kao CY. 2020a. Defeating antibiotic-resistant bacteria: Exploring alternative therapies for a post-antibiotic era. *Int J Mol Sci*. 21(3):1061.

Wang JW, Kuo CH, Kuo FC, Wang YK, Hsu WH, Yu FJ, Hu HM, Hsu PI, Wang JY, Wu DC. 2019. Fecal microbiota transplantation: Review and update. *J Formos Med Assoc*. 118(Suppl 1):S23–S31.

Wang L, Hu C, Shao L. 2017. The antimicrobial activity of nanoparticles: Present situation and prospects for the future. *Int J Nanomedicine*. 12:1227–49.

Wang LS, Gupta A, Rotello VM. 2016. Nanomaterials for the treatment of bacterial biofilms. *ACS Infect Dis*. 2(1):3–4.

Wang Q, Zhang P, Zhao D, Jiang Y, Zhao F, Wang Y, Li X, Du X, Yu Y. 2018. Emergence of tigecycline resistance in *Escherichia coli* co-producing MCR-1 and NDM-5 during tigecycline salvage treatment. *Infect Drug Resist*. 11:2241–48.

Wang S, Gao Y, Jin Q, Ji J. 2020b. Emerging antibacterial nanomedicine for enhanced antibiotic therapy. *Biomater Sci*. 8(24):6825–39.

Wang Y, Zhao Q, Han N, Bai L, Li J, Liu J, Che E, Hu L, Zhang Q, Jiang T, Wang S. 2015. Mesoporous silica nanoparticles in drug delivery and biomedical applications. *Nanomedicine*. 11(2):313–27.

Wang-Lin SX, Balthasar JP. 2018. Pharmacokinetic and pharmacodynamic considerations for the use of monoclonal antibodies in the treatment of bacterial infections. *Antibodies (Basel)*. 7(1):5.

Wanmakok M, Orrapin S, Intorasoot A, Intorasoot S. 2018. Expression in Escherichia coli of novel recombinant hybrid antimicrobial peptide AL32-P113 with enhanced antimicrobial activity in vitro. *Gene*. 671:1–9.

Wilcox MH, Gerding DN, Poxton IR, Kelly C, Nathan R, Birch T, Cornely OA, Rahav G, Bouza E, Lee C, Jenkin G, Jensen W, Kim YS, Yoshida J, Gabryelski L, Pedley A, Eves K, Tipping R, Guris D, Kartsonis N, Dorr MB; MODIFY I and MODIFY II Investigators. 2017. Bezlotoxumab for prevention of recurrent clostridium difficile infection. *N Engl J Med*. 376(4):305–17.

Wright GD. 2005. Bacterial resistance to antibiotics: Enzymatic degradation and modification. *Adv Drug Deliv Rev*. 57(10):1451–70.

Yasir M, Willcox MDP, Dutta D. 2018. Action of antimicrobial peptides against bacterial biofilms. *Materials (Basel)*. 11(12):2468.

Yazici A, Ortucu S, Taskin M, Marinelli L. 2018. Natural-based antibiofilm and antimicrobial peptides from microorganisms. *Curr Top Med Chem*. 18(24):2102–7.

Yenn TW, Arslan Khan M, Amiera Syuhada N, Chean Ring L, Ibrahim D, Tan WN. 2017. Stigmasterol: An adjuvant for beta lactam antibiotics against beta-lactamase positive clinical isolates. *Steroids*. 128:68–71.

Zhang R, Dong N, Huang Y, Zhou H, Xie M, Chan EW, Hu Y, Cai J, Chen S. 2018. Evolution of tigecycline- and colistin-resistant CRKP (carbapenem-resistant Klebsiella pneumoniae) *in vivo* and its persistence in the GI tract. *Emerg Microbes Infect*. 7(1):127.

Zharkova MS, Orlov DS, Golubeva OY, Chakchir OB, Eliseev IE, Grinchuk TM, Shamova OV. 2019. Application of antimicrobial peptides of the innate immune system in combination with conventional antibiotics–A novel way to combat antibiotic resistance? *Front Cell Infect Microbiol*. 9:128.

Zhou C, Cai H, Baruch A, Lewin-Koh N, Yang M, Guo F, Xu D, Deng R, Hazenbos W, Kamath AV. 2019. Sustained activity of novel THIOMAB antibody-antibiotic conjugate against Staphylococcus aureus in a mouse model: Longitudinal pharmacodynamic assessment by bioluminescence imaging. *PLoS One*. 14(10):e0224096.

Zschiedrich CP, Keidel V, Szurmant H. 2016. Molecular mechanisms of two-component signal transduction. *J Mol Biol*. 428(19):3752–75.

Zurawski DV, Reinhart AA, Alamneh YA, Pucci MJ, Si Y, Abu-Taleb R, Shearer JP, Demons ST, Tyner SD, Lister T. 2017. SPR741, an antibiotic adjuvant, potentiates the *In Vitro* and *In Vivo*activity of rifampin against clinically relevant extensively drug-resistant acinetobacter baumannii. *Antimicrob Agents Chemother*. 61(12):e01239–17.

3 Microbial Biofilms
Medical Importance and Management

Sayak Bhattacharya
Bijoy Krishna Girls' College, Howrah, India

CONTENTS

3.1 INTRODUCTION

Scientists have been working on biofilm formation by bacteria for several years. This term 'film' means adhesion and growth on surfaces. It was first applied on marine microbes to understand the differences between immobilised bacteria and free swimming 'planktonic' bacteria as early as 1933 (ZoBell and Allen, 1935; Henrici, 1933). Biofilm evolved almost 3.25 billion years ago. Both bacteria and archaea form biofilms. It is now a common realisation that the maximum number of

DOI: 10.1201/9781003345299-3

bacterial form biofilms than in isolated cells (Kolter and Greenberg, 2006). Microbial biofilms are nothing but the slimy layer of microorganisms that coat solid surfaces. The importance of biofilms is increased due to their capacity to cause several chronic diseases, antibiotic resistance, waste-water treatment, and biofouling (Heijnen et al., 1993). Cells can exchange secreted molecules, extracellular polymers, and siderophores among themselves in closely positions in biofilms. They are called 'cheater' because they consume resources from others but do not give to the group pool (West et al., 2006; Foster, 2005).

Biofilms are formed when they embed themselves within the extracellular polymeric substances (EPS). These polymers provide biofilm to maintain its proper structure (Boyd and Chakrabarty, 1994; Hughes et al., 1998). EPS protects against dehydration, ultraviolet radiation, and predator grazing (Crespi, 2001; Velicer, 2003; Keller and Surette, 2006;). It also facilitates signalling and enzymatic activity (West et al., 2006).

Several diseases are caused by microorganisms which form biofilm. These include *Mycoplasma pneumoniae*, *Pseudomonas aeruginosa*, *Mycobacterium tuberculosis*, *Candida albicans*, and *Mycobacterium abscessus*. All these microbes are able of causing serious health issues because they resist antimicrobial drugs and host immunity (Kumar et al., 2017). Both device- and tissue-related infections all are linked with formation of biofilm (Khatoon et al., 2018).

3.2 INFECTIONS ASSOCIATED WITH BIOFILM

Bacterial biofilm causes 65% of all bacterial infections (Lewis, 2001). Attachment occurs with host cells due to the presence of adhesive factors which are listed below.

3.2.1 Protein Adhesions

Microbial surface components recognising adhesive matrix molecules cause initial attachment with target cells. These surface molecules (mainly available in Gram-positive bacteria) detect and stick with certain target ECM components, like collagen, fibronectin, and lectin (Vengadesan and Narayana, 2011). The outer membrane protein (OMP) group acts as adhesive proteins in Gram-negative bacteria (Soto and Hultgren, 1999; Koebnik et al., 2000). These have reportedly been available in the ECM of *Pseudomonas aeruginosa*, *V. cholerae*, and *E. coli* (Orme et al., 2006; Fong and Yildiz, 2015; Toyofuku et al., 2012).

Several adhesive factors of *Staphylococcus aureus* aid to attach to host components (Gotz, 2002). Bacteria attach to implant surfaces by host plasma and other extracellular components. Virulence factors like fibrinogen-binding proteins, collagen-binding protein (Cna) (Switalski et al., 1993), and clumping factor A and B (ClfA and ClfB) are produced (Ni Eidhin et al., 1998) by *S. aureus*. These compounds bind with host plasma and extracellular matrix (ECM) components. Biofilm formation happens after attachment of bacteria to the surfaces. Then proliferation, aggregation, and recruitment of cells from those surrounding occur on those surfaces (Stoodley et al., 2002).

3.2.2 Adhesive Pili

Gram-positive and Gram-negative bacteria contain adhesive pili that help to bind to the cell surface for binding to abiotic surfaces and host cells. Characterisation of Pili in Gram-negative had already been carried out, but pili of Gram-positive was characterised only recently (Lauer et al., 2005). *Corynebacterium diptheriae* contains pilus system that is sortase-dependent (Ton-That and Schneewind, 2003). After translocation, a pilus-specific sortase enzyme cleaves the sorting signal. *Enterococcus faecalis* and *Streptococcus pneumoniae* contain sortase-dependent pili. *S. pneumonia* is adhered by the adhesin RrgA and the Rrg pili (Hilleringmann et al., 2008, 2009; Nelson et al., 2007). The chaperone-usher pathway (CUP) is utilised by *E. coli*, *Klebsiella* and *Pseudomonas* species to intricate multi subunit pili resolved by the tip adhesins (Fronzes et al., 2008; Busch and Waksman, 2012).

3.2.3 Amyloid Fibres

Amyloid fibres are a type of beta sheet fibre. These are encoded by several bacteria (Cohen and Kelly, 2003; Collinson et al., 1993).

3.2.4 Accumulation-Associated Protein

During biofilm formation on polymer surfaces, cell wall associated accumulation-associated protein (AAP) is expressed under sessile growth conditions (Hussain et al., 1997). Mutation studies showed that AAP played significant roles during biofilm production (Von Eiff et al., 2002). Device- and non-device-related infections were caused by bacterial biofilm. Breast implants (2%); Joint prostheses (2%); mechanical heart valves (4%); and about 40% for ventricular-assisted devices are caused by device-related infections (Darouiche 2004).

3.3 STRUCTURE OF BIOFILM

The structure of microbial biofilm has been investigated in detail.

3.3.1 Biofilm Formation of Living Tissues

Human adenoid tissues associated with mucosal biofilms were examined, and it was found that *Streptococcus salivarius*, *Corynebacterium argentoratense*, *Staphylococcus aureus*, and *Micrococcus luteus* form biofilm on mucosal surfaces (Kania et al., 2008). Similarly, FISH and SEM studies identified presence of polymicrobials like *Haemophilus influenzae*, *Moraxella catarrhalis*, and *Streptococcus pneumoniae*, respectively, in the middle ear (Hoa et al., 2009).

3.4 DEVICE-RELATED INFECTION

Surface-attached microbes face less challenge from biocides, antibiotics, and physical stress. These microbes can thrive at length on surfaces (Otter et al., 2015). Bacteria adopt different adhesion mechanisms which help them to attach to a surface after making contact. As biofilm was forms well on devices, it is difficult task to eradicate (Costerton et al., 2005). *E. coli*, *Staphylococcus epidermidis*, *P. aeruginosa*, species of *Serratia*, *Candida*, and *Proteus* can attach to contact lenses (Donlan and Costerton, 2002). It has been found that the architecture of biofilm formation can be changed depending on contact lenses (Imamura et al., 2008). A catheter is also a key device where biofilm formation occurs (Maki and Mermel, 1998). Biofilm formation differs on intravenous fluid. Gram-positive such as *Staphylococcus aureus*, *Staphylococcus epidermidis* cannot grow well there but Gram-negative bacteria like *Pseudomonas aeruginosa* and *Kleibsella* can grow well (Donlan et al., 1999). Strong biofilms in polystyrene microplates is formed by *Stenotrophomonas maltophilia*, while Borosilicate (BS) or polypropylene (PP) tubes contain strong, moderate, or weak biofilms (Passerini de Rossi et al., 2007). Explanted cardiac pacemaker devices (Okuda et al., 2018) and breast implants (Del Pozo et al., 2009) were covered by the biofilm of *Propionibacterium acne*. Chua et al., (1998) reported the endocarditis due to *Propionibacterium acnes*. Not only bacteria, but also fungi *Candida albicans* was able to form a drug-resistant biofilm on medical devices (Rajendran et al., 2010). *Candida* biofilms showed drug-resistant properties during the maturation process (Sardi et al., 2011). Genetic study of *Staphylococcus epidermidis* highlights the role of multiple physicochemical, protein, and polysaccharide factors (Heilmann et al., 1997). *Propionibacterium acnes form biofilm on lipid rich part of human skin* (Coenye et al., 2022).

3.5 NOSOCOMIAL INFECTION DUE TO PATHOGENIC BIOFILM

Biofilm formation is observed in clinically isolated *Acinetobacter baumannii* (Rodríguez-Baño et al., 2008). *Acinetobacter baumannii* has the capacity to form biofilm on hospital areas (Martí et al., 2011). Several nosocomial infections such as catheter-associated urinary tract infections (CAUTIs), lung infections, and prosthetic heart valve infections were caused in critical care units (CCU) associated by both bacterial and fungal biofilms (Koley and Mukherjee 2021). *A. baumannii* forms biofilm on ventilators in intensive care units (ICUs) (Pakharukova et al., 2018). However, the relation between biofilm formation and antibiotic resistance is still vague (Krzyściak et al., 2017).

Klebsiella pneumoniae causes infection with indwelling medical devices. *K. pneumonia* attaches to medical devices by type III fimbriae (Murphy and Clegg, 2012). Even hospital shower houses are suitable places for biofilm formation, by Mycobacterium like population (Soto-Giron et al., 2016). Enterococci are the most common nosocomial agents that infect the bloodstream, surgical sites, and urinary tract (Richards et al., 2000). Nosocomial infections caused by *Staphylococcus epidermidis* are among the five most frequent organisms. *S. epidermidis* protects itself by its own secretion of poly-gamma-DL-glutamic acid that gives protection from components of innate host defence (Kocianova et al., 2005). Significant research had been carried out to establish the mechanism behind biofilm formation and to control it (Mack et al., 2006).

Genetic factors of food-borne methicillin-resistant *Staphylococcus aureus* (MRSA) that are responsible for formation of biofilm were examined (Manago et al., 2006); 93.6% staphylococcal infections have been caused by MRSA (Nourbakhsh and Namvar, 2016); 53.5% of antibiotic resistant *Staphylococcus aureus* are high biofilm produces (ibid.). Pattern of biofilm formation by all pathogens is not the same. Confocal laser scanning microscopy pointed out the clear difference between the structures of biofilm of *S. haemolyticus* and *S. epidermidis* (Fredheim et al., 2009). Quorum-sensing (QS) system regulates the biofilm formation by *Serretia marcescens* (Rice et al., 2005).

3.6 MANAGEMENT OF BIOFILM FORMATION

From above discussions, it is clear that microorganisms deploy numerous virulent factors for biofilm formation. Management of biofilm formation can be achieved by targeting these factors.

3.6.1 Antipathogenic Agents—Targeting Quorum Sensing

Currently, new theories are emerging on antimicrobial chemotherapy and the origination of 'antipathogenic' agents. Quoram-sensing inhibitor drugs have more advantages than common drugs that either kill or retard microbial growth (Alksne and Projan, 2000). Previously, many experiments had been carried out to quench the quorum-sensing system of many Gram-negatives as therapeutic targets to prevent bacterial infection. Meta-bromo-thiolactone not only prevents the virulence factor and biofilm formation expression, but also gives defence to the host cells from quorum-sensing mediated killing by *P. aeruginosa* (O' Loughlin et al., 2013). A vast array of organisms including bacteria, fungi, plants, and animals produces quorum-sensing inhibitors (Musthafa et al., 2010; Koh et al., 2013; Natrah et al., 2011; Tello et al., 2011).

Aeromonas hydrophila and *Serratia liquefaciens* form abnormal biofilm due to targeted mutation of acylhomoserine lactone (AHL) synthesis enzymes (Lynch et al., 2002). Biofilm formation by *Vibrio* spp., *Streptococcus* spp., and *Staphylococcus* spp. was prevented by deletion of gene in AI-2 (Ahmed et al., 2009; Novick and Geisinger 2008). Antibiotics play crucial roles to prevent biofilm formation. Synthesis of C_4-homoserine lactone (C_4-HSL) and 3-oxo-C_{12}-HSL in *P. aeruginosa* were prevented by azithromycin. Thus reduction of bacterial adhesion happens to polystyrene surfaces.

Moreover, azithromycin cures chronic pulmonary infection caused by *P. aeruginosa* in a mouse model (Hoffmann et al., 2007).

RNAIII inhibiting peptide (RIP) acts as an QS inhibiting peptides. RIP and/or RIP analogues show antibiofilm effect against *Staphylococcus* spp, highlighted in several reports (Cirioni et al., 2013). Prevention of *Staphylococcus* spp biofilm formation is prevented by RIP loaded into polymethylmethacrylate beads (Kiran et al., 2008). In a study the quorum-sensing inhibitors were screened by automated docking program DOCK 5.3.0, and five compounds against *Pseudomonus aeruginosa* (Zeng et al., 2008).

3.6.2 Inhibition through Biofilm Disassembly

Breakdown of the extracellular matrix guides to prevention of biofilm formed and changes occur in the physiology of cell (Boles and Horswill, 2011). Several bacterial species secrete enzymes out of cells, and these enzymes degrade biofilm forming components, and thus disassembly happens.

A gene regulatory (*agr*) system contains small pore-forming toxins and several proteases also termed as phenol-soluble modulins (PSMs) (Boles and Horswill 2008). The extracellular proteases (e.g., sarA, Esp, sigB) disassemble the biofilm (Lauderdale et al., 2009; Tsang et al., 2008). Another mediator is nuclease that is responsible for biofilm disassembly (Mann et al., 2009).

3.6.3 Antibiofilm Antibiotics

The major drawback to treat device-related biofilm formation is that all the available antibiotics are only efficient against planktonic cells. Resistance against antimicrobial agents is better in biofilm than in planktonic cells. Diffusion limitation, change of metabolic activity, the phenotypic and genotypic conditions of biofilm cells are the features observed in biofilm (Dunman et al., 2001), and the adverse microenvironmental conditions is also present in the biofilms (Dunne 2002; Becker et al., 2001). Biofilm cells also are protected against antimicrobial agents that are harmful against planktonic cells. Biofilm formation is prevented by applying antibiotics in catheters in clinics (Casey et al., 2008; Hockenhull et al., 2009). *Pseudomonas aeruginosa* showed resistant against conventional antibiotics (Taylor et al., 2014). One likely target would be cell-to-cell communication that forms virulence phenotypes (Fitzpatrick et al., 2005).

3.6.4 Probiotic Products against the Different Pathogenic Biofilms

Lactobacillus sp. are the most famous probiotics as they manufacture several exometabolites such as EPSs, bacteriocins, (Sharma et al., 2018; Kaur et al., 2018) and antibiofilm biosurfactants (Yan et al., 2019). *Lactobacillus* showed antioxidant effects in presence of polysaccharide (Pan and Mei, 2010). The EPS of *Lactobacillus* spp. was functional against Gram-positive and negative bacteria. EPS concentration enables removal of biofilm (Mahdhi et al., 2017).

3.6.4.1 Probiotics Resists Biofilm Formation

Bacteriocin, one of the popular probiotics, demonstrates its antibiofilm activity. Bacteriocin isolated from *Lactobacillus brevis* DF01 resists biofilm formation, but does not maintain any effect on the *Escherichia coli* and *S. typhimurium* biofilms (Kim et al., 2019).

3.6.4.2 Probiotics Resists Biofilm Formation of Diarrhoea-causing Pathogens

It was found that the biofilm formation of *V. cholera was prevented by* media supernatant of seven isolates of *Lactobacillus* spp. (Kaur et al., 2018). *Clostridium difficile, another* diarrhoea causing microorganisms, was prevented by probiotics for example *Saccharomyces* and *Lactobacillus* (Lau and Chamberlain et al., 2016).

3.6.4.3 Prevention of Dental Biofilms by Probiotics

Dental diseases such as tooth decay and periodontal diseases are caused by multispecies biofilm (Kolenbrander et al., 2010). Comelli et al. (2002) studied that in presence of *Lactococcus lactis NCC2211* the biofilm formation of cancer-causing *S. sobrinus OMZ176 was disrupted* (Comelli et al., 2002).

3.6.4.4 Prevention of Wound Biofilm by Probiotics

P. aeruginosa, a Gram-negative bacterium, forms biofilm as well as cause chronic infections. The host immunity is changed for this infection (Rumbaugh et al., 2000). The biofilm formation of *P. aeruginosa* was disrupted by *L. plantarum*. It was observed that LPS impedes with the function of AHL and disrupts the general activity of *P. aeruginosa* QS (Ramos et al., 2012).

3.6.5 Matrix Targeting Enzymes

Disruption of the formation of biofilm happened due to degradation of the extracellular polymeric matrix of biofilms. Biofilm matrix comprises polysaccharide, eDNA and proteins, and a number of strategies have been employed to degrade these compounds (Kiedrowski and Horswill 2011). The oral bacterium *Actinobacillus actinomycetemcomitans* disrupts biofilm formed by other bacteria by producing dispersin B. Degradation of the extracellular matrix of *S. epidermidis* biofilm is caused by dispersin B (Kaplan et al., 2004). Extracellular genomic DNA is also present in extracellular matrix, released by the bacteria (Mann et al., 2009). As a result, biofilm of *Staphylococcus aureus* was dispersed by DNase I (Izano et al., 2008). *S. aureus* biofilms were disrupted effectively by the treatment of Proteinase K and trypsin (Chaignon et al., 2007). However, there are many restrictions with these approaches.

3.6.6 Biofilm Inhibition by Metal Ions

Zinc is used as an inhibitor against *Actinobacillus pleuropneumoniae* (Wu et al., 2013). Silver nanoparticles (Nag) and ionic silver (Ag^+) prevent *Pseudomonas aeruginosa* biofilm growth (Mann et al., 2021). Similarly Chakraborty et al. (2020) examined the function of caffeine to inhibit the biofilm formation of *Pseudomonas aeruginosa*. Biofilm growth of *Pseudomonas aeruginosa*, *Staphylococcus aureus*, and *E. coli* was prevented by copper, gallium and titanium (Gugala et al., 2017). Formation of biofilms is inhibited by the polyvalent metal ions (Brouse et al., 2017). Trimethylsilane (TMS) plasma nanocoatings and titanium were applied to coat surfaces and hence reduce the adherence of bacteria and biofilm formation. Novel Cu-bearing 304 type stainless steel (304CuSS) prevents biofilm forming *Staphylococcus aureus* by preventing activities of cell protein/enzymes (Nan et al., 2015). CUO NP is showing complete antibiofilm activity against marine *Staphylococcus lentus* (Padmavathi et al., 2019). It had been found that borate bioactive glass was showing antibiofilm activity and the effect was enhanced in the presence of copper and zinc (Jung et al., 2019). Olar et al. (2022) reviewed the role of 3D ions that have large-spectrum antimicrobial and antibiofilm activity. Rare metal ions are showing activity against *Pseudomonas aeruginosa* when applied with green tea-based polyphenol (catechin) (Liu et al., 2020).

3.6.7 Antimicrobial Mechanisms of Nanometals

Antimicrobial nanomaterials inhibit microbial growth or kill the invading organisms similar to antibiotics through one or more mechanisms (Khameneh et al., 2016). Sometimes, nanoparticles affect pathogens indirectly. Sharma et al. (2009) reported that biosynthesised silver nanoparticles consist of better antimicrobial effect than the conventional methods. Yu et al. (2016) showed that the application of AuNPs enhanced the activity of host immunity and thus can affect pathogens. The activity was enhanced when combination of biosynthesised silver nanoparticles and antibiotics such as

TABLE 3.1
List of Nanometals against Microbial Pathogens

Nanometals	Microbial Biofilm
AgNP	*Klebsiealla pneumoniae* (Siddique et al., 2020)
Gold and iron oxide nanoparticles	*Staphylococcus aureus* and *Pseudomonas aeruginosa* (Sathyanarayanan et al., 2013)
Ag-NPs and ZnO-NPs	*E.coli* and *S.aureus* (Fontecha-Umaña et al., 2020)
Silver nanoparticles	*S. aureus and marine biofilm* (Chaudhari et al., 2012; Inbakandan et al., 2013)
Copper and zinc oxides	Gram-positive and Gram-negative pathogens (Aswathanarayan and Vittal, 2017)
Silver (AgNPs), copper (CuNPs) and zinc oxide (ZnONPs)	*Enterococcus gallinarum*, *Staphylococcus haemolyticus*, *Enterobacter aerogenes*, and *Salmonella enterica* serovar *Typhi* (Ingle et al., 2020)

ampicillin and vancomycin were applied on to Gram-positive and Gram-negative microbial biofilm (Gurunathan et al., 2014). A list of nanoparticles against pathogenic bacteria produces biofilm is in Table 3.1.

3.7 CURRENT TREATMENT PROCESSES—BACTERICIDAL AND BACTERIOSTATIC AGENTS

There are limitations with traditional chemotherapy processes. Novel approaches are required to treat *Staphylococcus epidermidis* device-related infections that are not treated by common bactericidal or bacteriostatic mechanisms. These perspectives are generally on the progress of bioactive, anti-infective, or antimicrobial devices, that prevent bacterial adherence or growth by the presence of antimicrobial agents. Surfaces of medical devices are modified, hence making it inhospitable to bacterial colonisation. Low-energy surface acoustic waves from electrically activated piezo elements retard formation of microbial biofilm on medical devices. It had been shown that when elastic waves were applied in the nanometre range, the development of biofilms by four different bacteria and *Candida* species was prevented (Hazan et al., 2006).

3.8 TREATMENT BY BACTERIOPHAGES

Another popular method that appeared as a new alternative to antibiotics is phage therapy, and it is used to treat diseases. Anti-infectivity capacity of phages has already been tested in humans and animals (Kutateladze and Adamia 2010). Even phages that show lytic cycle, act as against pathogenic bacteria (Fischetti 2010). Over recent years, there has been a growing pursuit of phages able to eradicate biofilms (Donlan 2009). Though there is no such evidence, phages were used as a therapeutic agents against biofilm infections because they degrade the biofilm matrix and infect cells in this mode of growth (Hughes et al., 1998; Sillankorva et al., 2004). Fu et al. (2010) showed the effectiveness of a cocktail of phages against *Pseudomonas aeruginosa* responsible for biofilm formation on a catheter. Similarly *Pseudomonas aeruginosa* bacteriophages, Vb-Pa 4 and Vb-Pa 5 were effective against hospital strains of *P.aeruginosa* (Gabisoniya et al., 2016).

3.9 FUTURE CHALLENGES

Biofilm infections could be controlled by combined effect of common antibiotics with biofilm controlling compounds (Romilly et al., 2014). Generally, pathogenic cells could not be killed by biofilm dispersing medicines, and thus a mixture of antibiotics is required to bring a satisfactory outcome.

For example, patulin was ineffective on *P. aeruginosa* biofilm, but when it was treated in combination with tobramycin, it was highly effective (Rasmussen et al., 2005).

As antimicrobial agents are unable to treat biofilm infections, it remains a serious implication in healthcare services. Biofilms demonstrate resistance against antimicrobial agents, so it is urgent that we find effective treatment strategies against biofilm-associated infections. Several novel techniques exist, such as dispersal of formed biofilm or inhibition of quorum-sensing compounds, or application of antibiotics in combination to be carried out for effective removal of biofilms. However, many of these techniques still do not give promising outcomes, and more attention is required from the scientific community to develop novel and promising strategies.

REFERENCES

Ahmed, N.A., F.C. Petersen, and A.A. Scheie. 2009. AI-2/LuxS is involved in increased biofilm formation by *Streptococcus intermedius* in the presence of antibiotics. *Antimicrob Agents Chemother* 53(10): 4258–4263.

Alksne, L.E., and S.J. Projan. 2000. Bacterial virulence as a target for antimicrobial chemotherapy. *Curr Opin Biotechnol* 11: 625–636.

Antimicrobial Resistance n.d. National Institute of Allergy and Infectious Diseases. http://www3.niaid.nih.gov/about/overview/profile/fy2003/pdf/SSAR_ANTIMICROBIAL.pdf, accessed 17th January 2008.

Aswathanarayan, J.B., and R.R. Vittal. 2017. Antimicrobial, biofilm inhibitory and anti-infective activity of metallic nanoparticles against pathogens MRSA and *Pseudomonas aeruginosa* PA01. *Pharm Nanotechnol* 5(2): 148–153.

Becker, P., W. Hufnagle, G. Peters, and M. Herrmann. 2001. Detection of differential gene expression in biofilm-forming versus planktonic populations of *Staphylococcus aureus* using micro-representational-difference analysis. *Appl Environ Microbiol* 67: 2958–2965.

Boles, B.R., and A.R. Horswill. 2008. Agr-mediated dispersal of *Staphylococcus aureus* biofilms. *PLoS Pathog* 4: e1000052.

Boles, B.R., and A.R. Horswill. 2011. Staphylococcal biofilm disassembly. *Trends Microbiol* 19: 449–455.

Boyd, A., and A.M. Chakrabarty. 1994. Role of alginate lyase in cell detachment of *Pseudomonas aeruginosa*. *Appl Environ Microbiol* 60: 2355–2359.

Brouse, L., R. Brouse, and D. Brouse. 2017. Natural pathogen control chemistry to replace toxic treatment of microbes and biofilm in cooling towers. *Pathogens* 6(2): 14.

Busch, A., and G. Waksman. 2012. Chaperone-usher pathways: Diversity and pilus assembly mechanism. *Philos. Trans R Soc Lond B Biol Sci* 367: 1112–1122.

Casey, A.L., L.A. Mermel, P. Nightingale, and T.S. Elliott. 2008. Antimicrobial central venous catheters in adults: A systematic review and meta-analysis. *Lancet Infect Dis* 8: 763–776.

Chaignon, P., I. Sadovskaya, C. Ragunah, N. Ramasubbu, J.B. Kaplan, and S. Jabbouri. 2007. Susceptibility of staphylococcal biofilms to enzymatic treatments depends on their chemical composition. *Appl Microbiol Biotechnol* 75: 125–132.

Chakraborty, P., D.G. Dastidar, P. Paul, S. Dutta, D. Basu, S.R. Sharma, S. Basu, R.K. Sarker, A. Sen, A. Sarkar, and P. Prosun Tribedi. 2020. Inhibition of biofilm formation of *Pseudomonas aeruginosa* by caffeine: A potential approach for sustainable management of biofilm. *Arch Microbiol* 202: 623–635.

Chaudhari, P.R., S.A. Masurkar, V.B. Shidore, and S.P. Kamble. 2012. Effect of biosynthesized silver nanoparticles on *Staphylococcus aureus* biofilm quenching and prevention of biofilm formation. *Nano-Micro Lett* 4: 34–39.

Chua, A.G., J. Ding, P.E. Schoch, and B.A. Cunha. 1998. Pacemaker-induced endocarditis due to *Propionibacterium acnes*. *Clin Infect Dis* 27: 1541–1542.

Cirioni, O., F. Mocchegiani, I. Cacciatore, J. Vecchiet, C. Silvestri, L. Baldassarre, C. Ucciferri, E. Orsetti, P. Castelli, M. Provinciali, M. Vivarelli, E. Fornasari, and A. Giacometti. 2013. Quorum sensing inhibitor FS3-coated vascular graft enhances daptomycin efficacy in a rat model of staphylococcal infection. *Peptides* 40: 77–81.

Coenye, T., K.J. Spittaels, and Y. Achermann. 2022. The role of biofilm formation in the pathogenesis and antimicrobial susceptibility of *Cutibacterium acnes*. *Biofilm* 4: 100063.

Cohen, F.E., and Kelly, J.W. 2003. Therapeutic approaches to protein-misfolding diseases. *Nature* 426: 905–909.

Collinson, S.K., P.C. Doig, J.L. Doran, S. Clouthier, T.J. Trust, and W.W. Kay. 1993. Thin, aggregative fimbriae mediate binding of *Salmonella enteritidis* to fibronectin. *J Bacteriol* 175: 12–18.

Comelli, E.M., B. Guggenheim, F. Stingele, and J.R. Neeser. 2002. Selection of dairy bacterial strains as probiotics for oral health. *Eur J Oral Sci* 110(3): 218–224.

Costerton, J.W., L. Montanaro, and C.R. Arciola. 2005. Biofilm in implant infections: Its production and regulation. *Int J Artif Organs* 28(11): 1062–1068.

Crespi, B.J. 2001. The evolution of social behavior in microorganisms. *Trends Ecol Evol* 16: 178–183.

Darouiche, R.O. 2004. Treatment of infections associated with surgical implants. *N Engl J Med* 350: 1422–1429.

Del Pozo, J.L., N.V. Tran, P.M. Petty, C.H. Johnson, M.F. Walsh, U. Bite, R.P. Clay, J.N. Mandrekar, K.E. Piper, J.M. Steckelberg, and R. Patel. 2009. Pilot study of association of bacteria on breast implants with capsular contracture. *J Clin Microbiol* 47: 1333–1337.

Donlan, R., R. Murga, and L. Carson. 1999. Growing Biofilms in Intravenous Fluids. In Wimpenny, J., Gilbert, P., Walker, J., Brading, M., and Bayston, R., editors. *Biofilms, the Good, the Bad, and the Ugly. Presented at the Fourth Meeting of the Biofilm Club* (23–29) Powys, UK.

Donlan, R.M., and J.W. Costerton. 2002. Biofilms: Survival mechanisms of clinically relevant microorganisms. *Clin Microbiol Rev* 15: 167–193.

Dunman, P.M., E. Murphy, S. Haney, D. Palacios, G. Tucker-Kellogg, S. Wu, E.L. Brown, R.J. Zagursky, D. Shlaes, and S.J. Projan. 2001. Transcription profiling-based identification of *Staphylococcus aureus* genes regulated by the *agr* and/or *sarA* loci. *J Bacteriol* 183: 7341–7353.

Dunne, W.M. 2002. Bacterial adhesion: Seen any good biofilms lately? *Clin Microbiol Rev* 15: 155–166.

Fischetti, V.A. 2010. Bacteriophage endolysins: A novel anti-infective to control Gram-positive pathogens. *Int J Med Microbiol* 300: 357–362.

Fitzpatrick, F., H. Humphreys, and J.P. O'Gara. 2005. The genetics of staphylococcal biofilm formation—will a greater understanding of pathogenesis lead to better management of device-related infection? *Clin Microbiol Infect* 11(12): 967–973.

Fong, J.N., and F.H. Yildiz. 2015. Biofilm matrix proteins. *Microbiol Spectr* 3. 10.1128/microbiolspec.MB-0004-2014.

Fontecha-Umaña, F., A.G. Ríos-Castillo, C. Ripolles-Avila, and J.J. Rodríguez-Jerez. 2020. Antimicrobial activity and prevention of bacterial biofilm formation of silver and zinc oxide nanoparticle-containing polyester surfaces at various concentrations for use. *Foods* 9(4): 442.

Foster, K.R. 2005. Hamilotonian medicine: Why the social lives of pathogen matters. *Science* 308: 1269–1270.

Fredheim, E.G.A., C. Klingenberg, H. Rohde, S. Frankenberger, P. Gaustad, T. Flægstad, and J.E. Sollid. 2009. Biofilm formation by *Staphylococcus haemolyticus*. *J Clin Microbiol* 47(4): 1172–1180.

Fronzes, R., H. Remaut, and G. Waksman. 2008. Architectures and biogenesis of non-flagellar protein appendages in Gram-negative bacteria. *EMBO J* 27: 2271–2280.

Fu, W., T. Forster, O. Mayer, J.J. Curtin, S.M. Lehman, and R.M. Donlan. 2010. Bacteriophage cocktail for the prevention of biofilm formation by *Pseudomonas aeruginosa* on catheters in an in vitro model system. *Antimicrob Agents Chemother* 54(1): 397–404.

Gabisoniya, T.G., M.Z. Loladze, M.M. Nadiradze, N.K. Chakhunashvili, M.G. Alibegashvili, N.G. Tamarashvili, V.A. Pushkina. 2016. Effects of bacteriophages on biofilm formation by strains of *Pseudomonas aeruginosa*. *Prikl Biokhim Mikrobiol* 52(3): 312–317.

Gotz, F. 2002. Staphylococcus and biofilms. *Mol Microbiol* 43: 1367–1378.

Gugala, N., J. Lemire, and R. Turner. 2017. The efficacy of different anti-microbial metals at preventing the formation of, and eradicating bacterial biofilms of pathogenic indicator strains. *J Antibiot* 70: 775–780.

Gurunathan, S., J.W. Han, D.N. Kwon, and J.H. Kim. 2014. Enhanced antibacterial and anti-biofilm activities of silver nanoparticles against Gram-negative and Gram-positive bacteria. *Nanoscale Res Lett* 9(1): 1–17.

Hazan, Z., J. Zumeris, H. Jacob, H. Raskin, G. Kratysh, M. Vishnia, N. Dror, T. Barliya, M. Mandel, and G. Lavie. 2006. Effective prevention of microbial biofilm formation on medical devices by low-energy surface acoustic waves. *Antimicrob Agents Chemother* 50(12): 4144–4152.

Heijnen, J.J., M.C.M. van Loosdrecht, R. Mulder, R. Weltevrede, and A. Mulder. 1993. Development and scale-up of an aerobic biofilm air-lift suspension reactor. *Water Sci Technol* 27: 253–261.

Heilmann, C., M. Hussain, G. Peters, and F. Gotz. 1997. Evidence for autolysin-mediated primary attachment of *Staphylococcus epidermidis* to a polystyrene surface. *Mol Microbiol* 24: 1013–1024.

Henrici, A. 1933. Studies of freshwater bacteria. 1. A direct microscopic technique. *J Bacteriol* 25: 277–286.

Hilleringmann, M., F. Giusti, B.C. Baudner, V. Masignani, A. Covacci, R. Rappuoli, M.A. Barocchi, and I. Ferlenghi. 2008. Pneumococcal pili are composed of protofilaments exposing adhesive clusters of Rrg A. *PLoS Pathog* 4: e1000026.

Hilleringmann, M., P. Ringler, S.A. Müller, G. De Angelis, R. Rappuoli, I. Ferlenghi, and A. Engel. 2009. Molecular architecture of *Streptococcus pneumoniae* TIGR4 pili. *EMBO J* 28: 3921–3930.

Hoa, M., S. Tomovic, L. Nistico, L. Hall-Stoodley, P. Stoodley, L. Sachdeva, R. Berk, and J.M. Coticchia. 2009. Identification of adenoid biofilms with middle ear pathogens in otitis-prone children utilizing SEM and FISH. *Int J Pediatr Otorhinolaryngol* 73(9): 1242–1248.

Hockenhull, J.C., K.M. Dwan, G.W. Smith, C.L. Gamble, A. Boland, T.J. Walley, and R.C. Dickson. 2009. The clinical effectiveness of central venous catheters treated with anti-infective agents in preventing catheter-related bloodstream infections: A systematic review. *Crit Care Med* 37: 702–712.

Hoffmann, N., B. Lee, M. Hentzer, T.B. Rasmussen, Z. Song, H.K. Johansen, M. Givskov, and N. Høiby. 2007. Azithromycin blocks quorum sensing and alginate polymer formation and increases the sensitivity to serum and stationary-growth-phase killing of *Pseudomonas aeruginosa* and attenuates chronic *P. aeruginosa* lung infection in Cftr(-/-) mice. *Antimicrob Agents Chemother* 51(10): 3677–3687.

Hughes, K.A., I.W. Sutherland, and M.V. Jones. 1998. Biofilm susceptibility to bacteriophage attack: The role of phage-borne polysaccharide depolymerise. *Microbiology* 144: 3039–3047.

Hussain, M., M. Herrmann, C. Von Eiff, F. Perdreau-Remington, and G. Peters. 1997. A 140-kilodalton extracellular protein is essential for the accumulation of *Staphylococcus epidermidis* strains on surfaces. *Infect Immun* 65: 519–524.

Imamura, Y., J. Chandra, P.K. Mukherjee, A.A. Lattif, L.B. Szczotka-Flynn, E. Pearlman, ... M.A. Ghannoum. 2008. *Fusarium* and *Candida albicans* biofilms on soft contact lenses: Model development, influence of lens type, and susceptibility to lens care solutions. *Antimicrob Agents Chemother* 52(1): 171–182.

Inbakandan, D., C. Kumar, L.S. Abraham, R. Kirubagaran, R. Venkatesan, and S.A. Khan. 2013. Silver nanoparticles with anti microfouling effect: A study against marine biofilm forming bacteria. *Colloids Surf B Biointerfaces* 111: 636–643.

Ingle, A.P., S. Wagh, J. Biswas, M. Mondal, C.M. Feitosa, and M. Rai. 2020. Phyto-fabrication of different nanoparticles and evaluation of their antibacterial and anti-biofilm efficacy. *Curr Nanosci* 16(6): 1002–1015.

Izano, E.A., M.A. Amarante, W.B. Kher, and J.B. Kaplan. 2008. Differential roles of poly-N-acetylglucosamine surface polysaccharide and extracellular DNA in *Staphylococcus aureus* and *Staphylococcus epidermidis* biofilms. *Appl Environ Microbiol* 74: 470–476.

Jung, S., T. Day, T. Boone, B. Buziak, and A. Omar. 2019. Anti-biofilm activity of two novel, borate based, bioactive glass wound dressings. *Biomed Glasses* 5(1): 67–75.

Kania, R.E., G.E. Lamers, M.J. Vonk, E. Dorpmans, J. Struik, P. Tran Ba Huy, P. Hiemstra, G.V. Bloemberg, and J.J. Grote. 2008. Characterization of mucosal biofilms on human adenoid tissues. *Laryngoscope* 118(1): 128–134.

Kaplan, J.B., C. Ragunath, K. Velliyagounder, D.H. Fine, and N. Ramasubbu. 2004. Enzymatic detachment of *Staphylococcus epidermidis* biofilms. *Antimicrob Agents Chemother* 48: 2633–2636.

Kaur, S., P. Sharma, N. Kalia, J. Singh, and S. Kaur. 2018. Anti-biofilm properties of the fecal probiotic *lactobacilli* against *Vibrio* spp. *Front Cell Infect Microbiol* 8: 120.

Keller, L., and M.G. Surette. 2006. Communication in bacteria: An ecological and evolutionary perspective. *Nat Rev Microbiol* 4: 249–258.

Khameneh, B., R. Diab, K. Ghazvini, and B.S.F. Bazzaz. 2016. Breakthroughs in bacterial resistance mechanisms and the potential ways to combat them. *Microb Pathog* 95: 32.

Khatoon, Z., C.D. McTiernan, E.J. Suuronen, T.F. Mah, and E.I. Alarcon. 2018. Bacterial biofilm formation on implantable devices and approaches to its treatment and prevention. *Heliyon* 4: e01067.

Kiedrowski, M.R., and Horswill, A.R. 2011. New approaches for treating staphylococcal biofilm infections. *Ann N Y Acad Sci* 1241: 104–121.

Kim, N.N., W.J. Kim, and S.S.J.F.C. Kang. 2019. Anti-biofilm effect of crude bacteriocin derived from *Lactobacillus brevis* DF01 on *Escherichia coli* and *Salmonella Typhimurium*. *Food Control* 98: 274–280.

Kiran, M.D., N.V. Adikesavan, O. Cirioni, A. Giacometti, C. Silvestri, G. Scalise, R. Ghiselli, V. Saba, F. Orlando, M. Shoham, and N. Balaban. 2008. Discovery of a quorum sensing inhibitor of drug-resistant staphylococcal infections by structure-based virtual screening. *Mol Pharmacol* 73: 1578–1586.

Kocianova, S., C. Vuong, Y. Yao, J.M. Voyich, E.R. Fischer, F.R. DeLeo, and M. Otto. 2005. Key role of poly-gamma-DL-glutamic acid in immune evasion and virulence of *Staphylococcus epidermidis*. *J Clin Invest* 115: 688–694.

Koebnik, R., K.P. Locher, and P. Van Gelder. 2000. Structure and function of bacterial outer membrane proteins: Barrels in a nutshell. *Mol Microbiol* 37: 239–253.

Koh, C.L., C.K. Sam, W.F. Yin, L.Y. Tan, T. Krishnan, Y.M. Chong, and K.G. Chan. 2013. Plant-derived natural products as sources of anti-quorum sensing compounds. *Sensors* 13(5): 6217–6228.

Kolenbrander, P.E., R.J. Palmer, S. Periasamy, and N.S. Jakubovics. 2010. Oral multispecies biofilm development and the key role of cell–cell distance. *Nat Rev Microbiol* 8(7): 471–480.

Koley, S., and M. Mukherjee. 2021. Genetic Basis of Biofilm Formation and Spread of Nosocomial Infections. In Nag, M., and Lahiri, D., editors. *Analytical Methodologies for Biofilm Research* (pp. 269–298). New York, NY: Springer.

Kolter, R., and E.P. Greenberg. 2006. Microbial sciences: The superficial life of microbes. *Nature* 441: 300–302.

Krzyściak, P., A. Chmielarczyk, M. Pobiega, D. Romaniszyn, J. Wójkowska-Mach. 2017. *Acinetobacter baumannii* isolated from hospital-acquired infection: Biofilm production and drug susceptibility. *APMIS* 125: 1017–1026.

Kumar, A., A. Alam, M. Rani, N.Z. Ehtesham, and S.E. Hasnain. 2017. Biofilms: Survival and defense strategy for pathogens. *Int J Med Microbiol* 307(8): 481–489.

Kutateladze, M., and Adamia, R. 2010. Bacteriophages as potential new therapeutics to replace or supplement antibiotics. *Trends Biotechnol* 28: 591–595.

Lau, C.S. and R.S. Chamberlain. 2016. Probiotics are effective at preventing Clostridium difficile-associated diarrhea: a systematic review and meta-analysis. *International Journal of General Medicine*, *9*, 27–37.

Lauderdale, K.J., B.R. Boles, A.L. Cheung, and A.R. Horswill. 2009. Interconnections between Sigma B, agr, and proteolytic activity in *Staphylococcus aureus* biofilm maturation. *Infect Immun* 77: 1623–1635.

Lauer, P., C.D. Rinaudo, M. Soriani, I. Margarit, D. Maione, R. Rosini, A.R. Taddei, M. Mora, R. Rappuoli, G. Grandi, and J.L. Telford. 2005. Genome analysis reveals pili in Group B *Streptococcus*. *Science* 309: 105.

Lewis K. 2001. Riddle of biofilm resistance. *Antimicrob Agents Chemother* 45: 999e1007.

Liu, L., X. Xiao, K. Li, X. Li, B. Shi, and X. Liao. 2020. Synthesis of catechin-rare earth complex with efficient and broad-spectrum anti-biofilm activity. *Chem Biodivers* 17(3): e1900734.

Lynch, M.J., S. Swift, D.F. Kirke, C.W. Keevil, C.E. Dodd, and P. Williams. 2002. The regulation of biofilm development by quorum sensing in *Aeromonas hydrophila*. *Environ Microbiol* 4(1): 18–28.

Mack, D., H. Rohde, L.G. Harris, A.P. Davies, M.A. Horstkotte, and J.M. Knobloch. 2006. Biofilm formation in medical device-related infection. *Int J Artif Organs* 29(4): 343–359.

Mahdhi, A., N. Leban, I. Chakroun, A.M. Chaouch, J. Hafsa, K. Fdhila, K. Mahdouani, and H. Majdoub. 2017. Extracellular polysaccharide derived from potential probiotic strain with antioxidant and antibacterial activities as a prebiotic agent to control pathogenic bacterial biofilm formation. *Microb Pathog* 109: 214–220.

Maki, D.G., and L.A. Mermel. 1998. Infections due to Infusion Therapy. In Bennett, J.V., and Brachman, P.S., editors. *Hospital Infections*, 4th ed. (689–724). Philadelphia: Lippincott-Raven.

Manago, K., J. Nishi, N. Wakimoto, H. Miyanohara, J. Sarantuya, K. Tokuda, M. Iwashita, K. Yamamoto, M. Yoshinaga, I. Maruyama, and Y. Kawano. 2006. Biofilm formation by and accessory gene regulator typing of methicillin-resistant *Staphylococcus aureus* strains recovered from patients with nosocomial infections. *Infect Control Hosp Epidemiol* 27(2): 188–190.

Mann, E.E., K.C. Rice, B.R. Boles, J.L. Endres, D. Ranjit, L. Chandramohan, L.H. Tsang, M.S. Smeltzer, A.R. Horswill and K.W. Bayles. 2009. Modulation of eDNA release and degradation affects *Staphylococcus aureus* biofilm maturation. *PloS One* 4: e5822.

Mann, R., A. Holmes, O. McNeilly, R. Cavaliere, G.A. Sotiriou, S.A. Rice, C. Gunawan. 2021. Evolution of biofilm-forming pathogenic bacteria in the presence of nanoparticles and antibiotic: Adaptation phenomena and cross-resistance. *J Nanobiotechnol* 19: 291.

Martí, S., J. Rodríguez-Baño, M. Catel-Ferreira, T. Jouenne, J. Vila, H. Seifert, and E. Dé. 2011. Biofilm formation at the solid-liquid and air-liquid interfaces by *Acinetobacter* species. *BMC Res Notes* 4: 5.

Murphy, C.N., and S. Clegg. 2012. *Klebsiella pneumoniae* and type 3 fimbriae: Nosocomial infection, regulation and biofilm formation. *Future Microbiol* 7(8): 991–1002.

Musthafa K.S., A.V. Ravi, A. Annapoorani, I.S.V. Packiavathy, and S.K. Pandian. 2010. Evaluation of anti-quorum-sensing activity of edible plants and fruits through inhibition of the N-acyl-homoserine lactone system in *Chromobacterium violaceum* and *Pseudomonas aeruginosa*. *Chemotherapy* 56: 333–339.

Nan, L., K. Yang, and G. Ren. 2015. Anti-biofilm formation of a novel stainless steel against *Staphylococcus aureus*. *Mater Sci Eng C* 51: 356–361.

Natrah, F.M.I., M.M. Kenmegne, W. Wiyoto, P. Sorgeloos, P. Bossier, and T. Defoirdt. 2011. Effects of micro-algae commonly used in aquaculture on acyl-homoserine lactone quorum sensing. *Aquaculture* 317: 53–57.

Nelson, A.L., J. Ries, F. Bagnoli, S. Dahlberg, S. Fälker, S. Rounioja, J. Tschöp, E. Morfeldt, I. Ferlenghi, M. Hilleringmann, D.W. Holden, R. Rappuoli, S. Normark, M.A. Barocchi, and B. Henriques-Normark. 2007. RrgA is a pilus-associated adhesin in *Streptococcus pneumoniae*. *Mol Microbiol* 66: 329–340.

Ni Eidhin, D., S. Perkins, P. Francois, P. Vaudaux, M. Hook, and T.J. Foster. 1998. Clumping factor B (ClfB), a new surface-located fibrinogen-binding adhesin of *Staphylococcus aureus*. *Mol Microbiol* 30: 245–257.

Nourbakhsh, F., and A.E. Namvar. 2016. Detection of genes involved in biofilm formation in *Staphylococcus aureus* isolates. *GMS Hyg Infect Control* 22: 11.

Novick, R.P., and E. Geisinger. 2008. Quorum sensing in *staphylococci*. *Ann Rev Genetics* 42: 541–564.

O'Loughlin, C.T., L.C. Miller, A. Siryaporn, K. Drescher, M.F. Semmelhack, and B.L. Bassler. 2013. A quorum-sensing inhibitor blocks *Pseudomonas aeruginosa* virulence and biofilm formation. *Proc Natl Acad Sci* 110(44): 17981–17986.

Okuda, K.I., R. Nagahori, S. Yamada, S. Sugimoto, C. Sato, M. Sato, T. Tadayuki Iwase, K. Hashimoto, and Y. Mizunoe. 2018. The composition and structure of biofilms developed by *Propionibacterium acnes* isolated from cardiac pacemaker devices. *Front Microbial* 9: 182.

Olar, R., M. Badea, and M.C. Chifiriuc. 2022. Metal complexes—a promising approach to target biofilm associated infections. *Molecules* 27(3): 758.

Orme, R., C.W. Douglas,, S. Rimmer, and M. Webb. 2006. Proteomic analysis of *Escherichia coli* biofilms reveals the overexpression of the outer membrane protein OmpA. *Proteomics* 6: 4269–4277.

Otter, J.A., K. Vickery, J.D. Walker, E.D. Pulcini, P. Stoodley, S.D. Goldenberg, J.A. Salkeld, J. Chewins, S. Yezli, and J.D. Edgeworth. 2015. Surface-attached cells, biofilms and biocide susceptibility: Implications for hospital cleaning and disinfection. *J Hosp Infect* 89(1): 16–27.

Pakharukova, N., M. Tuittila, S. Paavilainen, H. Malmi, O. Parilova, S. Teneberg, S.D. Knight, and A.V. Zavialov. 2018. Structural basis for *Acinetobacter baumannii* biofilm formation. *Proc Natl Acad Sci U S A* 115: 5558–5563.

Pan, D., and X. Mei. 2010. Antioxidant activity of an exopolysaccharide purified from *Lactococcus lactis* subsp. lactis 12. *Carbohydr Polym* 80(3): 908–914.

Passerini de Rossi, B., M. Calenda, C. Vay, and M. Franco. 2007. Biofilm formation by *Stenotrophomonas maltophilia* isolates from device-associated nosocomial infections. *Rev Argent Microbial* 39(4): 204–212.

Rajendran R., D.P. Robertson, P.J. Hodge, D.F. Lappin, and G. Ramage. 2010. Hydrolytic enzyme production is associated with *Candida albicans* biofilm formation from patients with type 1 diabetes. *Mycopathologia* 170: 229–235.

Ramos, A.N., M.E. Sesto Cabral, D. Noseda, A. Bosch, O.M. Yantorno, and J.C. Valdez,. 2012. Antipathogenic properties of *Lactobacillus plantarum* on *Pseudomonas aeruginosa*: The potential use of its supernatants in the treatment of infected chronic wounds. *Wound Repair Regen* 20(4): 552–562.

Rasmussen, T.B., M.E. Skindersoe, T. Bjarnsholt, R.K. Phipps, K.B. Christensen, P.O. Jensen, J.B. Andersen, B. Koch, T.O. Larsen, M. Hentzer, L. Eberl, N. Høiby, and M. Givskov. 2005. Identity and effects of quorum-sensing inhibitors produced by *Penicillium* species. *Microbiology (Reading)* 151(Pt 5): 1325–1340.

Rice, S.A., K.S. Koh, S.Y. Queck, M. Labbate, K.W. Lam, and S.J. Kjelleberg. 2005. Biofilm formation and sloughing in Serratia marcescens are controlled by quorum sensing and nutrient cues. *Bacteriol* 187: 3477–3485.

Richards, M.J., J.R. Edwards, D.H. Culver, and R.P. Gaynes. 2000. Nosocomial infections in combined medical-surgical intensive care units in the United States. *Infect Control Hosp Epidemiol* 21: 510–551.

Rodríguez-Baño, J., S. Martí, S. Soto, F. Fernández-Cuenca, J.M. Cisneros, J. Pachón, A. Pascual, L. Martínez-Martínez, C. McQueary, L.A. Actis, and J. Vila. 2008. Biofilm formation in *Acinetobacter baumannii*: Associated features and clinical implications. *Clin Microbiol Infect* 14: 276–278.

Romilly, C., C. Lays, A. Tomasini, I. Caldelari, Y. Benito, P. Hammann, T. Geissmann, S. Boisset, P. Romby, and F. Vandenesch. 2014. A non-coding RNA promotes bacterial persistence and decreases virulence by regulating a regulator in *Staphylococcus aureus*. *PLoS Pathog* 10(3): e1003979.

Rumbaugh, K.P., J.A. Griswold, and A.N. Hamood. 2000. The role of quorum sensing in the in vivo virulence of *Pseudomonas aeruginosa*. *Microbes Infect* 2(14): 1721–1731.

Sardi, J.C., A.M. Almeida, and M. Mendes Giannini. 2011. New antimicrobial therapies used against fungi present in subgingival sites—a brief review. *Arch Oral Biol* 56: 951–959.

Sathyanarayanan, M.B., R. Balachandranath, Y. Genji Srinivasulu, S.K. Kannaiyan, and G. Subbiahdoss. 2013. The effect of gold and iron-oxide nanoparticles on biofilm-forming pathogens. *ISRN Microbiol* 2013: 272086.

Sharma, A.N., S. Kumar, and A.K. Tyagi. 2018. Effects of mannan-oligosaccharides and Lactobacillus acidophilus supplementation on growth performance, nutrient utilization and faecal characteristics in Murrah buffalo calves. *Journal of Animal Physiology and Animal Nutrition*, 102(3), 679–689.

Sharma, V.K., R.A. Yngarda, and Y. Lin. 2009. Silver nanoparticles: Green synthesis and their antimicrobial activities. *Adv Colloid Interface Sci* 145: 83–89.

Siddique, M.H., B. Aslam, M. Imran, A. Ashraf, H. Nadeem, S. Hayat, M. Khurshid, M. Afzal, I.R. Malik, M. Shahzad, U. Qureshi, Z.U.H. Khan, and S. Muzammil. 2020. Effect of silver nanoparticles on biofilm formation and EPS production of multidrug-resistant *Klebsiella pneumoniae*. *Biomed Res Int* 2020: 6398165.

Sillankorva S., R. Oliveira, M. Vieira, I.W. Sutherland, and J. Azeredo. 2004. Bacteriophage ξ-S1 infection of *Pseudomonas fluorescens* planktonic cells versus biofilms. *Biofouling* 20: 133–138.

Soto, G.E., and S.J. Hultgren. 1999. Bacterial adhesins: Common themes and variations in architecture and assembly. *J Bacteriol* 181: 1059–1071.

Soto-Giron, M.J., L.M. Rodríguez-R, C. Luo, M. Elk, H. Ryu, J. Hoelle, J.W. Santo Domingo, T. Konstantinos, and K.T. Konstantinidis. 2016. Biofilms on hospital shower hoses: Characterization and implications for nosocomial infections. *Appl Environ Microbial* 82: 2872–2883.

Stoodley, P., K. Sauer, D.G. Davies, and J.W. Costerton. 2002. Biofilms as complex differentiated communities. *Annu Rev Microbiol* 56: 187–209.

Switalski, L.M., J.M. Patti, W. Butcher, A.G. Gristina, P. Speziale, and M. Hook. 1993. A collagen receptor on *Staphylococcus aureus* strains isolated from patients with septic arthritis mediates adhesion to cartilage. *Mol Microbiol* 7: 99–107.

Taylor, P.K., A.T. Yeung, and R.E. Hancock. 2014. Antibiotic resistance in Pseudomonas aeruginosa biofilms: Towards the development of novel anti-biofilm therapies. *J Biotechnol* 191: 121–130.

Tello, E., L. Castellanos, C. Arevalo-Ferro, J. Rodríguez, C. Jimenez, C. Duque. 2011. Absolute stereochemistry of antifouling cembranoid epimers at C-8 from the Caribbean octocoral *Pseudoplexaura flagellosa*. Revised structures of plexaurolones. *Tetrahedron* 67: 9112–9121.

Ton-That, H., and O. Schneewind. 2003. Assembly of pili on the surface of *Corynebacterium diphtheriae*. *Mol Microbiol* 50: 1429–1438.

Toyofuku, M., B. Roschitzki, K. Riedel, and L. Eberl. 2012. Identification of proteins associated with the *Pseudomonas aeruginosa* biofilm extracellular matrix. *J Proteome Res* 11: 4906–4915.

Tsang, L.H., J.E. Cassat, L.N. Shaw, K.E. Beenken, and M.S. Smeltzer. 2008. Factors contributing to the biofilm-deficient phenotype of *Staphylococcus aureus* sarA mutants. *PloS One* 3: e3361.

Velicer, G.J. 2003. Social strife in the microbial world. *Trends Microbiol* 11: 330–337.

Vengadesan, K., and S.V. Narayana. 2011. Structural biology of gram-positive bacterial adhesins. *Protein Sci* 20: 759–772.

Von Eiff, C., G. Peters, and C. Heilmann. 2002. Pathogenesis of infections due to coagulase-negative *staphylococci*. *Lancet Infect Dis* 2: 677–685.

West, S.A., A.S. Griffin, A. Gardner, and S.P. Diggle. 2006. Social evolution theory for microorganisms. *Nat Rev Microbiol* 4: 597–607.

Wu, C., J. Labrie, Y.D.N. Tremblay, D. Haine, M. Mourez, and M. Jacques. 2013. Zinc as an agent for the prevention of biofilm formation by pathogenic bacteria. *J Appl Microbiol* 115(1): 30–40.

Yan, X., S. Gu, X. Cui, Y. Shi, S. Wen, H. Chen, and J. Ge. 2019. Antimicrobial, anti-adhesive and anti-biofilm potential of biosurfactants isolated from Pediococcus acidilactici and Lactobacillus plantarum against Staphylococcus aureus CMCC26003. *Microb Pathog* 127: 12–20.

Yu, Q., J. Li, Y. Zhang, Y. Wang, L. Liu, and M. Li. 2016. Inhibition of gold nanoparticles (AuNPs) on pathogenic biofilm formation and invasion to host cells. *Sci Rep* 6: 26667.

Zeng, Z., L. Qian, L. Cao, H. Tan, Y. Huang, X. Xue, Y. Shen, and S. Zhou. 2008. Virtual screening for novel quorum sensing inhibitors to eradicate biofilm formation of *Pseudomonas aeruginosa*. *Appl Microbiol Biotechnol* 79: 119–126.

ZoBell, C.E., and E. Allen. 1935. The significance of marine bacteria in the fouling of submerged surfaces. *J Bacteriol* 29: 239–251.

4 Bacterial Persister Cells and Medical Importance

Mahima S. Mohan, Archana Priyadarshini Jena, Simi Asma Salim, and Siddhardha Busi
Pondicherry University, Puducherry, India

CONTENTS

4.1 INTRODUCTION

The evolution, dissemination, and appearance of antimicrobial resistance (AMR) has become one of the most serious healthcare issues with global scope. Understanding the causes and hotspots involved in the diffusion and development of novel resistance patterns, particularly multidrug resistance is urgently required. Antimicrobial resistance is caused by a wide variety of biochemical and physiological mechanisms (Davies and Davies, 2010). It can be classified into four categories: limiting drug uptake, altering drug target, inactivating a drug, and active drug efflux. Limiting drug uptake includes decreased cell permeability; i.e., modified cell wall proteins. Altering drug target is mediated by mutational changes within the drug target. Similarly, inactivation of a drug can be due to the phosphorylation of antibiotic which causes reduction in its bacterial ribosomal binding capacity, while active drug efflux means increasing active efflux of drugs out of the cell (Moradali

DOI: 10.1201/9781003345299-4

et al., 2017). Among all the microbial resistance mechanisms, persistence is one of the major public health concerns in the current scenario. It provides strength to microbes for their survival even in the presence of higher antimicrobial concentrations. 'Persistence' is the term used to describe the process in which a small population of bacteria resists the drugs fatal effects (Cohen et al., 2013). The survival mechanism ultimately supports enhanced microbial infections leading to higher mortality and morbidity rate due to the generation of persister cells.

Soon after the discovery of antibiotics, the term persistence was first discovered by Joseph Bigger in 1944 during the sterilisation process of *Staphylococcus aureus* cultures with penicillin treatment (Bigger, 1944). Even during the period of administration of antibiotics, numerous bacterial cells can persist inside the host for an extended period. Antimicrobial resistance and immunosuppression may contribute to the establishment and maintenance of persister cells in the host (Fisher et al., 2017). Besides the growth inhibition and retarded growth, the persister phenotype is frequently accompanied by metabolic activity reduction (Amato et al., 2014). The researchers suggested that the persister cells are not always metabolically inactive and can be activated under the influence of favourable conditions, which leads to the resistance development (Helaine et al., 2014; Levin-Reisan et al., 2017). Contradictory to the mutants, neither the growth was observed in the proximity of drugs; alternatively, the regrowth of persister cell population begins only after the exclusion of drugs, which is a significant difference between persistence and resistance phenomenon (Cohen et al., 2013).

4.2 BACTERIAL PERSISTENCE AND PERSISTERS

As a consequence of transient switch, multidrug-resistant bacterial cells become persister cells associated with antibiotic treatment failure (Helaine and Kugelberg, 2014), which can be exemplified in case as bacterial phenotypic diversity (Dhar and McKinney, 2007). Particular switching rates regulates the conversion of regular bacterial cell to persister cells and vice versa (Figure 4.1). These switching rates and cell types are directing the persisters level in a bacterial population (Patra and Klumpp, 2013). Although switching can occur autonomously, the growth phase and environmental factors also have a significant impact (Keren et al., 2004). Existence of persistence has been well explained in ESKAPE pathogens and several bacterial species such as *Mycobacterium tuberculosis*, *Escherichia coli*, and *Salmonella typhimurium* (Harms et al., 2016; Michiels et al., 2016). However, this phenomenon is not only present in bacteria, but has also been reported in eukaryotes such as *Candida albicans*, *Saccharomyces cerevisiae* and cancerous cells.

The experimental signature of persistence can be determined by the 'biphasic killing' phenomenon, by the administration of fatal dosage of antimicrobial drug into a bacterial community; further, the remaining live cells were enumerated and monitored eventually (Figure 4.2). In the initial phase,

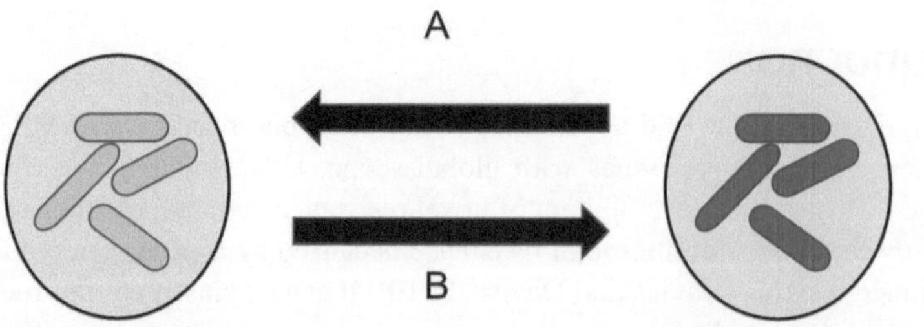

FIGURE 4.1 Basic principles of persistence. Interconversion of normal and persister cells at definite shifting rates denoted as A and B in an isogenic bacterial population.

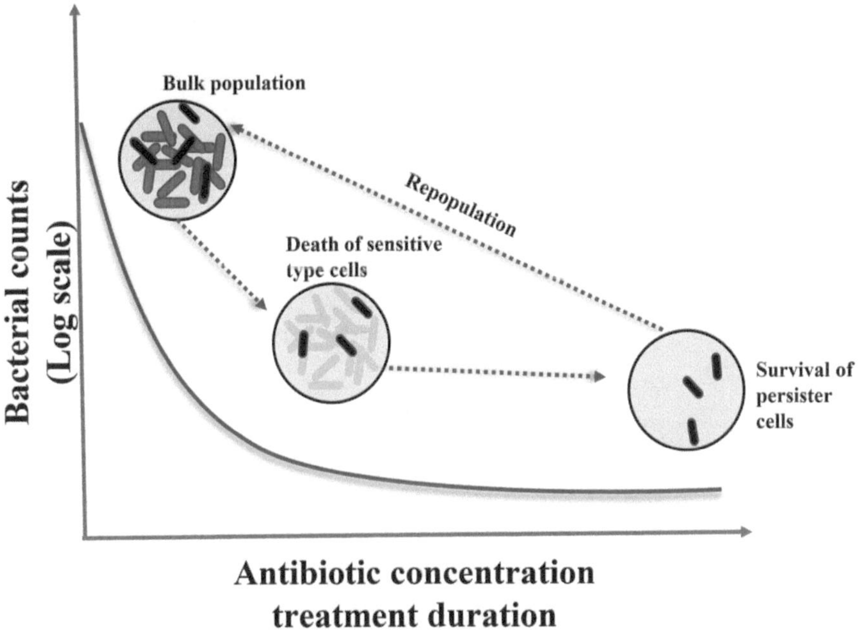

FIGURE 4.2 Biphasic killing curve in response to bactericidal antibiotics. Bacterial cell counts are plotted against the antibiotic concentration and/or treatment duration and depicted by a red curve. The above graphical interpretation shows the rapid killing of bacterial cells in the first phase that occurs due to the increased concentration of antibiotics and/or time duration. In the second phase, the slower killing rate indicates the persistent resuscitation rate. When antibiotic therapy is stopped, the population can be regenerated by active viable persisters.

there will be rapid declination of sensitive type cells. The latter state is followed by gradual death rate, which is due to the existence of extremely antibacterial drug-tolerant subpopulation of cells, which indicates the poor evacuation of persister population (Figure 4.2) (Balaban, 2012).

In recent years, much research has challenged the emergence of multidrug-resistant persisters, and provided the deeper perceptions about the mechanism that revert the persisters. Advanced technologies have been established for the study of persister cells, and various proposals are suggested for complete elimination. It was observed that the intrinsic potential of persister cells to withstand antimicrobial therapy, that can lead to the development of persistent infections without any genetic resistance phenomenon. In this chapter, we are addressing the emerging clinical problems of bacterial persistence and providing a summary of our knowledge about the current fundamental mechanisms. Specifically, we outline how persistence is correlated with resistance, the reappearance of persister population after the treatment has stopped, and the clinical relevance of persister cells. Furthermore, this chapter concludes multiple prospective methods for the control, prevention, and therapies against recurrent persister infections.

4.2.1 Resistence and Persistence

Antimicrobial resistance (AMR) and persistence are interconnected to an increased chance of treatment failure and chronic infections. They ultimately play a substantial role expanding the rates of morbidity and mortality that imparts high health expenditures.

Resistance can be obtained either by innate variation or by gene transfer (through a horizontal process) between bacterial species to the respective antimicrobial drugs which were susceptible before, resulting in enhanced chronic bacterial infection enabling severe illness and death (Smith,

2017). Whereas, persistence is the potential of bacteria to survive for a long time inside their host in the presence of bactericidal antibiotics without acquiring resistance-conferring genetic mutations (Cohen et al., 2013; Fisher et al., 2017). Contrary to antibiotic resistance, referred to as a bacteria's inherited potential to thrive in the proximity of an antibacterial drug, while persistence is related to the emergence of distinctive persister cells that tentatively exhibit phenotypic endurance to antibacterial therapies. These specialised adapted cells enter a physiologically dormant state to prevent being killed by antibiotics (Harms et al., 2016) (Figure 4.3).

Bacterial persistence is a kind of diversity in the phenotypes, where certain isogenic bacterial cells can withstand tolerance towards higher doses of antibiotics. The mechanism of antibiotic resistance is gene-based, which permits the bacterial cells for its continuous growth when exposed to antibiotics (Maclean et al., 2016) (Figure 4.3). The contrast between persistence and resistance is emphasised in Table 4.1.

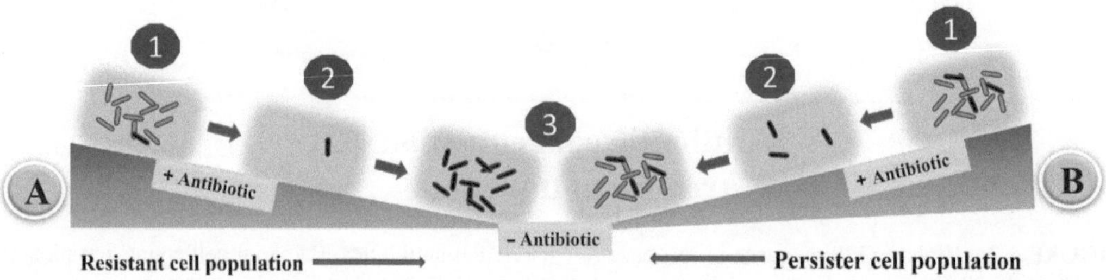

FIGURE 4.3 Resistance vs persistence. (A) (Adams et al., 2011) Initially bacterial population consists of both sensitive and resistant cells. (Amato et al., 2014) after antibiotic treatment (+ Antibiotic), only resistant cells remain unaffected. (Balaban, 2012) upon regrowth, (– Antibiotic), the entire population consists of resistant cells. (B) (Adams et al., 2011) Initially, bacterial population composed of both sensitive and persister cells. (Amato et al., 2014) Alternatively, bilateral phenotypic transformation to a lenient state may result in persister cells in the population. As a result, after exposure to antimicrobial drug (+Antibiotic), the remaining cells became persister type. (Balaban, 2012) After the termination of antibiotic treatment (– Antibiotic), these specialised cells revert to sensitive cells and starts multiplying hence exhibits the original population.

TABLE 4.1
Distinction between Resistance and Persistence

Resistance	Persistence
It is a congenital capacity of microbes that can grow at higher concentrations of antimicrobial drugs.	It is the capability of bacterial community to thrive temporary exposure to antibiotics.
It is based on genetic changes.	It is based on phenotypic heterogeneity.
Resistant cells are irreversible: they cannot revert to normal cells after the termination of treatment.	Persister cells are reversible.
These cells have the capacity to divide when exposed to medication.	These cells are unable to divide in the presence of bactericidal drugs.
Resistant cells are not genetically identical.	These cells are genetically similar to bacteria.
It can be quantified by MIC (Minimum Inhibitory Concentration) which is defined as the least amount of antibacterial drug necessary for the inhibition of bacterial population.	With MDK (minimal duration of killing), the tolerance level of antibiotics among bacterial strains can be correlated.
Resistant cells can be detected by antibiotic sensitivity tests.	Persistent cells can be evaluated easily by the biphasic killing curve.

4.2.2 Persistence and the Emergence of Drug Resistance

Persisters might serve as a 'catalyst' for the emergence of resistance that aids in the direct contribution to the failure of clinical therapies. In fact, persisters are a group of viable cells while antibiotics are present, and mathematical modelling predicts the increased rate of resistance via persistence (Levin and Rozen, 2006). In the persister subpopulation, growth usually does not occur as the cells are non-dividing hence the non-growing stressed cells cause mutations by a unique mechanism known to be stationary phase mutation or stress-induced mutation which is independent of DNA replication (Loewe et al., 2003). In growth-restricted conditions, it has been observed that a number of stress responses related to persistence increases the process of mutation in persister cells (Cohen et al., 2013).

Whilst the persistence and resistance are basically different mechanisms, persister cells and antibiotics resistant mutants sometimes appear to rely on similar strategies to overcome the detrimental effects of antibiotic treatment. Efflux pumps are crucial antibiotic resistance mechanism and have also been associated with intracellular persistence in *Mycobacterium* (Adams et al., 2011).

Even though the compelling theoretical considerations, the relationship between persistence and resistance evolution remains largely unexplored experimentally. However, it is uncertain if both the mechanisms are different or complementary to evolutionary responses to antibiotics. Understanding both microbial persistence and resistance is crucial in order to prevent AMR.

4.3 PERSISTER CELL DORMANCY

The major cause for the biofilm tolerance to antibiotics is because of the dormancy and reduced growth of the persister cells. Persister cells of *S. aureus* were initially described by Hobby et al. after treatment with penicillin, 1% of the cells were unaffected and showed that it was resistant to penicillin (Hobby et al., 1942). Bigger confirmed in 1944 the dormancy of persister cells by (i) reducing the optimal culture temperature to generate persister formation; (ii) decanting the culture media; and (iii) growth inhibition by the addition of boric acid, that causes the generation of persisters (Kim and Wood, 2016). Later, studies confirmed that they are metabolically inactive cells; for instance, it was shown that these dormant cells are resistant to antibiotic, ofloxacin. Studies reported that these cells become resistant towards antibiotics due to lack of protein synthesis and rise in persistence was confirmed by pretreatment of cells with (i) Carbonyl cyanide m-chlorophenyl hydrazone, which inhibits synthesis of ATP; (ii) rifampicin, which inhibits transcription of mRNA; and (iii) tetracycline (stops translation of proteins). After the treatments, a rise of around 10,000-fold was reported in the bacterial persister population (Wood, 2016). Extracellular factors, ciprofloxacin that trigger the persistence by the induction of TisB toxin at subinhibitory concentration, and also due to interspecies and the interkingdom signalling molecule, indole (Wood et al., 2013). Transcriptome analysis revealed comparison between active gene of persisters and exponential phase cells showed reduction in protein synthesis related to energy production and chemotaxis, which indicates the dormant state of persister cells. There is little knowledge available on the dormant state of persister cells and also the genes involved in the activation of resuscitation in bacterial community (Wainwright et al., 2021).

4.4 MECHANISM OF PERSISTER FORMATION

The formation of persister cells are enhanced by several environmental factors such as cellular stress, nutrient stress, change in the carbon sources, pH changes, DNA damage, and also by the biofilm and macrophages (Fisher et al., 2017). There are two main types of persisters: spontaneously generating persisters and stress-induced persisters. Spontaneous persisters are formed at very low rates, while stress-induced are formed often and can cause antibiotic resistance through numerous independent pathways (Gollan et al., 2019). Multiple mechanism of persistence was reported and the best studied is described below:

4.4.1 Toxin-Antitoxin (TA) System

The main components of TA system are: (i) toxin, which hinder with the cellular mechanisms such as translation and (ii) antitoxin, which is an RNA or protein, that counteract with the toxins (Wood et al., 2013). These antitoxin molecules are primarily function for the maintenance of plasmid and phage resistance, and it reversibly activates the bacterial dormancy, leads to persistence (Gollan et al., 2019). Toxin is suppressed by the antitoxin, which affects the cell metabolic process such as replication, translation, etc. The antitoxin becomes unstable and degraded under stress condition because of the activity of guanosine tetraphosphate (ppGp) signalling cascade, thereby toxin get accumulated. The cells undergo dormancy via TA system and the persister cells generation helps to overcome the antibiotics (Jung et al., 2019). The connection between the persister formation and the TA system is first identified in hip mutant of *Escherichia coli*, where the proportion of persisters are higher when compared with the wild strain. Mutation in hipAB is responsible for the formation of persisters, due to the phosphorylation of glutamyl tRNA synthetase GltX by hipA toxin, thereby accumulation of ppGpp occurs (Fisher et al., 2017). Other toxins engaged in persister formation are RelE, MazF, YafQ, MqsR, and DinJ (Jung et al., 2019). In *E. coli*, contact dependent growth inhibition, CdiA toxins also involved in the persister formation. CDI at higher concentration causes the formation of persisters because there is proportionally less antitoxin to neutralise toxin, resulting in elevated levels of ppGpp, leading to persister formation (Gollan et al., 2019).

4.4.2 (p)ppGpp

The activation of TA system and persister formation is highly dependent upon the alarmone signalling, which is mainly composed of guanosine pentaphosphate and tetraphosphate. These signal molecules are synthesised and degenerated by RelS-SpoT homolog family enzyme (Zou et al., 2022). Evidence shown that there is reduction in the proportion of persisters when there is a deletion of TA modules or the decreased levels of (p)ppGpp. Increased levels of (p)ppGpp hinder with the transcription of mRNA in amino acid lacking bacteria and elevates the expression of stress genes. In *E. coli* under stress conditions, the expression of (p)ppGpp is controlled by RelA and SpoT, where RelA with the large ribosomal subunit provokes the (p)ppGpp, whereas under normal conditions, (p) ppGpp is degraded by SpoT (Jung et al., 2019).

4.4.3 Phosphate Metabolism

Phosphate metabolism is essential for the generation of persister cells. In persister cells, Phosphate is necessary for the ATP synthesis and ppGpp formation in persister cells. Mutation in the negative regulator of phosphate metabolism, PhoU causes the generation of metabolically hyperactive cell. Mutants are more susceptible towards antibiotics, when compared to the non-mutant cells (Wood et al., 2013). Under stress condition, HipA activates ppGpp synthetase RelA, leads to the increased concentration of ppGpp, which eventually induces the deterioration of antitoxin molecule of the TA system. An example is *HipBA* module is regulated by phosphate metabolism by liberation of HipA, leads to persistence (Jung et al., 2019).

4.4.4 SOS Response

DNA damage causes the activation of regulator, RecA through the binding of regulator to the single strand DNA, which forms a nucleoprotein that induces the self-degradation of LexA, causing the inactivation of SOS genes. In cells, normally an SOS response is in inactive conditions, but during stress conditions it gets activated. ROS are formed due to agents such as UV, drugs, and stress, which can cause damage to the DNA and thereby induce the SOS response. Another method of SOS response formation is iron and nutrient starvation, activating the ppGpp via induction of *RelA* and

SpoT genes. The growth rate of the *E. coli* was reduced when there was an increase in the levels of ppGpp, due to the alteration of transcription (Podlesek and Žgur Bertok, 2020).

4.5 FACTORS TRIGGERING PERSISTER FORMATION

Several factors interfere with the formation of persister cells such as nutritional stress, extracellular signals, and effector pathways are depicted in Figure 4.4 (Gollan et al., 2019).

4.6 PERSISTER CELL INFECTION

Previously, it was thought that the immune system clears the bacterial cells that cause infections, but later infections due to persisters in immunocompromised individuals (such as cancer and HIV patients) has been reported. Patients suffering from cystic fibrosis also reported the high incidence of *P. aeruginosa* persister cell infection after periodic treatment with antibiotics at higher doses (Jung et al., 2019). Persister cells are extremely resistant to traditional antimicrobial drugs, while it is efficacious for the elimination of metabolically active bacteria. In order to overcome the dormancy and resistance in persister cells, there needs to be more focus on compounds that target the cells by passive diffusion and killing the cells without using the cell's machinery, which are mostly inactive. Because of the low population of persister cells, studies with compounds are difficult, so pretreatment with rifampicin would increase the persister population, and thereby, compounds can be tested directly (Wood, 2016). Persister cells are adapted with various mechanisms to overcome the harsh environmental conditions, antibiotics, etc. One such mechanism is the formation of biofilm, which is highly recalcitrant towards the antibacterial treatment and very difficult to eradicate. Another mechanism of drug tolerance is the expression of efflux pump proteins to overcome the antibiotic effect; other way such as SOS responses also increases the drug tolerance of persister cells. In *P. aeruginosa*, increase in persister levels causes the loss of function mutation in the genes of MexXY-OprM regulator such as *mucA*, *mutL*, *mutS*, and *mexZ*, leads to generation of mucoid phenotype (increased alginate production), and also efflux pump inactivation, observed in patients suffering from cystic fibrosis and respiratory infections. This is mainly because of the hypermutation in bacteria, there is high incidence in the level of persisters with antibiotic resistance and finally failure of the antimicrobial therapy (Khan et al., 2020).

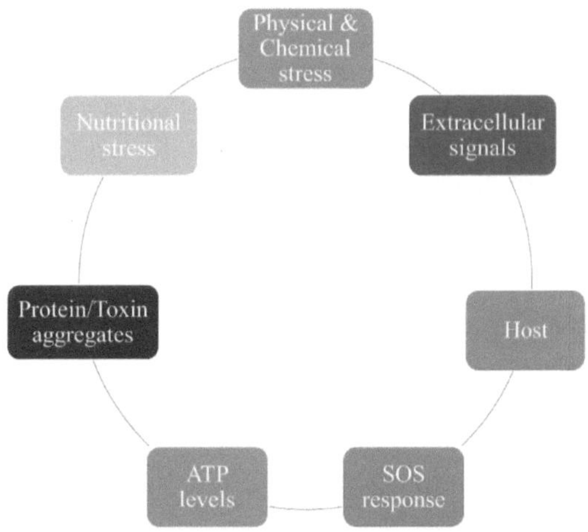

FIGURE 4.4 The main triggers for the formation of persister cells.

4.7 RESUSCITATION OF PERSISTER CELLS

Once the stress is over, the metabolically dormant persister cells will revert to the growth phase, where they actively grow and cause infections. The mechanism behind the resuscitation is not fully understood, but according to Yamasaki et al. (2020), reversion of *E. coli* persister cells into functional states depends upon the sensing the nutrients.

4.7.1 Sensing of Nutrients

Nutrients induce the awakening of persister cells, which makes them to grow in an exponential phase. Resuscitation of persistence can be triggered via the presence of a single amino acid, which was studied in *E. coli* in four groups containing five amino acids, acting as a carbon source. After incubation, reversion of persisters was observed in some treatment plates. Later, they studied amino acids individually, finally identifying alanine as the major reason for the reversion of persisters. In control groups without any amino acids, no growth was identified. Then they tested whether the persister cells wake in the absence of alanine,with alanine, ampicillin is added, which kills the normal cells. It was observed that after the waking of the persister, lysis happened due to ampicillin; in a control group without ampicillin no death happened, indicating that the persister cells resuscitate by recognising the nutrients rather than spontaneously. For further confirmation, whether alanine acts as an exact signal for the reversion of persisters, *dadA* isogenic mutation on *E. coli* were studied. Mutation in the *dadA* gene caused inactivation of D-amino acid dehydrogenase, unable to metabolise alanine. Around 4% of the cells wake after the treatment, but no cells wake without alanine. In the same study, instead of alanine, pyruvate is used, but no waking of cells was observed, indicating alanine to be the proper signal for the reversion of persister cells (Yamasaki et al., 2020).

4.7.2 Chemotaxis

Motility helps bacterial cells to move towards the nutrients or chemicals using chemotaxis system, are inhibited when the bacteria undergo dormant state. The chemotaxis protein affects the persister resuscitation was described by Yamasaki et al., 2020. In this study, they used pCA24N derived plasmids, synthesises *E. coli* protein (ASKA) which was electroporated into *E. coli* BW25113 and persister cells formed were plated onto alanine + chloramphenicol containing plates for the retention of plasmids. After eight days, the colonies were extracted and sequenced to find the proteins thought to stimulate persister resuscitation. Single cells containing ASKA plasmid are grown on agarose gel pads containing Ala and after 18 h incubation, it was compared with persisters containing empty plasmid. Those cells that expressed the regulators, CheA and CheY were resuscitated with an enhancement of 27-fold and 33-fold. Further, deletion mutants such as ΔCheY and ΔCheA were studied, which showed the resuscitation were totally inhibited. This depicted the necessity of chemotaxis system in persister reversion and sensing of nutrients (Wainwright et al., 2021).

4.7.3 Ribosomal Resuscitation

Elevated levels of cAMP can cause the decreased rate of resuscitation in persister cells by the hibernation of ribosomes. When they produce the cAMP inhibitor CpdA, the rate of resuscitation of the persisters were increased, where greater effect was observed when treated with drug atropine and CpdA in combination, decreases the levels of cAMP. Cells that produce the ribosome resuscitation factor (HfX) are capable of faster resuscitation than the cells expressing ribosome rescue factor (ArfB), which regulates cAMP and the control cells with empty plasmid (Wainwright et al., 2021).

4.8 STRATEGIES TO CONTROL PERSISTENCE

The significance of persisters in clinical settings are increased and the treatment strategies for the eradication of persisters are in developing stage. The biphasic growth and antibiotic resistance of persisters makes the eradication process harder. Reports shown that the definition of persistence and anti-persister strategy was initially developed in 1994, but sensitive cells are killed directly by the antibiotics, not the persister cells. These persister cells undergo a dormant state, and wake once the treatment stops (Van den Bergh et al., 2017). Dosage dependent treatment practice is efficacious in the eradication of revived persister cells, and it was reported with *P. aeruginosa*. The limitation to this strategy is the development of resistance towards antimicrobial agents and the treatment duration varied according to patient (Van den Bergh et al., 2017). Antiseptic usage can kill the persister cells, but due to its toxicity when in contact with the tissue, it cannot used for systemic application. To prevent contact of antiseptics with the cells, nontoxic antiseptics in covalent bonding with the antimicrobial entity to a surface are developed recently (Lewis, 2007). Possible strategies to eradicate persister cells is provided in Table 4.2.

TABLE 4.2
Action of Various Drugs and Combination against Various Persister Cells of Pathogenic Organisms

Sl. No	Drug	Action	Organism	References
Killing of Persister Cells				
1	Gentamicin	Metabolic stimuli cause PMF generation via sugars	*S. aureus*	Conlon (2014)
2	Fluoroquinolone HT61	Enhance antibiotic efficacy	*S. aureus*	Defraine et al. (2018)
3	Mitomycin C	DNA crosslinking	*E. coli* *P. aeruginosa* *S. aureus* *Borrelia burgdorferi*	Wood (2016)
4	Antibiotic acyldepsipeptide ADEP4	Protein degradation	*S. aureus*	Wood (2016)
5	SPI009 (1-((2,4-dichlorophenethyl) amino)-3-phenoxypropan-2-ol)	Membrane damage	Gram-postitive and Gram-negative bacteria	Defraine et al. (2018)
6	Bedaquiline	ATP Synthase	*Mycobacterium tuberculosis*	Kaldalu et al. (2020)
7	Delamanid	Blocks mycolic acid production	*Mycobacterium tuberculosis*	Kaldalu et al. (2020)
8	TCA1	Cell membrane	*Mycobacterium tuberculosis*	Van den Bergh et al. (2017)
9	Cisplastin and anthracycline	DNA crosslinker	*E. coli* *P. aeruginosa* *S. aureus* *B. burgdorferi*	Defraine et al. (2018)
10	NH125	Membrane rupture	Methicillin resistant *S. aureus*	Van den Bergh et al. (2017)
11	Bacteriophage produced LysH5 and CF-301	Peptidoglycan degradation	*S. aureus*	Defraine et al. (2018)
12	Peptide-lysin conjugate Art-175 and lipidated lysin 9	Membrane damage	*E. coli* *P. aeruginosa* *A. baumannii*	Defraine et al. (2018)

(*Continued*)

TABLE 4.2 (Continued)

Sl. No	Drug	Action	Organism	References
Persister Cells Sensitisation				
13	C10 and *cis*-2-decanoic acid	Metabolic stimulation	*E. coli* *P. aeruginosa*	Defraine et al. (2018)
14	Tobramycin and antimicrobial peptide conjugation	Upgrades the antibiotic activity	*S. aureus* *E. coli*	Van den Bergh et al. (2017)
15	Aminoglycosides and sugars	Enhanced killing efficiency	*Listeria monocytogenes*	Van den Bergh et al. (2017)
16	C10 and FQ Norfloxacin	Respiration	*M. tuberculosis*	Kaldalu et al. (2020)
17	Silver ions and antibiotics	Membrane permeability	*E. coli*	Defraine et al. (2018)
18	Penetratin protein sequence and tobramycin	Membrane disruption	*E. coli* *S. aureus*	Defraine et al. (2018)
19	Pyrazinamide	Inhibit persister formation	*E. coli* *M. tuberculosis*	Van den Bergh et al. (2017)
20	Pyrazinamide, Daptomycin, Cefoperazone, Doxycycline	Killing persister cells	*B. burgdorferi*	Wood (2016)
21	Brominated furanone (Z)-4-bromo-5-(bromomethylene)-3-methylfuran-2(5H)-one	Waking of persister cells	*P. aeruginosa*	Wood (2016)
22	*Cis*-2-decanoic acid and Ciprofloxacin	Protein synthesis inhibition and waking of persisters	*P. aeruginosa*	Wood (2016)

4.8.1 Killing of Persister Cells

The traditional practice in controlling the persister cells is by applying the conventional antibiotics to inactivate the target, but some metabolic functions remain active in the cells. Modern insights into pathways and development of more potent antimicrobials, or finding alternative targets with conventional antibiotics to overcome the tolerance towards antimicrobial agents, is gaining importance (Van den Bergh et al., 2017). An alternative strategy is to eradicate persister cells, which cause bacterial membrane destruction, DNA damage, inhibition of enzymes, and free radical generation. Another study in persister cells reported that addition of pyruvate, glycolysis intermediate, shown success in killing of persisters through the formation of proton motive force (pmf), which makes them susceptible to aminoglycosides (Wood et al., 2013).

Studies that target the bacterial membrane and the associated proteins were reported, which are considered as an ideal target against persister cells. Fluoroquinolone HT61 enhances the efficiency of antibiotics by destroying the cell wall of *Staphylococcus aureus* and thereby kill the dormant and non-dormant cells (Defraine et al., 2018). Some compounds interact with the lipid bilayer directly by disrupting the membrane function, which may be lipophilic, hydrophobic, or ionic in nature. Through arginine and tryptophan containing cationic membrane penetration peptides, the eradication of *E. coli* persister cells is possible by targeting cell wall component, lipopolysaccharide, finally causing membrane alteration and cell deaths, reported recently by Khan et al. (2020). It was also reported that persister cell and biofilm formation in *Mycobacterium tuberculosis* was inhibited by TCA1 by targeting the cell membrane (Van den Bergh et al., 2017).

Repurposing of anticancer drug also reported to eradicate the persister cells by targeting DNA. Mitomycin C, an anticancer drug was effective against both planktonic and biofilm form of several pathogens were reported *in vivo*. Mitomycin acts as a DNA crosslinking agent (5′-GC-3′), enters the cell through passive diffusion and gets activated by the cytoplasm reduction (Van den Bergh et al., 2017). Another study indicated that Mitomycin C was active against pathogens such as

E. coli, P. aeruginosa, S. aureus, Borrelia burgdorferi and *Acinetobacter baumannii* (Defraine et al., 2018). Another anticancer DNA crosslinker, cisplastin, and the antibiotic anthracycline was reported to kill the persisters of *E. coli, P. aeruginosa* and *B. burgdorferi* respectively (Defraine et al., 2018).

Another way to kill the persister cells is the use of phage or phage components. It has been reported that using phage derived or synthesised lysins that destroys persister cells via peptidoglycan degradation (Gollan et al., 2019). In a study, phage part bacteriophage encoded endolysin KZ144 was conjugated with the recombinant protein for transportation via outer membrane to form artilysin, having the potential to eradicate *P. aeruginosa* as well as *Acinetobacter baumannii* persister cells (Defraine et al., 2018). Staphylococcal bacteriophage sb-1 is reported to lyse the *S. aureus* persister cells and also disperse the biofilm by destroying the exopolysaccharide in combination with several antibiotics. LexA3, an engineered phage, increased the killing efficiency of persister cells of *E. coli* (Khan et al., 2020). The phage-derived enzyme, Lysin CF-301, has demonstrated the ability to degrade the peptidoglycan layer of Gram-positive organisms. Another bacteriophage enzyme, LysH5 was active against *S. aureus* and also *S. epidermidis*, killing cells and inhibiting biofilm formation (Khan et al., 2020).

4.8.2 Persister Cell Sensitisation

Antibiotics mainly targets the cellular metabolism but due to the dormant state of the persister makes them resistant to antibiotics. Another way to kill persisters is the resuscitation of persister cells, which could enhance the activity of antibiotics. Fatty acid cis-2-decenoic, and silver, triggers the persister cells for the uptake of disinfectant, triclosan, and aminoglycosides respectively. Another study reported that the *S. aureus* and also *E. coli* persister cells were eliminated when treated with an antimicrobial peptide, conjugated with the antibiotic tobramycin, through the transformation of antibiotic into self-transporting aminoglycoside. Sugars such as mannitol and fructose enhance eradication of *E. coli* and *P. aeruginosa* persister cells by increasing the proton motive force, thereby inducing the aminoglycosides (van den Bergh et al., 2017). Another study reported that sugars like glucose, fructose, mannitol, and pyruvate in combination with gentamicin are effective against *E. coli* and *S. aureus* persister cells by reverting the persistence, via the formation of pmf to increase the uptake of aminoglycosides (Khan et al., 2020). Stimulation of antibiotic influx is another strategy to eradicate persistence, which is accomplished by permeation or disrupting the membrane, helps in the antibiotics transport into the cell. The membrane porosity of *E. coli* (in both active as well as dormant cells) are enhanced by silver ions, which helps in killing cells with different antibiotics. The addition of an engineered amino acid sequence from penetratin, a cell-permeating peptide into tobramycin, was reported to eradicate the persisters of *E. coli* as well as *S. aureus*, by circumventing the active transport through membrane disruption (Defraine et al., 2018).

4.8.3 Combination Therapy

The development of biofilm and persister cells makes the antibiotic more resistant and treatment with single antibiotic is became ineffective. Treatment with monotherapy drugs requires frequent application of antibiotics, and careful monitoring is needed. In order to avoid such problems, an alternative strategy of using a combination of drugs has recently been considered for treatment. The main aims of combination therapy are: (i) enhancing the additive nature of drug on numerous targets; and (ii) the controlled and stable drug delivery (Khan et al., 2020). Combination of antibiotics with other different classes of antibiotics enhances the treatment via multi-target prolonged attack and leads to the eradication of subpopulation of cells. Combination of drugs such as doxycycline and daptomycin showed activity against proliferating cells of drug resistant *Borrelia burgdorferi*, while doxycycline and cefoperazone combination showed activity against non-proliferating cells, confirming the clinical application in Lyme disease treatment (Defraine et al., 2018). The activity of antibiotics such as gentamicin and ofloxacin against uropathogenic *E. coli* was enhanced in combination with colistin, which causes membrane damage to the bacterial cells. Moreover, *in vitro* and *in vivo* studies on combination of colistin with erythromycin were reported to eradicate the *P. aeruginosa* persister cells (Defraine et al., 2018). In the presence of mannitol, nisin and ampicillin combination can kill persister cells of *Salmonella* by removing 78% of the biofilms (Khan et al., 2020).

4.9 CONCLUSION

Tolerance towards antibiotics and the dormant state of the persister cells makes the treatment much harder. Due to the dormant condition of the persister cells, use of antibiotics for the treatment is unsuccessful, and the exact mechanism behind such dormancy and resistance patterns is unknown. Deeper insights into the persister formation as well as the resuscitation helps to recognise the mechanism behind, and also for the development of novel strategies to overcome the dormancy, resistance towards antibiotics, etc. Several studies had developed combination of various drugs which awakes the persisters from dormancy and another drug causes deleterious effects to the cells. But still, the connection between the persistence and the resistance mechanism is unambiguous, where the prolonged exposure to the antibiotic treatment causes increase in the formation of persisters. Persister formation and the development of mutation in the pathogenic species causes more problematic situations to the healthcare system; therefore, it is essential to develop anti-persister therapies.

REFERENCES

Adams, K. N., K. Takaki, L. E. Connolly, H. Wiedenhoft, K. Winglee, O. Humbert, P. H. Edelstein, C. L. Cosma, and L. Ramakrishnan. 2011. Drug Tolerance in Replicating Mycobacteria Mediated by a Macrophage-Induced Efflux Mechanism. *Cell* 145: 39–53. https://doi.org/10.1016/j.cell.2011.02.022

Amato, S. M., C. H. Fazen, T. C. Henry, W. W. K. Mok, M. A. Orman, E. L. Sandvik, K. G. Volzing, and M. P. Brynildsen. 2014. The Role of Metabolism in Bacterial Persistence. *Front Microbiol* 5: 1–9. https://doi.org/10.3389/fmicb.2014.00070

Balaban, N. Q. 2012. *Phenotypic Switch*, 1622. https://doi.org/10.1126/science.1099390

Bigger, J. W. 1944. Treatment of Staphylococcal Infections with Penicillin by Intermittent Sterilisation. *The Lancet* 244: 497–500. https://doi.org/10.1016/S0140-6736(00)74210-3

Cohen, N. R., M. A. Lobritz, and J. J. Collins. 2013. Microbial Persistence and the Road to Drug Resistance. *Cell Host Microbe* 13: 632–642. https://doi.org/10.1016/j.chom.2013.05.009

Conlon, B. P. 2014. Staphylococcus Aureus Chronic and Relapsing Infections: Evidence of a Role for Persister Cells. *Bioessays*, 36: 991–996.

Davies, J., and D. Davies. 2010. Origins and Evolution of Antibiotic Resistance. *Microbiol Mol Biol Rev* 74: 417–433. https://doi.org/10.1128/MMBR.00016-10

Defraine, V., M. Fauvart, J. Michiels. 2018. Fighting Bacterial Persistence: Current and Emerging Anti-Persister Strategies and Therapeutics. *Drug Resist Updat* 38: 12–26. https://doi.org/10.1016/j.drup.2018.03.002

Dhar, N., and J. D. McKinney. 2007. Microbial Phenotypic Heterogeneity and Antibiotic Tolerance. *Curr Opin Microbiol* 10: 30–38. https://doi.org/10.1016/j.mib.2006.12.007

Fisher, R. A., B. Gollan, and S. Helaine. 2017. Persistent Bacterial Infections and Persister Cells. *Nat Rev Microbiol* 15: 453–464. https://doi.org/10.1038/nrmicro.2017.42

Gollan, B., G. Grabe, C. Michaux, and S. Helaine. 2019. Bacterial Persisters and Infection: Past, Present, and Progressing. *Annu Rev Microbiol* 73: 359–385. https://doi.org/10.1146/annurev-micro-020518-115650

Harms, A., E. Maisonneuve, and K. Gerdes. 2016. Mechanisms of Bacterial Persistence during Stress and Antibiotic Exposure. *Science* 354. https://doi.org/10.1126/science.aaf4268

Helaine, S., and E. Kugelberg. 2014. Bacterial Persisters: Formation, Eradication, and Experimental Systems. *Trends Microbiol* 22(7): 417–424. https://doi.org/10.1016/j.tim.2014.03.008

Hobby, G. L., Meyer, K., and Chaffee, E. 1942. Observations on the Mechanism of Action of Penicillin. *Exp Biol Med* 50: 281–285. https://doi.org/10.3181/00379727-50-13773

Jung, S. H., C. M. Ryu, and J. S. Kim. 2019. Bacterial Persistence: Fundamentals and Clinical Importance. *J Microbiol* 57(10): 829–835. https://doi.org/10.1007/s12275-019-9218-0

Kaldalu, N., V. Hauryliuk, K. J. Turnbull, A. La Mensa, M. Putrinš, and T. Tenson. 2020. In Vitro Studies of Persister Cells. *Microbiol Mol Biol Rev* 84. https://doi.org/10.1128/MMBR.00070-20

Keren, I., N. Kaldalu, A. Spoering, Y. Wang, and K. Lewis. 2004. Persister Cells and Tolerance to Antimicrobials. *FEMS Microbiol Lett* 230(1): 13–18. https://doi.org/10.1016/S0378-1097(03)00856-5

Khan, F., D. T. N. Pham, N. Tabassum, S. F. Oloketuyi, and Y. M. Kim. 2020. Treatment Strategies Targeting Persister Cell Formation in Bacterial Pathogens. *Crit Rev Microbiol* 46(6): 665–688. https://doi.org/10.1080/1040841X.2020.1822278

Kim, J. S., and T. K. Wood. 2016. Persistent Persister Misperceptions. *Front Microbiol* 7: 1–7. https://doi.org/10.3389/fmicb.2016.02134

Levin-Reisman, I., I. Ronin, O. Gefen, I. Braniss, N. Shoresh, and N. Q. Balaban. 2017. Antibiotic Tolerance Facilitates the Evolution of Resistance. *Science* 355: 826–830. https://doi.org/10.1126/science.aaj2191

Levin, B. R., and D. E. Rozen. 2006. Non-Inherited Antibiotic Resistance. *Nat Rev Microbiol* 4(7): 556–562.

Lewis, K. 2007. Persister Cells, Dormancy and Infectious Disease. *Nat Rev Microbiol* 5: 48–56. https://doi.org/10.1038/nrmicro1557

Loewe, L., V. Textor, and S. Scherer. 2003. High Deleterious Genomic Mutation Rate in Stationary Phase of Escherichia coli. *Science* 302: 1558–1561.

Maclean, R. C., T. Vogwill, A. C. Comfort, and V. O. Furi. 2016. Persistence and Resistance as Complementary Bacterial Adaptations to Antibiotics. *J Evol Biol* 29: 1223–1233. https://doi.org/10.1111/jeb.12864

Michiels, J. E., B. Van den Bergh, N. Verstraeten, and J. Michiels. 2016. Molecular Mechanisms and Clinical Implications of Bacterial Persistence. *Drug Resist Updat* 29: 76–89. https://doi.org/10.1016/j.drup.2016.10.002

Moradali, M. F., S. Ghods, and B. H. A. Rehm. 2017. Pseudomonas Aeruginosa Lifestyle: A Paradigm for Adaptation, Survival, and Persistence. *Front Cell Infect Microbiol* 7: 39. https://doi.org/10.3389/fcimb.2017.00039

Patra, P., and S. Klumpp. 2013. Population Dynamics of Bacterial Persistence. *PloS One* 8. https://doi.org/10.1371/journal.pone.0062814

Podlesek, Z., and D. Žgur Bertok. 2020. The DNA Damage Inducible SOS Response Is a Key Player in the Generation of Bacterial Persister Cells and Population Wide Tolerance. *Front Microbiol* 11: 1–8. https://doi.org/10.3389/fmicb.2020.01785

Smith, M. 2017. *Antibiotic Resistance Mechanisms. Journeys in Medicine and Research on Three Continents Over 50 Years*, 95–99. https://doi.org/10.1142/9789813209558_0015

Van den Bergh, B., M. Fauvart, and J. Michiels. 2017. Formation, Physiology, Ecology, Evolution and Clinical Importance of Bacterial Persisters. *FEMS Microbiol Rev* 41: 219–251. https://doi.org/10.1093/femsre/fux001

Wainwright, J., G. Hobbs, and I. Nakouti. 2021. Persister Cells: Formation, Resuscitation and Combative Therapies. *Arch Microbiol* 203: 5899–5906. https://doi.org/10.1007/s00203-021-02585-z

Wood, T. K. 2016. Combatting Bacterial Persister Cells. *Biotechnol Bioeng* 113: 476–483. https://doi.org/10.1002/bit.25721

Wood, T. K., S. J. Knabel, and B. W. Kwan. 2013. Bacterial Persister Cell Formation and Dormancy. *Appl Environ Microbiol* 79: 7116–7121. https://doi.org/10.1128/AEM.02636-13

Yamasaki, R., S. Song, M. J. Benedik, and T. K. Wood. 2020. Persister Cells Resuscitate Using Membrane Sensors that Activate Chemotaxis, Lower cAMP Levels, and Revive Ribosomes. *IScience* 23: 100792. https://doi.org/10.1016/j.isci.2019.100792

Zou, J., B. Peng, J. Qu, and J. Zheng. 2022. Are Bacterial Persisters Dormant Cells Only? *Front Microbiol* 12: 1–10. https://doi.org/10.3389/fmicb.2021.708580

5 Alternative Strategies for Combating Antibiotic Resistance in Microorganisms

Subhaswaraj Pattnaik, Monika Mishra, and Pradeep Kumar Naik
Sambalpur University, Sambalpur, India

CONTENTS

5.1 INTRODUCTION

The antibiotic drug discovery is an important breakthrough in the history of biomedical sectors in the 20th century to tackle the chronic microbial infections. The emergence of antibiotics has revolutionised not only the conventional treatment strategies, but also critically improved life expectancies. Hence, antibiotics are considered 'Magical Bullets' in the history of mankind in providing

DOI: 10.1201/9781003345299-5

widespread public health benefits for disease prevention and development of therapeutic regimens against several infectious diseases (Hutchings et al., 2019; Cook and Wright, 2022). After penicillin's discovery, an extensive repertoire of antibiotics were reported from several sources, and are considered effective measures against infectious diseases, including microbial infections. Hence, the period from the 1950s to the 1970s is considered the 'Golden age of antibiotics' (Aminov, 2010; Hutchings et al., 2019). This golden period undoubtedly brings a paradigm shift in our research infrastructure for future drug discovery pipelines. However, the irrational and overuse of antibiotics, with inappropriate healthcare settings has substantially led to the antimicrobial resistance (AMR) (Aminov, 2010; Miethke et al., 2021). The resistance to antibiotics becomes a global health problem with greatest impact on healthcare infrastructure and the global economy.

5.1.1 Antibiotic Resistance: A Global Scenario

Antibiotics are the most effective drug that are delivered to improve human health. However, pathogenic microorganisms have developed an adaptive strategy of developing resistance to survive against therapeutic regimens. This adaptive transformation also facilitates several other secondary infections in hospital-acquired settings causing severe health burdens (Cameron et al., 2022). In addition, the recent incidence of severe acute respiratory syndrome Corona virus-2 (SARS-CoV-2) pandemics also created severe health hazards among immunocompromised patients. The immunocompromised patients in healthcare settings also face the serious concern of several other secondary infections associated with multidrug-resistant (MDR) bacteria which ultimately aided the ongoing AMR crisis worldwide. As per estimates, if we fail to address the issues with AMR, it could result in the death of 10 million people worldwide every year by 2050 (Diallo et al., 2020; Miethke et al., 2021). Based upon the healthcare hazards of AMR crisis, priority pathogens list was issued World Health Organisation (WHO) to mark their resistance to a multitude of next-generation antibiotics and that needs urgent attention from the scientific community to discover novel antimicrobial agents (Taconelli et al., 2018; Lewis, 2020).

Several anthropological and biological factors are responsible for the emergence of antibiotics. The anthropological factors such as the indiscriminate and irrational use of antibiotics, inappropriate prescription of antibiotics to treat microbial infections, extensive use in agricultural and livestock sectors, non-availability of new generation antibiotics in the market and regulatory approval issues in antibiotic discovery pipelines are the causes of AMR crisis (Ventola, 2015). Similarly, the generation of selective pressure owing to extensive use of antibiotics, restricted penetration ability of antibiotics, target modifications, the evolution of efflux machinery, target switching, target sequestration, and development of biofilms are notable factors in the generation of AMR crisis (Lewis, 2020). The priority pathogens as listed by WHO, particularly ESKAPE group of microorganisms (*Enterococcus faecium*, *Staphylococcus aureus*, *Klebsiella pneumoniae*, *Acinetobacter baumannii*, *Pseudomonas aeruginosa*, and *Enterobacter* sp.) are highly resistant towards next-generation antibiotics and create a global health hazard with characteristic influence on infection severity, increased mortality, and failure of therapeutic regimens. Hence, these pathogens require a coordinated global response for antimicrobial resistance surveillance (De Oliveira et al., 2020). ESKAPE pathogens also undergo several adaptive strategies to withstand environmental stress, including selective pressure associated with antibiotic misuse. These include production of specific enzymes which irreversibly modify and inactivate antibiotic dosage, mutation in antibiotic binding sites, reduced intracellular drug uptake and increased drug elimination, and ability to form recalcitrant biofilm matrix (Santajit and Indrawattana, 2016). ESKAPE microorganisms mediated nosocomial infections in hospital-acquired settings not only promote AMR phenomena but also pose a challenging economic crisis with the highest impact on low- and middle-income countries particularly in Asia and African continents (Dadgostar, 2019; Zhen et al., 2019).

5.1.2 Impact of Antibiotic Resistance on Public Health and Global Economy

The occurrence of resistance to antibiotics and the evolution of antibiotic-resistant genes (ARGs) in pathogenic microorganisms pose a serious health concern. As per the evidence-based studies, hospital-acquired settings with immunocompromised patients are more prone to nosocomial pathogen infections and drug resistance. Over the past 30 years, a characteristic evolutionary trend in the emergence of drug resistant microorganisms impose a severe threat to animal health. According to the report published in 2019, the bacterial AMR crisis played a pivotal role in the death of 4.95 million people with one or more infections in 2019. Among the deceased individuals, 1.27 million (approximately 25%) deaths occurred by direct influence of bacterial resistance. This staggering mortality rate is even higher than that of malaria and acquired immunodeficiency syndrome (AIDS) (Murray et al., 2022). From historical perspectives, resistance to available antibiotics significantly affected individuals' physical, mental, and psychological wellbeing and also controlled the socioeconomic wellbeing of human society globally. In the post-antibiotic era, the AMR crisis has critically transformed into globally acclaimed 'Silent Pandemics', which greatly affected human health with several other secondary infections in the majority of public health issues. Thus, the deaths associated with AMR pose a serious threat to human health and our health infrastructure (Michael et al., 2014; Founou et al., 2017). No doubt, India has become the manufacturing hub of the antibiotic market owing to cost efficiency. However, improper waste management skills and excessive and irrational emission of antibiotic residue to the environment have become a serious health issue in the form of environmental pollution (Zheng et al., 2021; Larsson and Flach, 2022).

As mentioned earlier, the rise of resistance to several classes of antibiotics and antimicrobial agents has significantly affected the socioeconomic wellbeing of low- and middle-income countries. This could be interpreted from a recent study which conferred an interesting finding where the overall defined daily doses (DDDs) of consuming antibiotics has been increased by 65%, whereas a 39% increase was observed in the antibiotic consumption rate (i.e., DDDs per 1,000 inhabitants) in the last 15 years. The excessive burden of the consumption of antibiotics in low- and middle-income countries thus critically affected their economy (Klein et al., 2018; Malik and Bhattacharyya, 2019; Allel et al., 2020). The incidence of AMR in the Research and Development sectors, critically increases the costs associated with therapeutic regimens under the current scenario. As per the estimates, if we fail to address the issues related to AMR phenomena, the global GDP could be 3–4 times lower than that with a low-AMR scenario by the year 2050 (World Bank report on "Drug Resistant Infections: A threat to our economic future", 2017). If this condition persists, by 2050, millions of deaths due to the AMR crisis will create an extra burden on the global economy ranging between US$60 Trillion to US$100 Trillion with the cost of healthcare settings would rise up to US$ 1.2 Trillion (Dadgostar, 2019). The increased prevalence of AMR could also eventually create an aided economic burden even on developed countries like United States of America (USA) as mentioned in Centers for Disease Control and Prevention (CDC) report. In the USA and on a global platform, the incidence of SARS-CoV-2 pandemics has inferred the resurgence of antimicrobial-resistant infections by more than 15% within the 2019–2020 financial year. Based upon the socioeconomic influence of AMR, it is important to design novel and effective strategic plans to mitigate the consequences of AMR. No doubt, several action plans were under exploration to minimise the severity of AMR on public health and the economy; we need to understand the basic biological cues associated with drug resistance patterns in pathogenic microorganisms. Based upon the insight understanding, we could formulate the strategic plan to develop effective diagnostic and therapeutic regimens.

5.1.3 Mechanism of Antibiotic Resistance

The most interesting aspect of antibiotic-resistant microorganisms is the adaptive ability to withstand antibiotic treatment and consequently survive in the antibiotic stress environment followed by reproductive mechanisms to transfer the ARGs to next-generation microbes through horizontal gene transfer. The resistance to antibiotics as evidenced from microbial pathogens has been categorised into four principal classes such as (i) natural or intrinsic resistance; (ii) acquired or adaptive resistance; (iii) cross resistance; and (iv) multidrug resistance (MDR).

- **Natural resistance**: In the case of intrinsic resistance, microorganisms naturally exhibit promising genetic plasticity to survive under antibiotic stress either by mutation in genes associated with antibiotic action or by acquiring foreign gene coding for resistance determinants through horizontal gene transfer (Munita and Arias, 2016). In addition, Gram-positive and Gram-negative bacteria exert specific mechanisms that facilitate decreased drug permeability with special influence on drugs having intracellular targets. The classic example of Vancomycin resistance in Gram-negative bacteria is that Vancomycin is unable to penetrate the outer membrane; thus, Gram-negative bacterial pathogens exert resistance to Vancomycin. Hence, Vancomycin could specifically be used against Gram-positive pathogens. It can be said that bacteria do not have the target sites for the drug, so the treated drug does not affect the bacteria and is naturally resistant to those agents. Apart from blockage to influx of antibiotics, pathogenic microorganisms also develop highly specific efflux pump machinery to extrude the antibiotics administered during treatment. Apart from that, efflux pump machinery is also a classic example of microorganisms with the specific sophisticated mechanism of self-defence against their own antibiotics (Munita and Arias, 2016; Peterson and Kaur, 2018). As per the classic trends, Gram-positive bacteria exert inherent resistance to Aztreonam class of antibiotics whereas Gram-negative microorganisms show intrinsic resistance towards glycopetides and lipopeptides class of antibiotics (Reygaert, 2018).
- **Acquired resistance**: Acquired resistance is an adaptive strategy that occurs due to the microbes' genetic modification. In this case, the bacteria are earlier susceptible to certain drugs or antibiotics, and become resistant owing to the genetic modification in the later stage of the life. Acquired resistance occurs either by mutation in genes associated with antibiotic action or by horizontal transfer of acquired foreign genes encoding antibiotic-resistant elements (Blair et al., 2015; Munita and Arias, 2016). These types of mutations occur mainly due to certain chemical or physical factors. The mutated chromosomes can pass from one to another bacterium in various ways with the horizontal gene transfer through plasmid mediated transmission becoming the most rapid and effective process (Reygaert, 2018). Biofilm formation in adverse environmental conditions is a classic example of adaptive resistance where the recalcitrant biofilm matrix forms a diffusive barrier against antibiotic access into the bacterial system. In addition, the inert metabolic activities within the biofilm matrix were also found to be suitable in resistance by modifying the antibiotic physiological targets (Pang et al., 2019).
- **Cross resistance**: Cross resistance in microbial pathogens occurs against the same class of antibiotics slowly over a period of time because it follows a single mechanism. Similarly, in any biological niche several species of pathogenic microorganisms coexist and in due course of time, few species also develop resistance to certain antibiotics by acquiring ARGs of the coexisting bacterial population (Ishii et al., 2021). Biofilms are a classic example of biological niches harbouring several groups of microorganisms that eventually involved in cross-species cooperation, enabling them to survive the antibiotic treatment regimens (Uruen et al., 2021; Luo et al., 2022).
- **Multidrug resistance**: In the current scenario, the evolution of pathogenic microorganisms in exhibiting resistance to several classes of antibiotics becomes a global trend. The multidrug-resistant (MDR) bacteria pose inherent potential to produce several classes of β-lactam antibiotic degrading enzymes called, β-lactamases which act upon the next-generation β-lactam antibiotics including Carbapenems, Cephalosporins, ampicillins,

among others (Bahr et al., 2021; Mojica et al., 2022). The majority of nosocomial pathogens including, ESKAPE pathogens, have the inherent potential to produce several classes of hydrolysing enzymes including β-lactamases which often infer MDR patterns in hospital-acquired infections (Galani et al., 2021).

5.2 FACTORS ASSOCIATED WITH ANTIBIOTIC DRUG RESISTANCE

Several biological, pathophysiological, and anthropological factors are considered as the key drivers in the emergence, transmission, and establishment of antibiotic resistance. These factors also facilitate the increase in public health hazards and socioeconomic implications (Chatterjee et al., 2018). In this section, the most prominent biological and pathophysiological drivers responsible for imparting drug resistance patterns in pathogenic microorganisms were discussed with special reference to the insight into their mechanism of antibiotic resistance (Figure 5.1).

5.2.1 Role of Quorum-Sensing and Biofilm Mechanics on Antibiotic Resistance

In antibiotic resistance mechanisms, bacterial virulence factors are the key regulators of mediating several pathophysiological responses. Quorum-sensing (QS) infers the cell density dependent cell–cell regulatory signalling cascade process which involves the secretion, detection, and response to extracellular signals called autoinducers (AIs). The QS regulatory network manages several pathophysiological events including production of virulence phenotypes, acquisition of nutrients, modulation of cellular functions, sporulation, induction of competence, production of antibiotics, transfer of genetic materials and biofilm formation and development (Rutherford and Bassler, 2012; Mukherjee and Bassler, 2019; Wu and Luo, 2021). The QS communication in bacterial pathogens is highly species-specific on detecting and responding to specific signal transduction systems with specific AIs. Normally, oligopeptides are considered as AIs in Gram-positive bacteria, whereas Gram-negative bacteria use acyl homoserine lactones (AHLs) as autoinducing signals. In addition, the presence of AI-2 signalling network confers the inter-species crosstalk (Duddy and Bassler, 2021).

The small peptides as AIs utilise two-component signalling systems as in *Staphylococcus aureus* (AgrC/A). The two-component transduction cascade involves in secretion of various toxins and degradable exoenzymes. As the concentration of AI peptides increases, they bind with the histidine kinase sensor, which, upon phosphorylation, mediate gene expressions. Meanwhile, Gram-negative bacteria follow AHL-based QS system, homologues of LuxI/R prototype as it was first observed in *Vibrio harveyi*. In LuxI/R regulatory network, LuxI corresponds to LuxI synthase which produces AHL signals. At optimum cell density, the AHLs bind to the response regulator, LuxR and modulate expression of virulent genes (Lazdunski et al., 2004; Papenfort and Bassler, 2016). In ESKAPE pathogens, the QS regulatory network plays a critical role from the beginning of nosocomial infections to host and drug resistance patterns. Hence, QS is considered a viable therapeutic target in developing a potential therapeutic regimen against AMR (Chadha et al., 2022).

Biofilms are self-assemblage of surface-associated microbial cells encased within an extracellular matrix which ultimately forms the pathophysiological barrier during stress conditions including antibiotic therapy. The extracellular matrix comprises of secreted exopolysaccharides, proteins and nucleic acids (extracellular DNA). The adaptive biofilms promote social cooperation between the encased microbial communities, enhanced ability for resource utilisation and enhanced survival efficacy following exposure to antibiotics (Donlan, 2002; Flemming et al., 2016; Muhammad et al., 2020). The QS regulation is considered as niche-specific pathway which controls the formation and development of biofilms based on the bacterial cell density above a threshold level. In addition, several biofilms associated phenotypes such as exopolysaccharides (EPS), alginate, rhamnolipid, and extracellular DNA (eDNA) are concomitantly regulated by QS hierarchical network. Several other QS virulence phenotypes control these biofilm factors' expression (Kjelleberg and Molin, 2002; Zhao et al., 2020; Zhou et al., 2020; Peerzada et al., 2022).

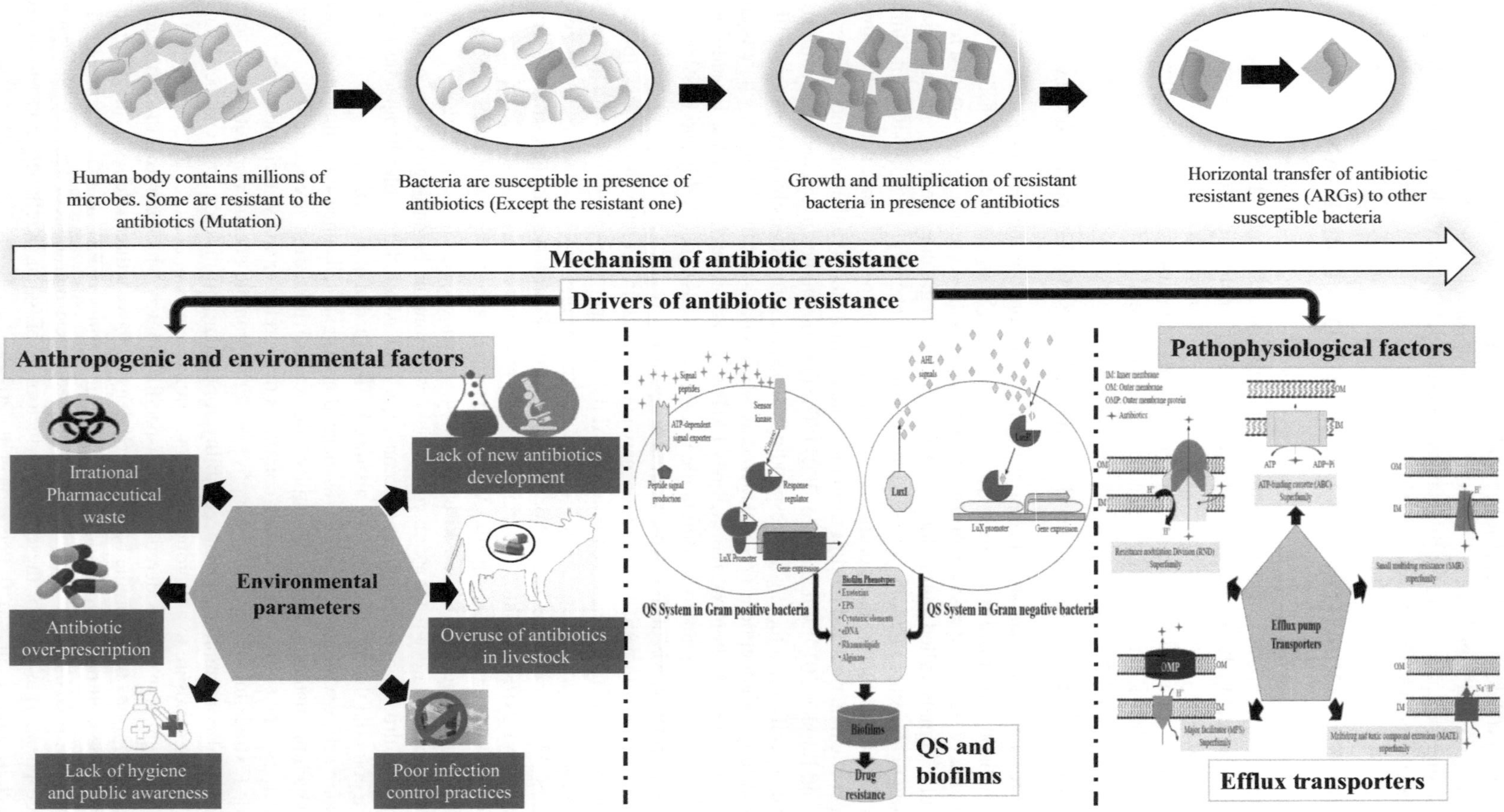

FIGURE 5.1 Schematic overview of the mechanism of antibiotic resistance and the key regulators (both environmental and pathophysiological factors) in the emergence of drug resistance patterns in pathogenic microorganisms.

Apart from that, the cellular transition of bacterial pathogens from planktonic state to sessile state (biofilm mode) also depends upon the modulation of signalling molecules and their signalling mechanisms (Dodson et al., 2022). The pathogenic microorganisms in the sessile biofilm state tend to release eDNA, that facilitate the remodelling of extracellular matrix and promote tethering of encased microbial community to form clusters by modulating acid base interactions. The release of eDNA also reported to facilitate the microbial competency by affecting the lateral gene transfer mechanisms. In addition, the release of eDNA involves in biofilm development and promotes antibiotic resistance by promoting the inactivation of cationic antibiotics (Das et al., 2013; Okshevsky and Meyer, 2015; Saxena et al., 2019). The QS-associated virulence phenotypes are critical in biofilm mechanics and facilitate the reduction of antibiotic penetration into the matrix, thereby promoting antibiotic resistance (Haque et al., 2021; Fan et al., 2022).

5.2.2 Effect of Multidrug Efflux Transport Machinery

The drug efflux transporter in microbial pathogens is an adaptive strategy to infer resistance to available antibiotics. These pump's main work is to remove the solutes present in the cell (Impey et al., 2020). The efflux transporters serve as the first line of defence against any foreign substances, including antibiotics, by decreasing the intracellular level of antibiotics (Randhawa et al., 2016; Rahman et al., 2017; Tran et al., 2020). Efflux pump permits the bacteria to adjust its internal environment by eradicating the toxic constituents including antimicrobial agents, certain secondary metabolites and quorum-sensing signalling fragments (Reygaert, 2018). The efflux pumps are composed of protein complexes localised on bacterial membrane and control the removal of toxins and antibiotics from the bacterial system to facilitate reduced sensitivity towards antibiotic treatment by promoting decreased cellular drug accumulation. Hence, efflux pump machinery are critical for inferring bacterial pathogenesis and antibiotic resistance (Venter et al., 2015; Blanco et al., 2016; Seukep et al., 2020; Thakur et al., 2021; Yoneda et al., 2022).

Five major classes of transporter superfamily such as adenosine triphosphate (ATP)-binding cassette (ABC), resistance-nodulation-division (RND), major facilitator superfamily (MFS), small multidrug resistance (SMR), and multidrug and toxic compound extrusion (MATE) are the major contributors of efflux transporters. The MFS efflux transporters are among the most predominant extruder system in pathogenic bacteria which are highly responsible for MDR phenomena (Blanco et al., 2016; Tambat et al., 2019). In addition, the efflux transporters also facilitate bacterial colonisation, QS controlled virulence and biofilms. As mentioned earlier, the QS regulatory network in pathogenic bacteria hovers around the release and sensing of signalling molecules, which depends not only upon the cell density, but also on the expression profiles of efflux transporters (Rezaie et al., 2018; Huang et al., 2022). Wolloscheck et al. (2018) also reported the differential kinetic properties of efflux transporters directly regulate the synthesis and pathophysiology of QS controlled virulence factors like cytotoxic phenazines (Wolloscheck et al., 2018). Hence, the key regulators of QS pathogenesis, biofilm mechanics and efflux transporters are inherently correlated and promote survival of microorganisms under antibiotic stress (Xu, 2016). Moreover, drug resistant factors like mutation in antibiotic targets and production of antibiotic degrading enzymes are concomitantly regulated by the presence of effective efflux transporters (Alibert et al., 2017). Since the efflux transporters and their regulated expression are critical in pathophysiological responses, these targeting the functional attributes of efflux transporters could be potential therapeutic regimens in the post-antibiotic era (Sharma et al., 2019).

5.2.3 Anthropogenic and Environmental Parameters Responsible for Drug Resistance

The indiscriminate and irrational exploitation of antibiotics to treat microbial infections has become a key anthropogenic factor in inferring drug resistance phenomena. These anthropogenic events facilitate the dissemination of ARGs from environmental and cellular resources and subsequently evolved to confer antibiotic resistance (Zhang et al., 2022). Several environmental factors such as disposal of excreted antibiotics to the soil microbiota, entry of genetic resistant elements into the soil through animal waste, irrational antibiotic waste management, lack of sanitation in drinking water linkages, uncontrolled hospital sewage disposal, inadvertent pharmaceutical waste disposal and transfer of antibiotic-resistant elements to human beings from food-producing animals (fed with inadvertent exposure of antibiotics for economic gain) through dedicated food chain lineage are influential in AMR crisis globally (Fletcher, 2015; Swift et al., 2019; Dominguez, Chacon, and Wallace, 2021; Truong et al., 2022). The transmission of ARGs across the environment, environmental microbiota and human gut environment are critically influenced by the indiscriminate anthropogenic activities as mentioned earlier. The dissemination of these ARGs with increased human intervention has received considerable attention as critical environmental contaminants with critical influence on human and environmental health (Zhu et al., 2017; Bhattacharyya et al., 2019; Lee et al., 2020; Scott et al., 2021).

Environmental factors significantly contribute in the mobilisation of ARGs by promoting the selective pressure on microbial community and negligible fitness costs. The environmental intervention promotes the emergence of resistance by several mechanisms including promotion of the transfer of ARGs between environmental, human, and animal associated bacteria, forming a biological niche for harbouring ARGs and drug resistant bacteria and endorsing evolutionary modification of environmental resistome profiles (Bengtsson-Palme et al., 2018; Bengtsson-Palme et al., 2021). Several spatiotemporal parameters like rainfall and intermittent sewage contamination also critically influenced the transmission of ARGs across different environments (Williams et al., 2022). Apart from these factors, several socioeconomic factors also significantly control the environmental resistome profiles and transmission of drug resistance. As per the report, socioeconomic factors including GDP per capita, primary education, rate of antibiotic consumption, healthcare facilities, and funding on public healthcare are inversely correlated with antibiotic resistance. Meanwhile, environmental pollution, high temperature, poor governance, and private to public health funding ratio are positively correlated with antibiotic resistance (Collignon et al., 2018). Hence, by only controlling antibiotic consumption is not the ultimate solution to mitigate resistance profiles. Instead, a logistic approach should be undertaken to develop alternative therapeutics in the fight against AMR phenomena.

5.3 EVIDENCE-BASED ALTERNATIVE THERAPEUTIC APPROACHES AGAINST ANTIBIOTIC RESISTANCE

In the quest for novel therapeutic agents against AMR and associated chronic infections, it is essential to consider three important points: (i) mitigation of infections; (ii) avoidance of resistance and control of resistance; and (iii) nurturing the natural microbiome (Brooks and Brooks, 2014). In this chapter, we will critically discuss several strategies under investigation in the infection control practices, with special reference to the management of drug resistance by targeting several pathophysiological targets. Based upon the different strategies being developed, targeting the antimicrobial-resistant enzymes, MDR and extensively drug resistant (XDR) bacteria with special reference to their pathophysiological responses, developing novel drug delivery platforms, designing potential physicochemical methods and advanced molecular tools received considerable attention (Murugaiyan et al., 2022).

5.3.1 Targeting Antimicrobial-resistant Enzymes

As mentioned earlier, pathogenic microorganisms tend to produce several enzymes which concomitantly function to target the administered antibiotics and make them non-functional (Annunziato, 2019). β-lactamases (Serine β-lactamase and metallo β-lactamase) are the highly defined drug resistant enzymes that critically affect the functional attributes of β-lactam group of antibiotics.

The resistance to β-lactam antibiotics could also be attributed to the mutation in penicillin binding proteins (PBPs; key regulator of peptidoglycan biosynthesis), hyperproduction of PBPs and decreased sensitivity of PBPs towards these antibiotics (Egorov et al., 2018). Boron-based compounds such as boric acid, boronic acid and its derivatives, phenylboronic acid and 3-aminophenylboronic acid were reported as the first class of compounds with β-lactamase inhibition potential (Beesley et al., 1983; Smoum et al., 2012; González-Bello et al., 2020). Similarly, Taniborbactam, boronic acid derivative used along with antibiotic, cefepime is under clinical Phase III trials (NCT03840148) as potential mitigating agents against serine β-lactamases and metallo β-lactamases produced by carbapenem-resistant bacterial infections associated with urinary tract infections (UTIs) (Liu et al., 2020; Mojica et al., 2022).

The FDA approved 1,6-Diazabicyclo[3,2,1]octanes derivative, Avibactam when used along with ceftazidime and metronidazole for treatment of drug resistance as non-β-lactam β-lactamase inhibitor (González-Bello et al., 2020). Recently, Zidebactam, derivative of 1,6-Diazabicyclo[3,2,1]octanes in combination with cefepime (WCK5222) is under the clinical phase I stage for its application in mitigation of chronic microbial infections and MDR patterns in microbial pathogens by binding to PBP2 and reported for the inhibition of serine β-lactamases (Livermore et al., 2017; Abuhussain et al., 2019; Bush and Bradford, 2019; Thomson et al., 2019).

Owing to the importance of the Penicillin-based sulfones as β-lactamase inhibitor, a Tazobactam derivative, Enmetazobactam along with cefepime is formulated and is under clinical Phase III trials to mitigate severe UTIs associated with nosocomial infections (Papp-Wallace et al., 2019; González-Bello et al., 2020). In an earlier study, clinically approved thiol derivatives, Captopril, Thiorphan, Tiopronin, and Dimercaprol were reported to be inhibitors of class B metallo-β-lactamases (New Delhi Metallo-βLactamase-1 (NDM-1), Verona-Integron-encoded Metallo-β-Lactamase 1, and Impenemase-7) of drug resistant bacterial pathogens (Klingler et al., 2015; González-Bello et al., 2020).

5.3.2 Targeting Antimicrobial-resistant Bacteria

In the quest for alternative therapeutic regimens against drug resistance and its complicacy, mitigation of drug resistant capability of pathogenic microorganisms proved to be influential in current scenario of antimicrobial drug discovery pipelines. Several strategic therapeutic programmes including phage therapy, mitigation of QS regulon, combating recalcitrant biofilm development, development of efflux pump inhibitors (EPIs) and design of antimicrobial peptides (specifically antibiofilm peptides) are currently under investigations in developing putative drug candidates against chronic microbial infections and drug resistance crisis (Figure 5.2). Apart from these approaches, nanotechnological intervention and development of physicochemical methods including antimicrobial photodynamic therapy (aPDT) have also provided novel avenues in dismantling the drug resistant patterns in pathogenic microorganisms.

5.3.2.1 Phage Therapy

The age-old phage therapy using bacteriophages could be considered as promising alternative therapeutic regimen in the current context of chronic microbial infections and drug resistance. The ubiquity and host specificity as shown by bacteriophages are the most important advantages in considering its candidature as potential alternative therapy (Kumar et al., 2021). The narrow host range is an important parameter of bacteriophages which enable them to selectively target the bacterial pathogens of interest without affecting the commensal bacterial community. Thus, these selective bacteriophages with narrow host ranges could be viable alternatives to traditional antibiotic treatment (Mattila et al., 2015). Since biofilms play pivotal roles in drug resistance incidence, targeting biofilms is considered as viable alternative therapeutics. In addition, the EPS matrix is the protective barrier for encased bacterial community against antibiotic treatment. Hence, bacteriophages encoded with EPS degrading enzymes could be used as suitable therapeutic regimen for biofilm

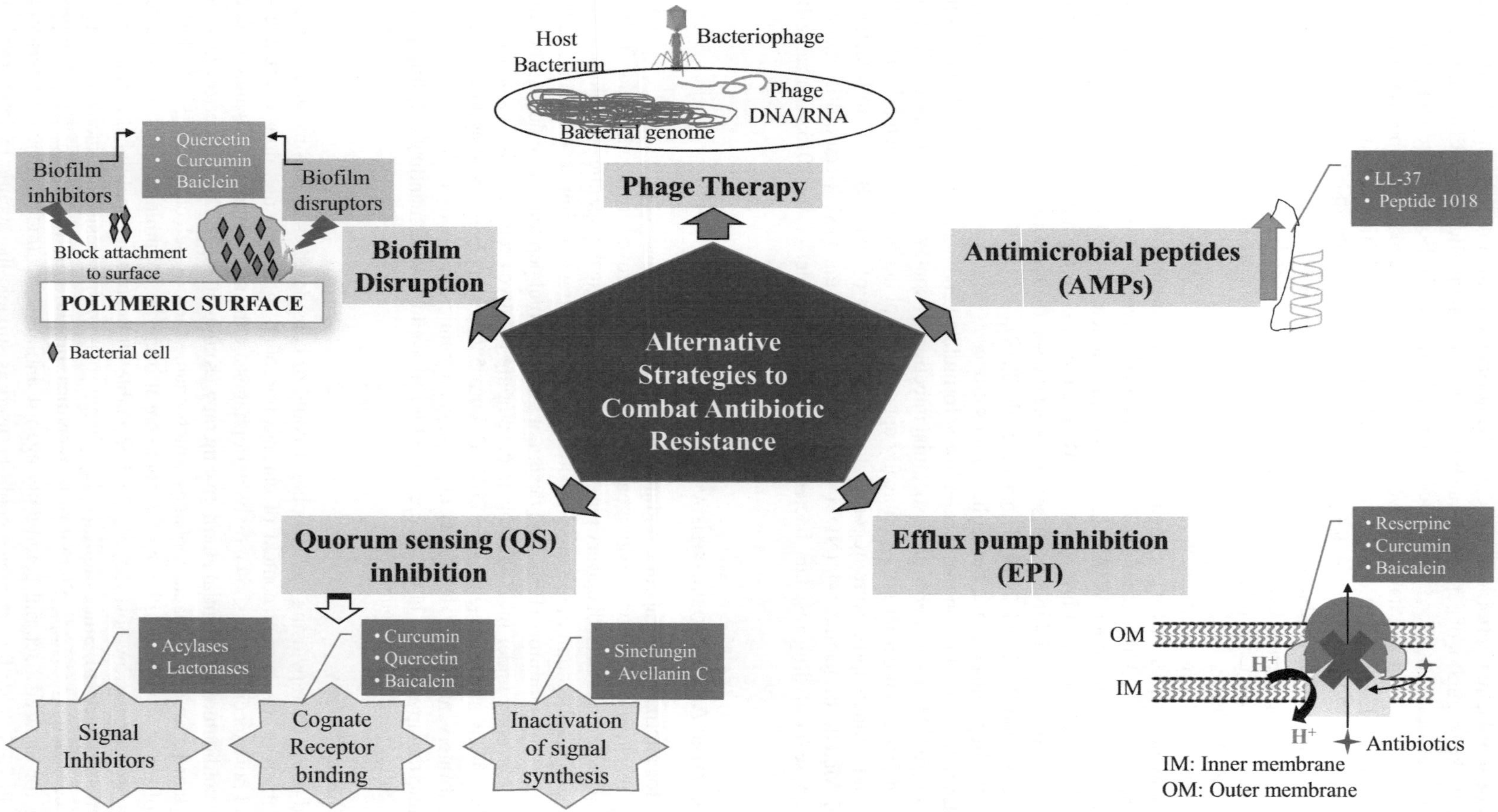

FIGURE 5.2 Schematic representation of the alternative therapeutic approaches under preclinical investigations for mitigation of multidrug resistance (MDR), pan drug resistance (PDR), and extensively drug resistance (XDR) in bacterial pathogens.

penetration and dispersal (Ferriol-González and Domingo-Calap, 2020). Bacteriophages, apart from lysing the bacterial hosts, also have the potential to prevent the biofilm mechanics in MDR bacteria and nosocomial infections (Moghadam et al., 2020; Chegini et al., 2021).

In this context, the combinatorial approach, called phage-antibiotic synergy (PAS) by combining the bacteriophages with sub-lethal concentrations of other antimicrobial agents could be an interesting therapeutic regimen. The lytic bacteriophage SAP-26 when used in combination with rifampicin, critically altered the biofilm-forming ability of methicillin resistant *S. aureus* (MRSA) (Rahman et al., 2011). Similarly, the combined treatment exhibited significant enhancement in the eradication of *Escherichia coli* ATCC 11303 biofilms as compared to the effect of bacteriophages and antibiotic used alone (Ryan et al., 2012). Another study showed the combination therapy of tobramycin with PB-1 phage significantly reduced the biofilm formation in ESKAPE pathogen, *P. aeruginosa* (Coulter et al., 2014). More recently, the combination of carbenicillin (at 0.25 MIC level) and Citrophage MRM57 (10^6 PFU/mL) resulted in the six-log reduction in growth count of *Citrobacter amalonaticus* (Enterobacteriaceae), the causative agents for UTIs, respiratory infections and other nosocomial infections (Manohar et al., 2022).

The concept of phage cocktails using several bacteriophages with widespread host ranges and targeting wide range of therapeutic targets could be engineered for targeting recalcitrant biofilms in *S. aureus* (Kelly et al., 2012; Pires et al., 2017). The advancement in biotechnological principles and molecular tools has given a suitable platform for the design and development of engineered bacteriophages. The engineered bacteriophages have the added advantage over the wild-type bacteriophages as the engineered therapeutic regimen could address the issues including resistance development, rapid elimination profiles and limited strain coverage, as evident from the application of wild-type bacteriophages (Czaplewski et al., 2016).

5.3.2.2 Attenuation of Quorum Sensing as a Therapeutic Module for Drug Resistance Management

As discussed earlier, QS in bacterial pathogens regulates the secretion and regulated modulation of several pathophysiological elements, biofilm determinants, biofilm development, and incidence of drug resistance. Hence, to counteract the transmission and nuisances of MDR and XDR in bacterial pathogens towards conventional antibiotics; QS could be considered as promising alternative therapeutic regimen. Basically, three prominent strategies: (i) inactivation of signal synthase, responsible for signal generation; (ii) enzymatic inactivation of QS signals; and (iii) disruption of cognate QS transcriptional receptor proteins (Yada et al., 2015; Haque et al., 2018; Martínez et al., 2019; Zhong and He, 2021). QS signalling molecules (AIs), synthesised by autoinducer synthase (AI synthase); inhibition of QS signal generation by targeting the AI synthase could be a viable alternative. Once the signal generation process is blocked, the secretion of AIs also concomitantly inhibited, which ultimately attenuate the QS regulated behaviour. Avellanin C, produced by *Hamigera ingelheimensis* significantly altered the signal molecule in *S. aureus* (Igarashi et al., 2015). Sinefungin, a structural analogue of S-adenosyl-L-methionine (SAM) (key regulator of signal biosynthesis) exhibited promising influence on the reduction of signal generation in *Streptococcus pneumoniae* (AI-2 signals) and AHL signal in *P. aeruginosa* (Parsek et al., 1999; Yadav et al., 2014). In earlier studies, at sub-lethal concentration, azithromycin significantly repressed the synthesis of AHL signals in nosocomial pathogen, *P. aeruginosa* (Tateda et al., 2001).

The second QS attenuation strategy infers the QS signal disruption by enzymatic mechanisms. The degradation of QS signals would work on the basis of creating a gap in communication among the bacterial community that concomitantly promote reduction in the production of virulence phenotypes. The enzymes responsible for the chemical degradation/inactivation of QS signals are called as quorum quenching (QQ) enzymes. The QQ enzymes are categorised into four catalytic classes such as (i) lactonases (target the homoserine lactone ring); (ii) amidases or acylases (target the amide bonds of AHL molecules); and (iii) oxidoreductases (catalyse oxidation of acyl chain or reduces 3-oxo-AHLs to their corresponding 3-hydroxy-AHL counterparts) (Grandclément et al., 2015; Sikdar and Elias, 2020). Based on the widespread occurrence, lactonases and acylases are highly explored in quenching QS regulatory network. Both QQ enzymes preferentially hydrolyse specific targets as acylases preferentially hydrolyse AHLs with long acyl chains, whereas lactonases prefer a broader spectrum of AHL substrates (Koch et al., 2014; Sikdar and Elias, 2020). One of the QQ enzymes, AHL lactonases (AiiA), encoded by *aiiA* isolated from *Bacillus* sp. and AiiM hydrolyse the ester bonds in HSL ring of QS signals in *P. aeruginosa* (Huma et al., 2011; Fong et al., 2018; Anandan and Vittal, 2019; López-Jácome et al., 2019). The AHL lactonases prevent the QS communication by degrading AHLs with different side chains which concomitantly attenuate the chronic infections in ESKAPE pathogens, *P. aeruginosa* and *A. baumannii* (Chow et al., 2014; Guendouze et al., 2017).

The AHL acylases also critically attenuated the QS machinery in ESKAPE pathogen, *P. aeruginosa* and also significantly altered the biofilm phenotypes (de Celis et al., 2021). In earlier studies, PvdQ, QQ acylases from *P. aeruginosa* not only hydrolyse AHL signals but also significantly improved the prolonged survival of pulmonary infected laboratory animals (Utari et al., 2017; Utari et al., 2018). Oxidoreductases also affect the AHLs signalling molecules by altering its side chain, which in turn interferes with the QS regulon expression (Uroz et al., 2005). Another QQ enzyme, dioxygenase tend to target the quinolone signalling of *P. aeruginosa* that results in the reduction of cytotoxic elements (pyocyanin, and lectins) and rhamnolipid. Therefore, anti-QS signalling enzymes could be used as alternatives to antibiotics to control bacterial infections mediated by QS regulated communication (Pustelny et al., 2009).

The third and the most widely explored QS attenuation strategy is the competitive inhibition of binding specific signal molecules to their cognate receptors. As a result, the transcription receptors will not activate the sequence of events associated with the regulatory virulence gene expression (Paczkowski et al., 2017). Certain plant-derived flavonoids or synthetic flavonoids can bind with the transcriptional receptors which ultimately reduces the chance of binding of specific signals resulting in the down regulation of QS regulatory network (Yang et al., 2012; Venkatramanan et al., 2020). Apart from these strategies, the QS inhibitory agents could also be combined with conventional antibiotics in order to provide effective management of QS regulated pathogenicity and biofilm associated infections (Furiga et al., 2016; Inoue et al., 2016; Kim et al., 2018). Since ancient times, natural products particularly plant sources have been used as folkloric medicines owing to the widespread pharmacological and therapeutic values. Hence, plant-derived natural products could also be engaged in the quest for novel anti-infective agents targeting QS controlled pathogenesis, biofilm-associated infections and drug resistance in pathogenic microorganisms. The widespread occurrence of bioactive phytoconstituents like alkaloids, polyphenols, terpenoids, tannins, organosulfur compounds and phenylpropanoids in the plant extracts has given novel avenues for establishing their role in QS and biofilm mitigating agents (Koh et al., 2013; Bacha et al., 2016). Similarly, plant-derived essential oils, being rich in bioactive phytoconstituents are prominently explored in the traditional pharmacopoeia. Hence, the pharmaceutical, pharmacological and medicinal properties of these essential oils could be considered in the management of AMR crisis by targeting the QS mechanism and biofilms (Camele et al., 2019; Cáceres et al., 2020; Mansuri et al., 2022). Table 5.1 depicts the inhibitory effect of medicinal plants, and plant-derived essential oils on modulation of QS regulatory network. Plant-derived phytochemicals, which belong to several chemical classes, are known for their pharmacological importance in the biomedical sectors. In the quest for anti-infective drugs, bioactive phytochemicals could thus be instrumental in providing novel avenues for potential drug candidates in the fight against QS regulated biofilms and drug resistance (Aswathanarayan and Vittal, 2018; Chadha et al., 2021). Table 5.2 depicts the inhibitory effect of bioactive phytocompounds on the production of QS regulated virulence phenotypes and modulation of QS regulatory network.

TABLE 5.1
Quorum-Sensing Inhibitory Effect of Medicinal Plants, Plant-Derived Essential Oils against Drug-Resistant Pathogens

A. Medicinal Plants as Anti-Quorum-Sensing Agents

Sl. No.	Plant Family	Plant Name	Plant Part Used	Target Microorganisms	Concentration (IC_{50}/MIC)	Effects	Reference
1.	**Acanthaceae**	*Andrographis paniculata* (Burm. f.) Wall. ex Nees	Whole plant	*Pseudomonas aeruginosa* PAO1	5 mg/mL	Inhibition of quorum-sensing (QS) regulated protease and elastase activity; reduction in production of pyocyanin and rhamnolipid; down regulation of expression of QS regulated genes	Banerjee et al. (2017)
2.	**Amaranthaceae**	*Achyranthes aspera*	Whole plant	*Streptococcus mutans*	125 µg/mL	Reduction in QS controlled biofilm formation	Murugan et al. (2013)
3.	**Amaryllidaceae**	*Allium sativum*	Bulbs	*Salmonella typhimurium*	0.66 µl/mL	Inhibited QS regulated biofilm formation	Alni et al. (2020)
4.	**Anacardiaceae**	*Pistacia atlantica*	Leaf	*Pseudomonas aeruginosa* PAO1	2 mg/mL	Showed anti-QS activity with the reduction in pyocyanin production	Kordbacheh et al. (2017)
5.		*Sclerocarya Birrea*	Bark	*P. aeruginosa* (MTCC 2453)	200 µg/mL	Inhibited QS dependent pathogenic factors (swarming, pyoverdin, and protease)	Sarkar et al. (2014)
6.		*Amphypterygium adstringens*	Stem bark	*Chromobacterium violaceum* CV 12472	55 µg/mL	Reduction in rhamnolipid and pyocyanin production	Castillo-Juárez et al., (2013)
7.	**Apiaceae**	*Anethum graveolens* L.	Aerial parts	*C. violaceum*	50 µl/mL	Inhibited the QS related behaviour and Violacein production	Makhfian et al. (2015)
8.		*Ferula asafetida*	Whole plants	*P. aeruginosa* ATCC 27853	4.4 mg/mL	Inhibited the QS controlled virulence genes	Tüfekci et al. (2020)
9.		*Carum copticum*	Seeds	*C. violaceum*	0.6 mg/mL	Inhibited the QS regulated virulence genes	Snoussi et al. (2018)
10.		*Cuminum cyminum*	Bulbs	*S. typhimurium*	2.18 µl/mL	Inhibited QS activity	Adonizio et al. (2006)
11.		*Apium graveolens*	Oleorasins	*C. violaceum* CV12472	10%	Inhibition of QS activity	Nagar et al. (2020)

(Continued)

TABLE 5.1 (Continued)

A. Medicinal Plants as Anti-Quorum-Sensing Agents

Sl. No.	Plant Family	Plant Name	Plant Part Used	Target Microorganisms	Concentration (IC_{50}/MIC)	Effects	Reference
12.	**Apocynaceae**	*Aspidosperma quebracho-blanco*	Bark	*P. aeruginosa* PAO1	12.5 μg/mL	Reduction in QS regulated bacterial motility	Ahmed et al. (2021)
13.	**Araliaceae**	*Panax ginseng*	Roots	*P. aeruginosa* PAO1	1.25%	Reduction in *lasA* and *lasB* mediated virulence factors production	Song et al. (2010)
14.	**Arecaceae**	*Cocos nucifera* Linn.	Husk	*Pseudomonas* sp.	1.46 mg/mL	Inhibition of QS regulated biofilm formation and biofilm phenotypes	Viju, Satheesh, and Vincent (2013)
15.		*Coreopsis tinctoria* Nutt	Flowers	*C. violaceum*	4 mg/mL	Reduction in violacein production by the presence of okanin	Mu, Zeng, and Chen (2020)
16.	**Bignoniaceae**	*Kigelia africana* (Lam.)	Fruits	*C. violaceum* ATCC12472	—	Violacein inhibition was observed	Chenia (2013)
17.	**Combretaceae**	*Conocarpus erectus*	Leaves	*C. violaceum* ATCC 12472	20 μl/mL	Strong QS activity was seen in water and ethanol extract	Adonizio et al. (2006)
18.		*Bucida burceras*	Leaf	*P. aeruginosa*	100 μl/mL	Inhibition of LasA protease, LasB elastase, pyoverdin production, and biofilm formation	Adonizio et al. (2008)
19.		*Terminalia catappa*	Whole plant	*P. aeruginosa* ATCC 10145	62.5 μg/mL	*lasA* activity was significantly reduced	Taganna et al. (2011)
20.		*Terminalia bellerica*	Leaves	*P. aeruginosa* PAO1	0.0625–0.5 mg/mL	Inhibition of QS controlled pyocyanin and exopolysaccharides (EPS) production	Ganesh and Rai (2018)

21.	**Fabaceae**	*Parkia javonica*	Fruit	*P. aeruginosa*	180 µg/mL	Inhibition of QS regulated virulence phenotypes, pyocyanin, pyoverdin production, reduction in biofilm formation	A. Das et al. (2017a)
22.		*Trigonella foenum*	Whole plant	*P. aeruginosa* PAO1	1200 µg/mL	Inhibition of AHL regulated virulence factors, protease, LasB elastase, pyocyanin production, chitinase, exopolysaccharides (EPS), and swarming motility	Husain et al. (2015a)
23.		*Glycyrrhiza Glabra*	Whole plant	*Acinetobacter baumannii*	2.0 mg/mL	Down regulation of auto-inducer synthase gene	Bhargava et al. (2015)
24.		*Quercus infectoria*	Galls	*P. aeruginosa*	2.5 mg/mL	Reduction in QS controlled *las*, *rhl*, and *exotoxin A* genes expression levels	Ahmed and Salih (2019)
25.	**Hypericaceae**	*Hypericum perforatum*	Aerial parts	*P. aeruginosa*	500 µg/mL	Significant modulation of QS regulatory pathway by interference in LasIR and RhlIR system	Doğan et al. (2019)
26.	**Lamiaceae**	*Satureja sahendica*	Whole plant in flowering stage	*S. typhimurium*	2.34 mg/mL	*sidA* gene expression is decreased	Sharchi et al. (2020)
27.		*Hyptis suaveolens* (L.) Poit	Leaves	*Escherichia coli, Proteus vulgaris, Proteus mirabilis, Klebsiella pneumoniae*, and *Serratia marcescens*	—	Inhibition of biofilm formation	Salini et al. (2015)
28.	**Lythraceae**	*Punica granatum*	Peel	*P. aeruginosa*	50 µg/mL	Inhibition of pyocyanin production	Joshi, Patel, and Kothari (2019)
29.	**Melastomataceae**	*Tetrazygia bicolor*	Leaf	*C. violaceum*	20 µl/mL	violacein production was inhibited	Adonizio et al. (2006)
30.	**Menispermaceae**	*Tinosporia cordifolia*	Stem	*P. aeruginosa*	IC_{50}: 4.79 mg/mL	Inhibition of QS regulated pyocyanin, production, and LasA protease and LasB elastase activities	Gala et al. (2016)

(*Continued*)

TABLE 5.1 (Continued)

A. Medicinal Plants as Anti-Quorum-Sensing Agents

Sl. No.	Plant Family	Plant Name	Plant Part Used	Target Microorganisms	Concentration (IC_{50}/MIC)	Effects	Reference
31.	**Melastomataceae**	*Melastoma malabathricum* (Indian Rhododendron)	Leaves, Flowers	*Staphylococcus aureus*	6.25 mg/mL	Regulation of QS mediated biofilm formation	Das et al. (2021)
32.	**Myrtaceae**	*Syzygium aromaticum*	Whole plant	*C. violaceum*	0.78 mg/mL	Violacein production significantly altered	Moradi et al. (2020)
33.		*Eucalyptus camaldulensis*	Whole plant	*Aeromonas veronii*	0.195 mg/mL	Signalling communication between the bacterial communities was affected	Moradi et al. (2020)
34.		*Callistemon viminalis*	Whole plant	*C. violaceum*	20 µl/mL	The QS activity is inhibited	Adonizio et al. (2006)
35.		*Syzygium aromaticum*	Bud	*P. aeruginosa* PA01	3.9 mg/mL	Inhibition of QS regulated swarming motility and pyocyanin production	Sagar and Singh (2021)
36.		*Syzygium jambos* (L.) Alston	Leaves	*P. aeruginosa* PAO1	1 mg/mL	Inhibition of QS controlled production of pyocyanin, EPS, rhamnolipid; reduction in protease, chitinase, and elastase activity	Rajkumari et al. (2018c)
37.		*Eucalyptus camaldulensis*	Leaves	*P. aeruginosa* PAO1	—	Reduction in QS signalling molecules	Moradi et al. (2020)
38.		*Callistemon viminalis*	Leaf	*C. violaceum*	20 µl/mL	Virulence genes are inhibited	Adonizio et al. (2006)
39.	**Nymphaeaceae**	*Nymphaea tetragona*	Whole plant	*C. violaceum* ATCC12472	5.0 mg/mL	Swarming motility was affected and decreased violacein production	Hossain et al. (2015)
40.	**Polygonaceae**	*Rumex alveolatus*	Leaf	*Staphylococcus aureus*, *P. aeruginosa*	62.5 mg/mL	Reduction in pyocyanin production, biofilm inhibition in *S. aureus*	Korkorian and Mohammadi-Sichani (2017)

41.	**Primulaceae**	*Dionysia revoluta* Boiss.	Whole plant	*P. aeruginosa*	1.56 mg/mL	QS related virulence gene is reduced	Moradi et al., (2020)
42.		*Dionysia revoluta*	Whole plant	*C. violaceum*	6.25 mg/mL	Exhibited anti-QS activity by inhibiting violacein production	Hadi et al. (2020)
43.	**Rosaceae**	*Rubus rosaefolius*	Fruit	*C. violaceum* ATCC 6357 and *Aeromonas hydrophila*	—	Inhibition of QS regulated violacein production, swarming motility, and biofilm formation	Oliveira et al. (2016)
44.	**Rutaceae**	*Zanthoxylum chalybeum*	Stem bark	*S. aureus*, *Enterococcus faecium*	16 µg/mL, 32 µg/mL	Exhibited promiing quorum quenching ability	Schultz et al. (2020)
45.	**Salvadoraceae**	*Salvador persica* (L.)	Whole plant	*P. aeruginosa*	20 mg/mL	Significant reduction in swarming motility	Noumi et al. (2017)
46.	**Solanaceae**	*Solanum torvum*	Root	*P. aeruginosa* jb07	3 mg/mL	Reduction in QS controlled virulence factors and biofilm formation	Vadakkan et al. (2019a)
47.	**Zygophyllaceae**	*Tribulus terrestris*	Root	*P. aeruginosa* SV1 and *Serratia marcescens* BJV5	0–2.5 mg/mL	Effective down regulation of QS controlled mechanisms including pigment production and biofilm formation	Vadakkan et al. (2019b)

B. Plant-Derived Essential Oils as Anti-Quorum-Sensing Agents

Sl. No.	Essential Oils	Plant Source	Plant Part Used	Target Microorganisms	Concentration (IC_{50}/MIC)	Effects	Reference
48.	**Clementina**	*Citrus clementina* Hort. ex Tan.	Peel	*Chromobacterium violaceum*	1.50 mg/mL	Inhibition of QS regulated violacein production and biofilm formation	Poli et al. (2018)
49.	**Coffee husk oil**	*Coffee arabica* L.	Coffee husk	*Pseudomonas aeruginosa* PAO1	50 µg/mL	Reduction in QS mediated virulence factors, elastase, pyocyanin, EPS, and biofilm	Al-Yousef and Amina (2018)
50.	**Lavender**	*Lavendula angustifolia*	Flowers	*Escherichia coli*	0.5% (v/v)	Inhibition of QS regulated virulence	Yap et al. (2014)
51.	**Clove bud**	*Eugenia caryophyllus*	Flower buds	*P. aeruginosa* PAO1 (ATCC15692)	0.1–1% (v/v)	Reduction in QS regulated exopolysaccharides production, decrease in motility patterns	Jayalekshmi et al. (2016)
52.	**Oregano**	*Origanum vulgare*	Leaf	*Streptococcus pneumoniae*	2.5–10 µl/mL	*luxS* and *pfs* genes were down regulated	Sharifi et al. (2018a)

(*Continued*)

TABLE 5.1 (Continued)

B. Plant-Derived Essential Oils as Anti-Quorum-Sensing Agents

Sl. No.	Plant Family	Plant Name	Plant Part Used	Target Microorganisms	Concentration (IC_{50}/MIC)	Effects	Reference
53.	**Mandarin**	*Citrus reticulata*	Fruit	*P. aeruginosa* ATCC 27853	>4 mg/mL	Regulation of QS controlled biofilm mechanics	Luciardi et al. (2016)
54.	**Cinnamon Bark oil**	*Cinnamomum veruim*	Bark	*P. aeruginosa*	5 µM	Down regulation of pyocyanin production, swarming motility and haemolytic activity	YG. Kim et al. (2015b)
55.	**Thyme**	*Thymus vulgaris*	Leaf	*E. coli* O157:H7, *E. coli* O33, and *Staphylococcus epidermidis* ATCC 12228	0.37–0.75 mg/mL	Inhibition of QS dependent biofilm production.	Cáceres et al. (2020)
56.	**Thymus essential oil**	*Thymus daenensis* L.	Leaves	*S. aureus*	0.0625 µl/mL	Inhibition of QS regulated biofilm formation	Sharifi et al. (2018b)
57.	**Lemon verbena**	*Aloysia triphylla*	Leaf	*E. coli*	25 µg/mL	Inhibition of QS mediated factors	Cervantes-Ceballos, Caballero-Gallardo, and Olivero-Verbel (2015)
58.	**Lemon oil**	*Citrus limon* L.	Fruit	*P. aeruginosa* ATCC 27853	0.1–4 mg/mL	Reduction in swarming motility, inhibition of QS signal synthesis	Luciardi et al. (2021)
59.	**Cinnamon leaf oil**	*Cinnamomum veruim*	Leaves	*E. coli*, *Klebsiella pneumoniae*, *S. aureus*, and *Acinetobacter baumannii*	8–16 µl/mL	Eradication of QS controlled biofilm matrix	Alibi et al. (2020)
60.	*Coridothymus capitatus* **essential oil**	*Coridothymus capitatus*	Leaves	*P. aeruginosa*	5% v/v	Significant reduction in swarming and swimming motility, inhibition of pyocyanin production and modification of extracellular matrix of bacterial biofilms	Vrenna et al. (2021)
61.	*Lippa alba* **essential oil**	*Lippa alba*	Aerial parts	*S. aureus* ATCC 6538	0.5 mg/mL	QS activity and biofilm formation was altered	Porfírio et al. (2017)

62.	**Cardamom essential oil**	*Elletlaria cardamomum*	Bark	*P. aeruginosa* PAO1	0.048–0.19 mg/mL	Reduction in elastase and protease activity, inhibition of swarming motility	Noumi et al. (2018)
63.	**Peppermint oil**	*Mentha piperita*	Aerial parts	*C. violaceum, P. aeruginosa* PAO1	0.6% v/v, 6.4% v/v	AHL regulated virulence factor is inhibited	Husain et al. (2015)
64.	**Swinglea essential oil**	*Swinglea glutinosa*	Aerial parts	*Escherichia. coli* O157:H7, *Escherichia coli* O33, and *Staphylococcus epidermidis* ATCC 12228	0.37–0.75 mg/mL	Moderate biofilm inhibition	Cáceres et al. (2020)
65.	**Coriandrum essential oil**	*Coriandrum sativum*	Aerial parts	*A. baumannii*	8 µl/mL	Inhibited the biofilm formation and interference in the QS activity	Alves et al. (2016)
66.	**Citrus essential oil**	*Citrus reticulata*	Aerial parts	*C. violaceum*	1 mg/mL	QS mediated violacein production was inhibited	Ngenge et al. (2021)

TABLE 5.2
Quorum-Sensing Inhibitory Effect of Plant-Derived Bioactive Phytocompounds against Drug-Resistant Pathogens

Sl. No.	Chemical Class	Phytocompounds	Source of Plants	Target Microorganisms	Concentration (IC_{50}/MIC)	Effects	Reference
1.	**Alkaloids**	Berberine	*Berberis* sp.	*Pseudomonas aeruginosa* PAO1	1.25 mg/mL	Modulation of biofilm dynamics	Aswathanarayan and Vittal (2018)
2.		Hordenine	*Hordeum vulgare*	*P. aeruginosa*	2.5 mg/mL	Down regulation of QS regulated virulence, biofilm mechanics	Zhou et al. (2018a, 2018b); Zhou et al. (2019)
3.	**Anthocyanins**	Proanthocyanidin	*Vaccinium macrocarpon*	*P. aeruginosa*	0.2 mg/mL	Inhibition of QS controlled virulence determinants	Maisuria et al. (2016)
4.		Petunidin	Pure	*Klebsiella pneumoniae*	200 µg/mL	Potential inhibitor of QS signalling molecules towards *LasR* protein activity	Gopu et al. (2016)
5.	**Anthraquinones**	Emodin	*Rheum palmatum*	*Staphylococcus aureus*	8 µg/mL	Down regulation of *cidA*, *icaA*, *dltB*, *agrA*, *sortaseA* and *sarA* gene	Yan et al. (2017)
6.		Aloe-emodin	*Rheum officinale*	*S. aureus*	16 µg/mL	Inhibition of initial adhesion and proliferation stages of biofilm development	Xiang et al. (2017)
7.	**Benzophenones**	7-Epiclusiaonone	*Rheedia brasiliensis*	*Streptococcus mutans*	>250 µg/mL	Inhibition of the synthesis of EPS and biofilm attenuation	Murata et al. (2010)
8.	**Carotenoids**	Zeaxanthin	Pure	*P. aeruginosa*	6 µM	QS related genes were down regulated	Gökalsın et al. (2017)
9.	**Coumarins**	Coumarins	—	*P. aeruginosa*	1–2 mM	Attenuation of QS controlled virulence, biofilm formation	Gutiérrez-Barranquero et al. (2015); Zhang et al. (2018)

10.	**Flavonoids**	Baicalin	*Scutellaria baicalensis*	*P. aeruginosa*	>1024 µg/mL	QS regulatory genes (*lasI*, *lasR*, *rhlI*, *rhlR*, *pqsR* and *pqsA*) were repressed	Luo et al. (2017)
11.		Quercetin	Pure	*P. aeruginosa*	>256 µg/mL	QS regulated genes, *lasI*, *lasR*, *rhlI*, and *rhlR* were significantly down regulated	Gopu, Meena, and Shetty (2015)
12.		Quercetin	Pure	*Staphylococcus epidermidis*	>500 µg/mL	Regulation of QS controlled biofilm formation and modulation of exopolysaccharides (EPS) composition	Mu et al., (2021)
13.		Baicalein	*Scutellaria baicalensis*	*P. aeruginosa*	>1024 µg/mL	Attenuation of QS regulated virulence factors, reduction in LasA protease and Las B elastase, biofilm inhibition	Luo et al. (2016)
14.		Baicalein	*Scutellaria baicalensis*	*S. aureus*	256 µg/mL	Destruction of biofilms, reduction in the Staphylococcal enterotoxin A and α-hemolysins	Chen et al. (2016)
15.		Catechin	*Azadirachta indica*	*Alcaligenes faecalis*, *Pseudomonas gingivalis*	180 µg/mL	Down regulation of *luxX* gene, inhibition of biofilm, and virulence factors affecting QS activity	Lahiri et al. (2021a)

(*Continued*)

TABLE 5.2 (Continued)

Sl. No.	Chemical Class	Phytocompounds	Source of Plants	Target Microorganisms	Concentration (IC_{50}/MIC)	Effects	Reference
16.		Epigallocatechin-3-gallate	*Camellia sinesis*	*P. aeruginosa*	512 µg/mL	Significant inhibition in QS regulated virulence phenotypes, biofilm formation and down regulation of QS regulatory genes (i.e., *las*, *rhl*, and *pqs*)	Hao et al. (2021)
17.		Naringenin	*Combretum albiflorium*	*P. aeruginosa* PAO1	4 mM	Reduced production of signalling molecules, down regulation of QS controlled genes	Vandeputte et al. (2011)
18.		Calycopterin	*Marcetia latifolia*	*P. aeruginosa* PA14	>32 µM	Modulation of QS mediated behaviours, motility, and biofilm dynamics	Froes et al. (2020)
19.		Vitexin	*Vitex* sp.	*P. aeruginosa*	260 µg/mL	Various quorum-sensing (QS) arbitrated phenomenon such as swarming motility, azocasein degrading protease activity, pyoverdin, and pyocyanin production, *LasA* and *LasB* activity	Das et al. (2016)
20.	**Furan derivatives**	5-Hydroxymethylfurfural	Pure	*P. aeruginosa* PAO1	2.5 µl/mL; 400 µg/mL	Biofilm inhibition and down regulation of QS regulatory genes (*lasI* and *rhlI*, *lasR* and *rhlR*)	Rajkumari et al. (2019); Vijayakumar and Ramanathan (2020)
21.	**Organosulfur compounds**	Allicin	*Alium sativum*	*P. aeruginosa*	8 µM	Down regulation of QS regulated virulence factors	Xu et al. (2019)
22.		Ajoene	*Alium sativum*	*P. aeruginosa*, *S. aureus*	0.02–0.08 mg/mL	Alteration in biofilm architecture, rhamnolipid and c-di-GMP inhibition	Jakobsen et al. (2012)
23.		Diallyl sufide	*Alium sativum*	*P. aeruginosa*	-	Alteration in QS regulon	Li et al. (2019)

24.	**Phenolic acids**	Salicylic acid	*Salix* sp.	*P. aeruginosa*	3.62 mM	Intereferene in QS regulatory network with inhibition at transcriptional and extracellular level	Ahmed et al. (2019)
25.	**Phenylpropanoids**	Methyl eugenol	*Cuminum cyminum*	*P. aeruginosa, Proteus mirabilis, Serratia marcescens*	4 mg/mL, 3 mg/mL	Significant alteration in QS regulated virulence phenotypes production	Packiavathy et al. (2012)
26.		6-Gingerol	*Zingiber officinale*	*P. aeruginosa*	0–100 μM	The QS related genes were down regulated	HS. Kim et al. (2015a)
27.		Ellagic acid	*Terminalia chebula*	*P. aeruginosa* PAO1	5 mg/ml	Inhibition of biofilm formation, downregulation of *lasI* and *rhlI* and their cognate receptor *lasR* and *rhlR*	Sarabhai et al. (2013)
28.		Curcumin	*Curcuma longa*	*Acinetobacter baumannii*	50 μg/mL	Down regulation of *bfmR* gene	Raorane et al. (2019)
29.		Cinnamic acid	*Averrhoa carambola*	*P. aeruginosa* PAO1	500 μg/mL	Attenuation of QS regulated virulence factors production, biofilm inhibition	Rajkumari et al. (2018b)
30.	**Tannins**	Punicalagin	*Punica granatum*	*S. typhimurium*	250 μg/mL	Inhibition of QS related gene *sdiA* and *srgE*	Li et al. (2014)
31.		Tannic acid	Pure	*Aeromonas hydrophila*	0.1 mg/mL	Down regulation of QS regulated genes	Patel et al. (2017)

(Continued)

TABLE 5.2 (Continued)

Sl. No.	Chemical Class	Phytocompounds	Source of Plants	Target Microorganisms	Concentration (IC_{50}/MIC)	Effects	Reference
32.	**Terpenes and Terpenoids**	Carvacrol & Thymol	*Thymus vulgare*	*Pseudomonas fluorescens, P. aeruginosa*	4.0 µl/mL, 7.9 mM	Reduced AHL related *flgA* gene expression, inhibition of QS virulence factors	Myszka et al. (2016); Tapia-Rodriguez et al. (2017)
33.		Hyperforin	*Hypercium perforatum*	*S. aureus*	8–32 µg/mL	Moderate inhibition of QS activity	Lyles et al., (2017)
34.		Menthol	*Menthe piperita*	*P. aeruginosa*	6.4% (v/v)	Decrease in LasB elastase activity, reduced in the activity of virulence factors and biofilm formation	Husain et al. (2015)
35.		Phytol	*Piper betle*	*S. marcescens*	800–1000 µg/mL	Inhibition of QS controlled virulence factors like prodigiosin, protease, biofilm, and hydrophobicity in *S. marcescens*	Srinivasan et al. (2016)
36.		Betulin and Betulinic acid	*Syzygium jambos*	*P. aeruginosa* PAO1	250 µg/mL	Inhibition of QS signalling pathway by targeting transcriptional regulators, LasR	Rajkumari et al. (2018a)
37.		Glycyrrhetinic acid	*Glycyrrhiza glabra*	*P. aeruginosa* ATCC 25619	0.08 mg/mL	Inhibition of QS controlled production of pathogenic determinants and biofilm formation	Kannan et al. (2019)
38.		Terpinen-4-ol	Plant-derived essential oils	*S. aureus* ATCC 25923	0.25% (v/v)	Inhibition of QS controlled biofilm formation	Cordeiro et al. (2020)
39.	Diarylheptanoids	Diarylheptanoids	*Alnus viridis*	*P. aeruginosa*	150–500 µg/mL	Reduced the production of signalling molecules	Ilic-Tomic et al. (2017)
40.	Vanillylacetone	Zingerone	*Zingiber officinale*	*P. aeruginosa*	6.4 µg/mL	Interference with QS transcriptional regulators, TraR, LasR, RhlR and PqsR	Kumar et al. (2015)

5.3.2.3 Role of Biofilm Blockers in Combating Antibiotic Resistance

There are many problems faced by people globally, owing to the increased incidence of AMR crisis associated with nosocomial pathogens, particularly ESKAPE pathogens. In the present scenario, due to stringent regulatory issues, new antibiotic discovery remains a statutory challenge to overcome this problem. In this context, biofilm blockers targeting the biofilm matrix could be influential in combating resistance profiles against antibiotics by facilitating the penetration of available antibiotics (Lopes et al., 2022). Since the bacterial community protect themselves from the antibiotics or drugs by developing a shield around them called biofilm matrix, the concept of developing potent biofilm blockers (i.e., antibiofilm agents) received considerable attention in 21st century (Li and Lee, 2017). The recent developments in the engineered antimicrobial peptides (AMPs) have provided novel avenues to target biofilms of several bacterial pathogens (Wang et al., 2016). There are several proposed mechanisms depicting an insight into the mode of action of engineered AMPs. One of the broadly used mechanisms involves the formation of transmembrane pores leading to extrusion of cellular contents and lead to the cell lysis which ultimately facilitate the functioning of conventional antibiotics (Yamaguchi et al., 2001). Since, EPS production and development has to play a critical role in biofilm formation, development, and maturation; targeting the enzymatic machinery responsible for development of EPS could provide novel avenues in identifying biofilm blockers (Alhobeira et al., 2021).

Nature has given us a reservoir of natural products and their derivatives of diverse chemical class. As compared to chemically synthesised sources, natural derived products on the basis of possible toxicity received considerable attention in the road to discover putative antibiofilm agents against recalcitrant biofilms (Mishra et al., 2020). For example, plants derived phytochemicals such as alkaloids, polyphenols, glycosides, terpenes, terpenoids, phenylpropanoids, organosulfur compounds, and other bioactive compounds are considered as the source of potential antibiofilm agents. The putative antibiofilm agents concomitantly facilitate our fight against drug resistance phenomena. Table 5.3 depicts the effect of compounds of natural and synthetic origin on mitigation of biofilm formation and disruption of biofilms in bacterial pathogens.

5.3.2.4 Efflux Pump Inhibitors (EPIS) as Promising Strategies to Combat Antimicrobial Resistance

As discussed earlier, the over expression efflux transporters in pathogenic microorganism enable the activation of several other secondary determinants of acquired resistance such as β-lactamases, metallo-β-lactamases and other resistant phenotypes (Choudhury et al., 2016). Hence, it is important to develop novel efflux pump inhibitors (EPIs) to counteract the efflux transporters mediated resistance. Several strategic plans like genetic regulation for efflux pumps expression, modulation of efflux pump functional assembly, blocking the efflux pumps, energy modulation to avoid the energy sources for activation of such efflux pumps and redesigning the antibiotics are the approaches to develop novel EPIs (Bhardwaj and Mohanty, 2012; Sharma et al., 2019). As per recent trends, the EPIs being developed at preclinical stage could be categorised into two groups based upon their mode of action. The EPIs under investigation are either designed with focus on the decoupling of energy source, or direct binding to the efflux pumps (Sharma et al., 2019).

Verapamil, though originally used in hypertension, due to its functional attributes as an ion channel blocker, gained considerable attention. Preclinical studies have reported the ability of Verpamil in potentiating the action of conventional antibiotics, Bedaquiline and Ofloxacin towards efflux mechanism in *Mycobacterium tuberculosis*. Verapamil also inhibited the functional activity of MATE pumps by targeting its prototype, DinF and NorM (Singh et al., 2011; Gupta et al., 2014; Radchenko et al., 2015). In terms of EPIs from natural sources, Reserpine, a bioactive alkaloid from the roots of *Rauwolfia serpentina*, first reported to modulate MFS and RND efflux pumps by directly altering the amino acid residues of efflux protein superfamily and thereby promotes inhibition of tetracycline efflux from Bmr efflux pump in *Bacillus subtilis*. Similarly, reserpine also exhibited

TABLE 5.3
List of Compounds (Natural/Synthetic) Reported for Mitigation of Biofilm Formation in Drug Resistant Pathogens

Sl. No.	Plant Extract/Compounds	Target Microorganisms	Concentration (IC_{50}/MIC)	Therapeutic Effects	Reference
1.	*Allium sativum*	STEC (Shiga toxin producing *Escherichia coli*)	30–140 μl/mL	Inhibited the biofilm formation by modulation of exopolysaccharides (EPS) composition	Bhatwalkar et al. (2019)
2.	*Azadirachta indica*	*Staphylococcus aureus* and *Vibrio cholerae*	75 μg/mL, 100 μg/mL	Antibiofilm activity was enhanced when ripe seed extract was given	Guchhait et al. (2022)
3.	*Bergenia crassifolia*	*Streptococcus mutans*	1.25 mg/mL	Reduction of EPS production and decrease in the adherence properties	Liu et al., (2017)
4.	*Camellia sinensis*	*Morganella morganii*	62.5 μg/mL	Has the ability to inhibit the biofilm formation by down regulation of QS regulated genes	Guzman et al. (2020)
5.	Cinnamaldehyde	*E. coli* and *Vibrio* sp.	1,750 μM, 150 μM	There is a reduction in the formation of biofilm, there is a decrease in the DNA-binding ability of *LuxR*	Niu and Gilbert (2004); Brackman et al. (2008)
6.	Ginkgolic acid	*E. coli 0157:H7*	5 μg/mL	Reduced in fimbriae production and biofilm formation	Lee et al. (2014)
7.	Hordenine	*Pseudomonas aeruginosa*	2.5 mg/mL	Presence of Hordenine reduced the effect of biofilm formation and reduced the effect of QS gene *lasI, lasR, rhlI*, and *rhlR*	Zhou et al. (2018b)
8.	Isolimonic acid	*Vibrio harveyi*	100 μl/mL	Interfere in biofilm formation with modulation of *luxO* expression	Vikram et al. (2011)
9.	*Mentha piperita*	*Aggregatibacter actinomycetemcomitans*	3.125–12.5 μl/mL	Prominent inhibition in biofilm formation	Karicheri and Antony (2016)
10.	N-(Heptyl sulfanylacetyl)-L-homoserine lactone	*P. aeruginosa*	35 mg/mL	Inhibition of biofilm formation	Harjai et al. (2010)
11.	*Ocimum tenuiflorum*	*P. aeruginosa*	300 μg/mL	Biofilm causing genes were down regulated	Lahiri et al. (2021b)
12.	*Piper betle* leaf	*P. aeruginosa* PAO1	50 μg/mL	Inhibition of Biofilm phenotype, swarming motility	Datta et al. (2016)
13.	*Herba patriniae*	*P. aeruginosa*	160 μg/mL	Decreased in the exopolysaccharide production	Fu et al. (2017)
14.	Phloretin	*E. coli O157:H7*	25 μg/mL	phloretin repressed *hlyE* and *stx2*, *lsrACDBF*, *csgA*, and *csgB* in *E. coli* O157:H7 biofilm cells	Lee et al. (2011)

15.	Polyphenols	*Streptococci*	25% v/v	Hydrophobicity, aprime factor in biofilm dynamics was attenuated	Yamanaka et al. (2004)
16.	Quercetin	*S. aureus, P. aeruginosa*	250 μg/mL, 8 μg/mL	Prevention of bacterial adhesions and QS pathways	Memariani et al. (2019)
17.	*Syzygium aromaticum*	*S. aureus*	85.46 μg/mL	It showed antibiofilm effects in various surfaces and detected by MALDI-TOF MS type detector	Kačániová et al. (2021)
18.	*Tridax procumbens*	*S. aureus*	100 μg/mL	The small peptides derived from *Tridax* has a potential to act as a biofilm blocker	Krishna and Kondapalli (2018)
19.	*Vitex negundo*	*E. coli*	75 μg/mL	The study revealed plant extracts inhibited biofilm formation, enhanced effect on biofilm inhibition was recorded in polymer coated extracts of the plants	Namasivayam and Roy (2013)
20.	*Zingiber officinale*	*S. mutans*	256 μg/mL	Biofilm inhibition by dispersion of the cell and reduction in the virulence gene	Hasan et al. (2015)

promising antagonistic activity against NorA-mediated drug resistance patterns in ESKAPE pathogen, *S. aureus* (Gibbons et al., 2003; Stavri et al., 2007; Sharma et al., 2019). Another report, depicting the inhibitory effect of dietary diterpenes, carnosic acid and carnosol (derived from *Rosmarinus officinalis*), against MsrA and TetK efflux pumps (ABC superfamily) of MDR *S. aureus* (Oluwatuyi et al., 2004; Sharma et al., 2019). The functional attributes of NorA efflux pump in S. *aureus* was also significantly altered in presence of an alkaloid, piperine and its derivative piperidine (Kumar et al., 2008). Conessine, a steroidal alkaloid from *Holarrhena antidysenterica* significantly inhibited over expressed MexAB-OprM efflux pump in ESKAPE pathogen, *P. aeruginosa*. In addition, conessine also critically restored the functional attributes of conventional antibiotics when used in a combinatorial study (Siriyong et al., 2017). Tannic acid, as reported earlier for its promising QS inhibitory activities and biofilm mitigation, also exhibited efflux pump inhibition in MDR *S. aureus* (Tintino et al., 2016, 2017). Thus, from the last few decades, a significant contribution has been made to the field of developing natural products as EPIs against MDR pathogens. Hence, to design novel plant-derived compounds in combination with traditional antibiotics for targeting the efflux pump assembly could provide novel avenues in our efforts to fight the drug resistance scenarios worldwide.

5.3.2.5 Antimicrobial Peptides against Drug Resistance

In the quest for alternative therapeutic strategies against AMR crisis, the evolution of antimicrobial peptides (AMPs) has received considerable attention, since they pose promising antimicrobial properties and widespread sequence space which promote low tendency of AMPs against drug resistance mechanisms. AMPs from natural sources are called host defence peptides (HDPs) or defensins, and these also exhibited promising antibacterial properties by targeting the bacterial membrane without being affected by AMR (Xu et al., 2021). The emerging concept of AMPs provides cues to the functional attributes of AMPs in rapid targeting and killing of the invading pathogens by membrane disruption without creating selective pressure on pathogens (Mishra et al., 2017; Di Somma et al., 2020). The most unique feature of AMPs is their ability to stimulate the protective immunological response against the invading pathogens and significant suppression of inflammatory responses (Dostert et al., 2021). In last few years, mesenchymal stem cells (MSCs) have been considered as a potential source of several secretomes, which could be used as AMPs in the fight against microbial infections mediated resistance to antibiotics (M. Kumar et al., 2021). One of the examples of AMPs is LL-37, a human cathelicidin peptide having antimicrobial properties against multiple antibiotic-resistant bacteria (Schittek et al., 2008). It also exhibited biofilm inhibition against *P. aeruginosa, S. aureus*, and *E. coli* (Aka, 2015).

In an earlier study, it was observed that the antibiofilm peptide (peptide 1018) exhibited promising synergistic effect with conventional antibiotics against multidrug-resistant ESKAPE pathogens (Reffuveille et al., 2014). The promising aspects of antibiofilm peptides in the fight against biofilm associated infections could also be observed from their antibiofilm efficacy in several *in vivo* models (Silveira et al., 2021). In addition, the widespread potential of AMPs and antibiofilm peptides lies in the fact that these peptides have the inherent ability to target several pathophysiological processes, and physical structures of pathophysiological importance (Hancock et al., 2021).

5.3.3 Nano-based Drug Delivery Systems for Effective Clearance of Drug Resistance

In the quest for novel and effective therapeutic regimens against biofilm associated chronic infections and drug resistance, the nanotechnological intervention, owing to their unique physicochemical properties, have received considerable attention in pharmaceutical sectors. In the last few decades, a revolutionised progression has been made in the field of nanotechnology-based sensing, diagnosis, and therapeutics of several diseases and disorders. The targeted and localised delivery are the most explored strategies for widespread biomedical applications (Subhaswaraj et al., 2020). Green nanotechnology considering the reducing properties of natural products for synthesis of nanoparticles, and nanocomposites has provided a new dimension to our efforts for the development of an effective, non-toxic and target-oriented approach to handle the drug resistance nuisances (X. Zhu et al., 2014; B. Das et al., 2017b). The unique physicochemical properties of synthesised nanomaterials not only exhibit therapeutic effects, but also complement the conventional antibiotics with several interesting aspects like sustained drug release profile, increased therapeutic index, and prolonged therapeutic effect (Zaidi et al., 2017).

The mycofabricated silver nanoparticles (AgNPs) synthesised by *Rhizopus arrhizus* BRS-07 exhibited promising influence on QS controlled virulence phenotypes by modulating the QS signalling network of LasI/R and RhlI/R system (Singh et al., 2015). Similarly, zinc oxide nanoparticles (ZnO NPs) synthesised using *Nigella sativa* seed extract also exhibited biofilm inhibitory effect on QS mediated biofilm formation in drug resistant food borne pathogens, *E. coli*, *P. aeruginosa* and *Listeria monocytogenes* (Al-Shabib et al., 2016). Selenium based nanoparticles (SeNPs) also considered as effective nanomaterials. In this context, polyphenols fabricated onto SeNPs for the mitigation of biofilm mechanics in *P. aeruginosa* (Prateeksha et al., 2017). The encapsulation of bioactive phytochemicals onto chitosan-based nanomaterials also critically influence the attenuation of QS regulated virulence factors production and biofilm mechanics in drug resistant *P. aeruginosa* PAO1 (Pattnaik et al., 2018; Subhaswaraj et al., 2018). The cinnamon bark extracts-loaded PLGA nanoparticles also showed inhibitory effect on *S. typhimurium* and *L. monocytogenes* biofilms (Tan et al., 2014). The curcumin-liposomal nanomaterials also showed promising aspects in mitigation of QS regulated virulence in *Aeromonas hydrophila* and *Serratia grimesii* by critically improving the therapeutic index of encapsulated curcumin (Ding, Li, and Li, 2018). No doubt, nanotechnology in the field of biomedical and pharmaceutical sectors has critically improved our drug discovery and development pipelines. Hence the engineered nanomaterials could be used rationally considering all the limitations. Apart from that more integrative approach considering all the regulatory issues is required to explore these engineered nanomaterials for clinical manifestations in drug resistance management.

5.3.4 Antimicrobial Photodynamic Therapy (APDT) against Drug Resistance

The established therapeutic modalities of photodynamic therapy (PDT) in cancer treatment and dermatological intervention has urged the scientific community for its application in tackling drug resistance mediated infectious diseases by photodynamic inactivation (Sharma et al., 2011). The recent advent of antibacterial photodynamic therapy (aPDT) received considerable attention to tackle multidrug-resistant pathogens without inferring selective pressure to microbial community thereby showing low tendency to induce drug resistance (Liu et al., 2015). The concept of aPDT hovers around three primary components: a light source (irradiation), photosensitiser (that absorbs light and excite at particular wavelength), and oxygen. The photosensitisers upon irradiation with a specific wavelength, undergo a transition from a stable and low energy ground state to an excited high energy singlet state. In the subsequent phase, the photosensitiser either return to its ground state (fluorescence emission) or may undergo further transition to a higher energy triplet state. At this high energy triplet state, the excited photosensitisers interact with the target sites via two pivotal molecular photosensitising pathways, type I and II. Type I reaction involves the production of highly reactive free radicals through dedicated electron transfer reactions. These reactive free radical species proved to be instrumental in dismantling cell membrane integrity leading to loss of membrane functions, and biological damage. Meanwhile, type II reaction involves photosensitisation of molecular oxygen by the interaction of triplet state photosensitiser to produce an electronically excited and highly reactive singlet oxygen ($^1O_2$). The short lived singlet oxygen further interact with biological entities causing severe oxidative damage (Rajesh et al., 2011; Vera et al., 2021; Songca and Adjei, 2022). The application of aPDT in tackling drug resistance in pathogenic microorganisms has several advantages over other conventional modalities. These advantages include broad-spectrum activities against several groups of pathogens, low possibility to develop photoresistance, high selectivity on host microbes, rapid bactericidal action unlike that of antibiotics, low probability to develop mutagenic effects, feasibility to be used in combination with other therapeutic modalities and inexpensive therapeutic setup (Sharma et al., 2011; Vera et al., 2021; Songca and Adjei, 2022).

Lasers, inorganic and organic light-emitting diodes (LEDs) and broad-spectrum discharge lamps are important light sources for emitting red light (wavelength range: 600–800 nm) during aPDT. The use of red light of longer wavelength conferred aided advantage of improved cell penetration efficacy which concomitantly improved aPDT mediated photoinactivation (Lian et al., 2019; Kim and Darafsheh, 2020). As discussed earlier, the use of appropriate photosensitisers critically affects the therapeutic outcome of aPDT mediated photoinactivation. Several classes of synthetic and natural molecules are used as photosensitisers. Among synthetic sources, synthetic dyes (Toluidine Blue O, Rose Bengal, Methylene blue, etc.), and porphyrins are regularly used for photoinactivation processes. Similarly, naturally derived photosensitisers (e.g., curcumin and hypericin) also gained considerable attention in aPDT mediated therapeutics (Ghorbani et al., 2018; Ludacka et al., 2022).

Recently, Cationic porphyrin conjugate, protoporphyrin IX-methyl ethylenediamine derivative (PPIX-MED) was used in combination with an antibiotic to determine the potential of designed aPDT against the burn wound infections manifested by drug resistant pathogens, *E. coli*, *P. aeruginosa*, and methicillin resistant *S. aureus* (Zhao et al., 2021). Prior to this, methylene blue based photoreceptor was used for wound healing activities by aPDT mechanisms against drug resistant *E. coli* K12 and *S. aureus* (Karner et al., 2020; Songsantiphap et al., 2022). The use of chelators in combination with photosensitisers also significantly improved the efficacy of aPDT. Curcumin, a natural photosensitiser when used in combination with chelating agents like ethylene diamine tetraacetic acid (EDTA) and hydroxyethylidene bisphosphate (HEBP); the cell viability and biofilm vitality of ESKAPE pathogen, *E. faecalis* significantly altered by the action of photoinactivation (Méndez et al., 2021). Several supramolecular architectures utilise aPDT based photoinactivation by non-covalent interaction of photosensitiser with nano-based materials like carbon-based nanomaterials (Graphene oxides, Carbon nanotubes), metallic nanoparticles (AgNPs, AuNPs), and lipid-based nanoparticles (Micelle, liposomes) (Vera et al., 2021). In this regard, chlorophyllin-phycocyanin mixture and toluidine blue O were used as photosensitisers in combination with propolis nanoparticles (PNPs) against *Streptococcus mutans* biofilms. As compared to the growth inhibitory effect of photosensitisers used alone; the growth inhibitory effect significantly improved when used in combination with PNPs suggesting the synergistic influence of nano-based materials in biofilm mitigation (Afrasiabi et al., 2020). In an earlier study, toluidine blue (TB) encapsulated in mesoporous silica nanoparticles (MSNs) were reported for promising biofilm inhibition in ESKAPE pathogens, *P. aeruginosa* and *S. aureus* by critically reducing the production of extracellular matrix (Parasuraman et al., 2019). Though, aPDT has revolutionised the scientific efforts in the management of drug resistance, more stringent and regulatory issues need to be understood properly. In addition, aPDT in combination with putative drug molecules and phytochemicals could also be considered in mitigating the severity of drug resistance.

5.4 CURRENT TRENDS AND FUTURE AVENUES IN ALTERNATIVE THERAPEUTIC MODULES IN COMBATING ANTIBIOTIC RESISTANCE

Understanding the severity of AMR phenomena on public health, environmental pollution and socioeconomic patterns globally; a few years back, the Transatlantic Taskforce on Antimicrobial Resistance (TATFAR) recommended several preventive and therapeutic measures in the form of antibiotic stewardship. These antibiotic stewardship strategies included rational use of antibiotics on humans as well as other animals, preventive measures in the health infrastructure and environmental setup in order to avoid drug resistant infections, and formulation of improving the drug discovery pipeline for development of novel antimicrobial drugs. This concept was also considered by the Infectious Diseases Society of America with additional surveillance mechanisms to control the AMR crisis (Brooks and Brooks, 2014). For combating the recalcitrant biofilm development, it is important to develop a well-established multidisciplinary collaboration between research and development (R&D) groups, academic institutions, hospital management, pharmaceutical sectors, local administration, funding agencies, and government undertaking. Such collaboration could pave the way for the identification of potential antibiofilm agents, their systemic investigations through clinical settings, their combinatorial studies, removal of the infected foreign bodies, management of biomedical devices in the healthcare settings and selective administration of anti-quorum-sensing or biofilm dispersal agents (H. Wu et al., 2015).

In last decade, the evolution of CRISPR-Cas9 genome editing tool has revolutionised the understanding of several pathophysiological responses and promote promising therapeutic regimens. In the fight against AMR crisis, it was evident from recent reports on the use of CRISPR-Cas9 in selective clearance of AMR by removal of ARGs by programmable and sequence-specific antimicrobials using RNA-guided nuclease Cas9 thereby promoting sensitisation of microbial community towards available antibiotics (Bikard et al., 2014; Citorik et al., 2014; Kumar et al., 2021). The concept of drug repurposing also gained recent recognition as the concept hovers around the use of clinically manifested drugs in the management of other diseases, which were originally used for other purposes. In the context of drug resistance severity and introduction of new and potent antibiotics into the market, drug repurposing seemed to be effective in terms of tackling the AMR crisis (Farha and Brown, 2019; Konreddy et al., 2019; Liu et al., 2021). Apart from that, several other physicochemical modules such as sonodynamic antimicrobial chemotherapy and photoinactivation could also be used in the efficient management of biofilm dispersion and drug resistance (Murugaiyan et al., 2022). The recent emergence of computational tools in biological manifestations of drug development also provides new dimension to the scientific society in the management of drug resistance. The next-generation sequencing (NGS) techniques using computational database also provide a platform to predict the drug resistant patterns in MDR and XDR pathogens., which ultimately provide cues to the development of potential drug candidates targeting the specific molecular and pathophysiological targets as predicted through computational approaches (Muzondiwa et al., 2020; Maryam et al., 2021).

5.5 CONCLUSION

In the last few decades we have achieved several advancements in biotechnological settings, the biomedical and pharmaceutical sectors. However, the incidence of resistance to antibiotics significantly increased owing to the irrational use of antibiotics, mismanagement in handling the drug resistant bacterial strains, and slowness in developing new antibiotics. In this context, a comprehensive approach should be taken into consideration. This comprehensive approach needs collaborative efforts at national and international levels to manifest regulatory and therapeutic guidelines in the management of AMR crisis. Similarly, collaborative efforts from several institutional bodies such as academic institutions, R&D sectors, research institutes, local administration, funding agencies and government regulatory bodies would provide a suitable platform to support funding and regulations in stringent antimicrobial stewardship policy. Another policy that needs urgent attention is the global "One Health" programme for scientific collaborations, innovations in therapeutic regimens, educational awareness, and leadership development with a common goal of tackling AMR crisis.

ACKNOWLEDGEMENT

The authors thank Department of Biotechnology and Bioinformatics, Sambalpur University, Sambalpur, Odisha, India for the extensive support provided during the manuscript preparation. The author, Subhaswaraj Pattnaik also acknowledges DST-SERB N-PDF program in Life Sciences for the Post-Doctoral fellowship (Reference No. PDF/2021/001260).

REFERENCES

Abuhussain, S.S.A., Avery, L.M., Abdelraouf, K., and Nicolau, D.P. 2019. *In vivo* efficacy of humanized WCK 5222 (cefepime-zidebactam) exposures against carbapenem-resistant *Acinetobacter baumannii* in the neutropenic thigh model. *Antimicrob. Agents Chemother.*, 63(1), e01931–18.

Adonizio, A., Kong, K.F., and Mathee, K. 2008. Inhibition of quorum sensing-controlled virulence factor production in *Pseudomonas aeruginosa* by South Florida plant extracts. *Antimicrob. Agents Chemother.*, 52(1), 198–203.

Adonizio, A.L., Downum, K., Bennett, B.C., and Mathee, K. 2006. Anti-quorum sensing activity of medicinal plants in southern Florida. *J. Ethnopharmacol.*, 105(3), 427–435.

Afrasiabi, S., Pourhajibagher, M., Chiniforush, N., and Bahador, A. 2020. Propolis nanoparticle enhances the potency of antimicrobial photodynamic therapy against *Streptococcus mutans* in a synergistic manner. *Sci. Rep.*, 10, 15560.

Ahmed, A.A., and Salih, F.A. 2019. *Quercus infectoria* gall extracts reduce quorum sensing-controlled virulence factors production and biofilm formation in *Pseudomonas aeruginosa* recovered from burn wounds. *BMC Complement. Altern. Med.*, 19(1), 1–11.

Ahmed, S.A.K.S., Rudden, M., Smyth, T.J., Dooley, J.S.G., Marchant, R., and Banat, I.M. 2019. Natural quorum sensing inhibitors effectively downregulate gene expression of *Pseudomonas aeruginosa* virulence factors. *Appl. Microbiol. Biotechnol.*, 103, 3521–3535.

Ahmed, S.O., Zedan, H.H. and Ibrahim, Y.M. 2021. Quorum sensing inhibitory effect of bergamot oil and *Aspidosperma* extract against *Chromobacterium violaceum* and *Pseudomonas aeruginosa. Arch. Microbiol.*, 203, 4663–4675.

Aka, S.T. 2015. Killing efficacy and antibiofilm activity of synthetic human cationic antimicrobial peptide cathelicidin hCAP-18/LL37 against urinary tract pathogens. *J. Microbiol. Infect. Dis.*, 5(01), 15–20.

Alhobeira, H.A., Al Mogbel, M., Khan, S., Khan, M., Haque, S., Somvanshi, P., et al. 2021. Prioritization and characterization of validated biofilm blockers targeting glucosyltransferase C of *Streptococcus mutans. Artif. Cells. Nanomed. Biotechnol.*, 49(1), 335–344.

Alibert, S., Diarra, J.N., Hernandez, J., Stutzmann, A., Fouad, M., Boyer, G., and Pages, J.M. 2017. Multidrug efflux pumps and their role in antibiotic and antiseptic resistance: A pharmacodynamic perspective. *Exp. Opin. Drug. Metabolism. Toxicol.*, 13(3), 301–309.

Alibi, S., Ben Selma, W., Ramos-Vivas, J., Smach, M.A., Touati, R., Boukadida, J., et al. 2020. Antioxidant, antibacterial, antibiofilm, and anti-quorum sensing activities of four essential oils against multidrug-resistant bacterial clinical isolates. *Curr. Res. Transl. Med.*, 68(2), 59–66.

Allel, K., Garcia, P., Labarca, J., Munita, J.M., Rendic, M., Grupo Colaborativo de Resistencia Bacteriana, and Undurraga, E.A. 2020. Socioeconomic factors associated with antimicrobial resistance of *Pseudomonas aeruginosa*, *Staphylococcus aureus*, and *Escherichia coli* in Chilean hospitals (2008–2017). *Rev. Panam. Salud Publica*, 44, e30.

Alni, R.H., Ghorban, K., and Dadmanesh, M. 2020. Combined effects of *Allium sativum* and *Cuminum cyminum* essential oils on planktonic and biofilm forms of *Salmonella typhimurium* isolates. *3 Biotech.*, 10(7), 315.

Al-Shabib, N.A., Husain, F.M., Ahmed, F., Khan, R.A., Ahmad, I., Alsaraheh, E., et al. 2016. Biogenic synthesis of Zinc oxide nanostructures from *Nigella sativa* seed: Prospective role as food packaging material inhibiting broad-spectrum quorum sensing and biofilm. *Sci. Rep.*, 6, 36761.

Alves, S., Duarte, A., Sousa, S., and Domingues, F.C. 2016. Study of the major essential oil compounds of *Coriandrum sativum* against *Acinetobacter baumannii* and the effect of linalool on adhesion, biofilms and quorum sensing. *Biofouling*, 32(2), 155–165.

Al-Yousef, H.M., and Amina, M. 2018. Anti-quorum and biofilm formation inhibition by coffee husk oil (*Coffee arabica* L.). *Biomed. Res.*, 29(10), 2096–2100.

Aminov, R.I. 2010. A brief history of the antibiotic era: Lessons learned and challenges for the future. *Front. Microbiol.*, 1, 134.

Anandan, K., and Vittal, R.R. 2019. Quorum quenching activity of AiiA lactonase KMMI17 from endophytic *Bacillus thuringiensis* KMCL07 on AHL- mediated pathogenic phenotype in *Pseudomonas aeruginosa. Microb. Pathog.*, 132, 230–242.

Annunziato, G. 2019. Strategies to overcome antimicrobial resistance (AMR) making use of non-essential target inhibitors: A review. *Int. J. Mol. Sci.*, 20(23), 5844.

Aswathanarayan, J.B., and Vittal, R.R. 2018. Inhibition of biofilm formation and quorum sensing mediated phenotypes by berberine in *Pseudomonas aeruginosa* and *Salmonella typhimurium. RSC Adv.*, 8, 36133–36141.

Bacha, K., Tariku, Y., Gebreyesus, F., Zerihun, S., Mohammed, A., Weiland-Brauer, N., et al. 2016. Antimicrobial and anti-Quorum Sensing activities of selected medicinal plants of Ethiopia: Implication for development of potent antimicrobial agents. *BMC Microbiol.*, 16, 139.

Bahr, G., Gonzalez, L.J., and Vila, A.J. 2021. Metallo-β-lactamases in the age of multidrug resistance: From structure and mechanism to evolution, dissemination, and inhibitor design. *Chem. Rev.*, 121(13), 7957–8094.

Banerjee, M., Moulick, S., Bhattacharya, K.K., Parai, D., Chattopadhyay, S., and Mukherjee, S.K. 2017. Attenuation of *Pseudomonas aeruginosa* quorum sensing, virulence and biofilm formation by extracts of *Andrographis paniculata*. *Microb. Pathog.*, 117, 85–93.

Beesley, T., Gascoyne, N., Knott-Hunziker, V., Petursson, S., Waley, S.G., Jaurin, B., and Grundstrom, T. 1983. The inhibition of class C β-lactamases by boronic acids. *Biochem. J.*, 209(1), 229–233.

Bengtsson-Palme, J., Johnsson, V., and Heß, S. 2021. What is the role of the environment in the emergence of novel antibiotic resistance genes? A modeling approach *Environ. Sci. Technol.*, 55(23), 15734–15743.

Bengtsson-Palme, J., Kristiansson, E., and Joakim Larsson, D.G. 2018. Environmental factors influencing the development and spread of antibiotic resistance. *FEMS Microbiol. Rev.*, 42(1), fux053.

Bhardwaj, A.K., and Mohanty, P. 2012. Bacterial efflux pumps involved in multidrug resistance and their inhibitors: Rejuvinating the antimicrobial chemotherapy. *Recent Pat. Antiinfect. Drug Discov.*, 7, 73–89.

Bhargava, N., Singh, S.P., Sharma, A., Sharma, P., and Capalash, N. 2015. Attenuation of quorum sensing-mediated virulence of *Acinetobacter baumannii* by *Glycyrrhiza glabra* flavonoids. *Future Microbiol.*, 10(12), 1953–1968.

Bhattacharyya, A., Haldar, A., Bhattacharyya, M., Ghosh, A. 2019. Anthropogenic influence shapes the distribution of antibiotic resistant bacteria (ARB) in the sediment of Sundarban estuary in India. *Sci. Total Environ.*, 647, 1626–1639.

Bhatwalkar, S.B., Gound, S.S., Mondal, R., Srivastava, R.K., and Anupam, R. 2019. Antibiofilm and antibacterial activity of *Allium sativum* against drug resistant shiga-toxin producing *Escherichia coli* (STEC) isolates from patient samples and food Sources. *Indian. J. Microbiol.*, 59(2), 171–179.

Bikard, D., Euler, C.W., Jiang, W., Nussenzweig, P.M., Goldberg, G.W., Duportet, X., et al. 2014. Exploiting CRISPR-Cas nucleases to produce sequence-specific antimicrobials. *Nat. Biotechnol.*, 32, 1146–1150.

Blair, J.M.A., Webber, M.A., Baylay, A.J., Ogbolu, D.O., and Piddock, L.J.V. 2015. Molecular mechanisms of antibiotic resistance. *Nat. Rev. Microbiol.*, 13, 42–51.

Blanco, P., Hernando-Amado, S., Reales-Calderon, J.A., Corona, F., Lira, F., Alcalde-Rico, M., et al. 2016. Bacterial multidrug efflux pumps: Much more than antibiotic resistance determinants. *Microorganisms*, 4(1), 14.

Brackman, G., Defoirdt, T., Miyamoto, C., Bossier, P., Van Calenbergh, S., Nelis, H., and Coenye, T. 2008. Cinnamaldehyde and cinnamaldehyde derivatives reduce virulence in *Vibrio* spp. by decreasing the DNA-binding activity of the quorum sensing response regulator LuxR. *BMC Microbiol.*, 8(1), 1–14.

Brooks, B.D., and Brooks, A.E. 2014. Therapeutic strategies to combat antibiotic resistance. *Adv. Drug Deliv. Rev.*, 78, 14–27.

Bush, K., and Bradford, P.A. 2019. Interplay between β-lactamases and new β-lactamase inhibitors. *Nat. Rev. Microbiol.*, 17, 295–306.

Cáceres, M., Hidalgo, W., Stashenko, E., Torres, R., and Ortiz, C. 2020. Essential oils of aromatic plants with antibacterial, antibiofilm and anti-quorum sensing activities against pathogenic bacteria. *Antibiotics*, 9(4), 147.

Camele, I., Elshafie, H.S., Caputo, L., and De Feo, V. 2019. Anti-quorum sensing and antimicrobial effect of mediterranean plant essential oils against phytopathogenic bacteria. *Front. Microbiol.*, 10, 2619.

Cameron, A., Esiovwa, R., Connolly, J., Hursthouse, A., and Henriquez, F. 2022. Antimicrobial resistance as a global health threat: The need to learn lessons from the COVID-19 pandemic. *Global Pol.*, 13(2), 179–192.

Castillo-Juárez, I., García-Contreras, R., Velázquez-Guadarrama, N., Soto-Hernández, M., and Martínez-Vázquez, M. 2013. *Amphypterygium adstringens* anacardic acid mixture inhibits quorum sensing-controlled virulence factors of *Chromobacterium violaceum* and *Pseudomonas aeruginosa*. *Arch. Med. Res.*, 44(7), 488–494.

Cervantes-Ceballos, L., Caballero-Gallardo, K., and Olivero-Verbel, J. 2015. Repellent and anti-quorum sensing activity of six aromatic plants occurring in Colombia. *Nat. Prod. Commun.*, 10(10), 1753–1757.

Chadha, J., Harjai, K., and Chhibber, S. 2021. Revisiting the virulence hallmarks of *Pseudomonas aeruginosa*: A chronicle through the perspective of quorum sensing. *Environ. Microbiol.*, 24(6), 2630–2656.

Chadha, J., Harjai, K., and Chhibber, S. 2022. Repurposing phytochemicals as anti-virulent agents to attenuate quorum sensing-regulated virulence factors and biofilm formation in *Pseudomonas aeruginosa*. *J. Microbial. Biotechnol.*, 15(6), 1695–1718.

Chatterjee, A., Modarai, M., Naylor, N.R., Boyds, S.E., Atun, R., Barlow, J., et al. 2018. Quantifying drivers of antibiotic resistance in humans: A systematic review. *Lancet Infect. Dis.*, 18(12), e368–e378.

Chegini, Z., Khoshbayan, A., Vesal, S., Moradabadi, A., Hashemi, A., and Shariati, A. 2021. Bacteriophage therapy for inhibition of multi drug-resistant uropathogenic bacteria: A narrative review. *Ann. Clin. Microbiol. Antimicrob.*, 20, 30.

Chen, Y., Liu, T., Wang, K., Hou, C., Cai, S., Huang, Y., et al. 2016. Baicalein inhibits *Staphylococcus aureus* biofilm formation and the quorum sensing system *in vitro*. *PloS One*, 11(4), e0153468.

Chenia, H.Y. 2013. Anti-quorum sensing potential of crude *Kigelia africana* fruit extracts. *Sensors*, 13(3), 2802–2817.

Choudhury, D., Paul, D., Ghosh, A.S., Das Talukdar, A., Dutta Choudhury, M., Maurya, A.P., et al. 2016. Effect of single-dose carbapenem exposure on transcriptional expression of bla_{NDM-1} and mexA in *Pseudomonas aeruginosa*. *J. Glob. Antimicrob. Resist.*, 7, 72–77.

Chow, J.Y., Yang, Y., Tay, S.B., Chua, K.L., and Yew, W.S. 2014. Disruption of biofilm formation by the human pathogen *Acinetobacter baumannii* using engineered quorum-quenching lactonases. *Antimicrob. Agents Chemother.*, 58(3), 1802–1805.

Citorik, R.J., Mimee, M., and Lu, T.K. 2014. Sequence-specific antimicrobials using efficiently delivered RNA-guided nucleases. *Nat. Biotechnol.*, 32, 1141–1145.

Collignon, P., Beggs, J.J., Walsh, T.R., Gandra, S., and Laxminarayan, R. 2018. Anthropological and socioeconomic factors contributing to global antimicrobial resistance: A univariate and multivariable analysis. *Lancet Planet. Health.*, 2(9), e398–e405.

Cook, M.A., and Wright, G.D. 2022. The past, present, and future of antibiotics. *Sci. Transl. Med.*, 14, eabo7793.

Cordeiro, L., Figueiredo, P., Souza, H., Sousa, A., Andrade-Junior, F., Medeiros, D., et al. 2020. Terpinen-4-ol as an antibacterial and antibiofilm agent against *Staphylococcus aureus*. *Int. J. Mol. Sci.*, 21(12), 4531.

Coulter, L.B., McLean, R.J.C., Rohde, R.E., and Aron, G.M. 2014. Effect of bacteriophage infection in combination with tobramycin on the emergence of resistance in *Escherichia coli* and *Pseudomonas aeruginosa* biofilms. *Viruses*, 6, 3778–3786.

Czaplewski, L., Bax, R., Clokie, M., Dawson, M., Fairhead, H., Fischetti, V.A., et al. 2016. Alternative to antibiotics: A pipeline portfolio review. *Lancet Infect. Dis.*, 16(2), 239–251.

Dadgostar, P. 2019. Antimicrobial resistance: Implications and costs. *Infect. Drug. Resist.*, 12, 903–3910.

Das, A., Das, M.C., Sandhu, P., Das, N., Tribedi, P., De, U.C., et al. 2017a. Antibiofilm activity of *Parkia javanica* against *Pseudomonas aeruginosa*: A study with fruit extract. *RSC Adv.*, 7, 5497.

Das, B., Dash, S.K., Mandal, D., Ghosh, T., Chattopadhyay, S., Tripathy, S., et al. 2017b. Green synthesized silver nanoparticles destroy multidrug-resistant bacteria via reactive oxygen species mediated membrane damage. *Arab. J. Chem.*, 10, 862–876.

Das, H., Samanta, A.K., Kumar, S., Roychoudhury, P., Sarma, K., Akter, F., Subudhi, P.K., and Dutta, T.K. 2021. *In vitro* antimicrobial, antibiofilm and antiquorum sensing activity of Indian Rhododendron (*Melastoma malabathricum*) against clinical isolates of *Escherichia coli* and *Staphylococcus aureus*. *Indian J. Animal Res.*, doi: 10.18805/ijar.b-4415.

Das, M.C., Sandhu, P., Gupta, P., Rudrapaul, P., De, U.C., Tribedi, P., et al. 2016. Attenuation of *Pseudomonas aeruginosa* biofilm formation by Vitexin: A combinatorial study with azithromycin and gentamicin. *Sci. Rep.*, 6, 23347.

Das, T., Sehar, S., and Manefield, M. 2013. The roles of extracellular DNA in the structural integrity of extracellular polymeric substance and bacterial biofilm development. *Environ. Microbiol. Rep.*, 5(6), 778–786.

Datta, S., Jana, D., Maity, T.R., Samanta, A., and Banerjee, R. 2016. *Piper betle* leaf extract affects the quorum sensing and hence virulence of *Pseudomonas aeruginosa* PAO1. *3 Biotech*, 6(1), 1–6.

de Celis, M., Serrano-Aguirre, L., Belda, I., Liebana-Garcia, R., Arroyo, M., Marquina, D., et al. 2021. Acylase enzymes disrupting quorum sensing alter the transcriptome and phenotype of *Pseudomonas aeruginosa*, and the composition of bacterial biofilms from wastewater treatment plants. *Sci. Total Environ.*, 799, 149401.

De Oliveira, D.M.P., Forde, B.M., Kidd, T.J., Harris, P.N.A., Schembri, M.A., Beatson, S.A., et al. 2020. Antimicrobial resistance in ESKAPE pathogens. *Clin. Microbiol. Rev.*, 33, e00181–19.

Diallo, O.O., Baron, S.A., Abat, C., Colson, P., Chaudet, H., and Rolain, J.M. 2020. Antibiotic resistance surveillance systems: A review. *J. Glob. Antimicrob. Resist.*, 23, 430–438.

Ding, T., Li, T., and Li, J. 2018. Impact of curcumin liposomes with anti-quorum sensing properties against foodborne pathogens *Aeromonas hydrophila* and *Serratia grimesii*. *Microb. Pathog.*, 122, 137–143.

Dodson, T.A., Carlson, E.A., Warner, N.C., Morse, C.N., Gadient, J.N., and Prestwich, E.G. 2022. Characterization of distinct biofilm cell subpopulations and implications in quorum sensing and antibiotic resistance. *MBio*, 13(3), e00191–22.

Doğan, S., Gokalsm, B., Senkardes, I., Dogan, A., and Sesal, N.C. 2019. Anti-quorum sensing and anti-biofilm activities of *Hypericum perforatum* extracts against *Pseudomonas aeruginosa*. *J. Ethnopharmacol.*, 235, 293–300.

Dominguez, D.C., Chacon, L.M., and Wallace, D. 2021. Anthropogenic activities and the problem of antibiotic resistance in Latin America: A water issue. *Water*, 13(19), 2693.

Donlan, R.M. 2002. Biofilms: Microbial life on surfaces. *Emerg. Infect. Dis*., 8(9), 881–890.

Dostert, M., Trimble, M.J., and Hancock, R.E.W. 2021. Antibiofilm peptides: Overcoming biofilm-related treatment failure. *RSC Adv*., 11, 2718–2728.

Duddy, O.P., and Bassler, B.L. 2021. Quorum sensing across bacterial and viral domains. *PLoS Pathog*., 17(1), e1009074.

Egorov, A.M., Ulyashova, M.M., and Rubtsova, M.Y. 2018. Bacterial enzymes and antibiotic resistance. *Acta Naturae*., 10(4), 33–48.

Fan, Q., Zuo, J., Wang, H., Grenier, D., Yi, L., and Wang, Y. 2022. Contribution of quorum sensing to virulence and antibiotic resistance in zoonotic bacteria. *Biotechnol. Adv*., 59, 107965.

Farha, M.A., and Brown, E.D. 2019. Drug repurposing for antimicrobial discovery. *Nat. Microbiol*., 4, 565–577.

Ferriol-González, C., and Domingo-Calap, P. 2020. Phages for biofilm removal. *Antibiotics*, 9(5), 268.

Flemming, H.C., Wingender, J., Szewzyk, U., Steinberg, P., Rice, S.A., and Kjelleberg, S. 2016. Biofilms: An emergent form of bacterial life. *Nat. Rev. Microbiol*., 14, 563–575.

Fletcher, S. 2015. Understanding the contribution of environmental factors in the spread of antimicrobial resistance. *Environ. Health. Prev. Med*., 20(4), 243–252.

Fong, J., Zhang, C., Yang, R., Boo, Z.Z., Tan, S.K., Nielsen, T.E., et al. 2018. Combination therapy strategy of quorum quenching enzyme and quorum sensing inhibitor in suppressing multiple quorum sensing pathways of *P. aeruginosa*. *Sci. Rep*., 8, 1155.

Founou, R.C., Founou, L.L., Essack, S.Y. 2017. Clinical and economic impact of antibiotic resistance in developing countries: A systematic review and meta-analysis. *PloS One*, 12(12), e0189621.

Froes, T.Q., Nicastro, G.G., de Oliveira Pereira, T., de Oliveira Carneiro, K., Reis, I.M.A., Conceição, R.S., et al. 2020. Calycopterin, a major flavonoid from *Marcetia latifolia*, modulates virulence-related traits in *Pseudomonas aeruginosa*. *Microb. Pathog*., 144, 104142.

Fu, B., Wu, Q., Dang, M., Bai, D., Guo, Q., Shen, L., and Duan, K. 2017. Inhibition of *Pseudomonas aeruginosa* biofilm formation by traditional Chinese medicinal herb *Herba patriniae*. *Biomed. Res. Int*., 2017, 9584703.

Furiga, A., Lajoie, B., El Hage, S., Baziard, G., and Roques, C. 2016. Impairment of *Pseudomonas aeruginosa* biofilm resistance to antibiotics by combining the drugs with a new quorum-sensing inhibitor. *Antimicrob. Agents Chemother*., 60(3), 1676–1686.

Gala, V.C., John, N.R., Bhagwat, A.M., Datar, A.G., Kharkar, P.S., and Desai, K.B. 2016. Attenuation of quorum sensing-regulated behaviour by *Tinospora cordifolia* extract & identification of its active constituents. *Indian J. Med. Res*., 144(1), 92–103.

Galani, I., Karaiskos, I., Giamarellou, H. 2021. Multidrug-resistant *Klebsiella pneumoniae*: Mechanisms of resistance including updated data for novel β-lactam-β-lactamase inhibitor combinations. *Exp. Rev. Anti-Infect. Ther*., 19(11), 1457–1468.

Ganesh, P.S., and Rai, V.R. 2018. Attenuation of quorum sensing-dependent virulence factors and biofilm formation by medicinal plants against antibiotic resistant *Pseudomonas aeruginosa*. *J. Tradit. Complement. Med*., 8, 170–177.

Ghorbani, J., Rahban, D., Aghamiri, S., Teymouri, A., and Bahador, A. 2018. Photosensitizers in antimicrobial photodynamic therapy: An overview. *Laser. Ther*., 27(4), 293–302.

Gibbons, S., Oluwatuyi, M., and Kaatz, G.W. 2003. A novel inhibitor of multidrug efflux pumps in *Staphylococcus aureus*. *J. Antimicrob. Chemother*., 51, 13–17.

Gökalsın, B., Aksoydan, B., Erman, B., and Sesal, N.C. 2017. Reducing virulence and biofilm of *Pseudomonas aeruginosa* by potential quorum sensing inhibitor carotenoid: Zeaxanthin. *Microb. Ecol*., 74(2), 466–473.

González-Bello, C., Rodriguez, D., Pernas, M., Rodriguez, A., and Colcon, E. 2020. β-Lactamase inhibitors to restore the efficacy of antibiotics against superbugs. *J. Med. Chem*., 63(5), 1859–1881.

Gopu, V., Meena, C.K., Murali, A., and Shetty, P.H. 2016. Petunidin as a competitive inhibitor of acylated homoserine lactones in *Klebsiella pneumoniae*. *RSC Adv*., 6(4), 2592–2601.

Gopu, V., Meena, C.K., and Shetty, P.H. 2015. Quercetin influences quorum sensing in food borne bacteria: *In vitro* and *in silico* evidence. *PloS One*, 10(8), e0134684.

Grandclément, C., Tannieres, M., Morera, S., Dessaux, Y., and Faure, D. 2015. Quorum quenching: Role in nature and applied developments. *FEMS Microbiol. Rev*., 40(1), 86–116.

Guchhait, K.C., Manna, T., Barai, M., Karmakar, M., Nandi, S.K., Jana, D., et al. 2022. Antibiofilm and anticancer activities of unripe and ripe *Azadirachta indica* (neem) seed extracts. *BMC Compl. Med. Ther*., 22(1), 1–18.

Guendouze, A., Plener, L., Bzdrenga, J., Jacquet, P., Rémy, B., Elias, M., et al. 2017. Effect of quorum quenching lactonase in clinical isolates of *Pseudomonas aeruginosa* and comparison with quorum sensing inhibitors. *Front. Microbiol*., 8, 227.

Gupta, S., Cohen, K.A., Winglee, K., Maiga, M., Diarra, B., and Bishai, W.R. 2014. Efflux inhibition with verapamil potentiates bedaquiline in *Mycobacterium tuberculosis*. *Antimicrob. Agents Chemother.*, 58, 574–576.

Gutiérrez-Barranquero, J.A., Reen, F.J., McCarthy, R.R., and O'Gara, F. 2015. Deciphering the role of coumarin as a novel quorum sensing inhibitor suppressing virulence phenotypes in bacterial pathogens. *Appl. Microbiol. Biotechnol.* 99(7), 3303–3316.

Guzman, J.P.M.D., De Las Alas, T.P.L., Lucban, M.C., and Sevilla, C.E.C. 2020. Green tea (*Camellia sinensis*) extract inhibits biofilm formation in acyl homoserine lactone-producing, antibiotic-resistant *Morganella morganii* isolated from Pasig River, Philippines. *Heliyon*, 6(10), e05284.

Hadi, N., Moradi, F., Rohi Jahromi, R., and Akbari, M. 2020. Antibacterial and anti-quorum sensing properties of *Dionysia revolute* Boiss against secondary bacterial infections of COVID-19 patients: An *in-vitro* study. *Herbal Med. J.*, 5(3), 91–99.

Hancock, R.E.W., Alford, M.A. and Haney, E.F. 2021. Antibiofilm activity of host defence peptides: Complexity provides opportunities. *Nat. Rev. Microbiol.*, 19, 786–797.

Hao, S., Yang, D., Zhao, L., Shi, F., Ye, G., Fu, H., et al. 2021. EGCG-mediated potential inhibition of biofilm development and quorum sensing in *Pseudomonas aeruginosa*. *Int. J. Mol. Sci.*, 22(9), 4946.

Haque, M., Islam, S., Sheikh, M.A., Dhingra, S., Uwambaye, P., Labricciosa, F.M., et al. 2021. Quorum sensing: A new prospect for the management of antimicrobial-resistant infectious diseases. *Exp. Rev. Antiinfect. Ther.*, 19(5), 571–586.

Haque, S., Ahmad, F., Dar, S.A., Jawed, A., Mandal, R.K., Wahid, M., et al. 2018. Developments in strategies for Quorum Sensing virulence factor inhibition to combat bacterial drug resistance. *Microb. Pathog.*, 121, 293–302.

Harjai, K., Kumar, R., and Singh, S. 2010. Garlic blocks quorum sensing and attenuates the virulence of *Pseudomonas aeruginosa*. *FEMS Immunol. Med. Microbiol.*, 58(2), 161–168.

Hasan, S., Danishuddin, M., and Khan, A.U. 2015. Inhibitory effect of *Zingiber officinale* towards *Streptococcus mutans* virulence and caries development: *In vitro* and *in vivo* studies. *BMC Microbiol.*, 15(1), 1–14.

Hossain, M.A., Lee, S.J., Park, J.Y., Reza, M.A., Kim, T.H., Lee, K.J., et al. 2015. Modulation of quorum sensing-controlled virulence factors by *Nymphaea tetragona* (water lily) extract. *J. Ethnopharmacol.*, 174, 482–491.

Huang, L., Wu, C., Gao, H., Xu, C., Dai, M., Huang, L., et al. 2022. Bacterial multidrug efflux pumps at the frontline of antimicrobial resistance: An overview. *Antibiotics*, 11(4), 520.

Huma, N., Shankar, P., Kushwah, J., Bhushan, A., Joshi, J., Mukherjee, T., et al. 2011. Diversity and polymorphism in AHL-lactonase gene (aiiA) of *Bacillus*. *J. Microbiol. Biotechnol.*, 21(10), 1001–1011.

Husain, F.M., Ahmad, I., Khan, M.S., Ahmad, E., Tahseen, Q., Khan, M.S., and Alshabib, N.A. 2015a. Sub-MICs of *Mentha piperita* essential oil and menthol inhibits AHL mediated quorum sensing and biofilm of Gram-negative bacteria. *Front. Microbiol.*, 6, 420.

Husain, F.M., Ahmad, I., Khan, M.S., and Al-Shabib, N.A. 2015b. *Trigonella foenum-graceum* (Seed) extract interferes with quorum sensing regulated traits and biofilm formation in the strains of *Pseudomonas aeruginosa* and *Aeromonas hydrophila*. *Evidence. Compl. Alt. Med.*, 2015, 879540.

Hutchings, M.I., Truman, A.W., and Wilkinson, B. 2019. Antibiotics: Past, present and future. *Curr. Opin. Microbiol.*, 51, 72–80.

Igarashi, Y., Gohda, F., Kadoshima, T., Fakuda, T., Hanafusa, T., Shojima, A., et al. 2015. Avellanin C, an inhibitor of quorum-sensing signaling in *Staphylococcus aureus*, from *Hamigera ingelheimensis*. *J. Antibiot.*, 68, 707–710.

Ilic-Tomic, T., Sokovic, M., Vojnovic, S., Ciric, A., Veljic, M., Nikodinovic-Runic, J., and Novakovic, M. 2017. Diarylheptanoids from *Alnus viridis* ssp. *viridis* and *Alnus glutinosa*: Modulation of quorum sensing activity in *Pseudomonas aeruginosa*. *Planta Med.*, 83(01/02), 117–125.

Impey, R.E., Hawkins, D.A., Sutton, J.M., and Soares da Costa, T.P. 2020. Overcoming intrinsic and acquired resistance mechanisms associated with the cell wall of Gram-negative bacteria. *Antibiotics*, 9(9), 623.

Inoue, Y., Togashi, N., and Hamashima, H. 2016. Farnesol-induced disruption of the *Staphylococcus aureus* cytoplasmic membrane. *Biol. Pharm. Bull.*, 39(5), 653–656.

Ishii, A., Shigemura, K., Kitagawa, K., Harada, M., Kan, Y., Hayashi, F., et al. 2021. Cross-resistance and the mechanisms of cephalosporin-resistant bacteria in urinary tract infections isolated in Indonesia. *Curr. Microbiol.*, 78, 1771–1777.

Jakobsen, T.H., Van Gennip, M., Phipps, R.K., Shanmugham, M.S., Christensen, L.D., Alhede, M., et al. 2012. Ajoene, a sulfur-rich molecule from garlic, inhibits genes controlled by quorum sensing. *Antimicrob. Agents Chemother.*, 56(5), 2314–2325.

Jayalekshmi, H., Omanakuttan, A., Pandurangan, N., Vargis, V.S., Maneesh, M., Nair, B.G., and Kumar, G.B. 2016. Clove bud oil reduces kynurenine and inhibits *pqs A* gene expression in *P. aeruginosa. Appl. Microbiol. Biotechnol.*, 100(8), 3681–3692.

Joshi, C., Patel, P., and Kothari, V. 2019. Anti-infective potential of hydroalcoholic extract of *Punica granatum* peel against gram-negative bacterial pathogens. *F1000Res*, 8, 70.

Kačániová, M., Galovičová, L., Borotová, P., Valková, V., Ďúranová, H., Kowalczewski, P.Ł., et al. 2021. Chemical composition, in vitro and in situ antimicrobial and antibiofilm activities of *Syzygium aromaticum* (Clove) essential oil. *Plants*, 10(10), 2185.

Kannan, S., Sathasivam, G., and Marudhamuthu, M. 2019. Decrease of growth, biofilm and secreted virulence in opportunistic nosocomial *Pseudomonas aeruginosa* ATCC 25619 by glycyrrhetinic acid. *Microb. Pathog.*, 126, 332–342.

Karicheri, R., and Antony, B. 2016. Antibacterial and antibiofilm activities of peppermint (*Mentha piperita* Linn) and menthol mint (*Mentha arvensis* Linn) essential oils on *Aggregatibacter actinomycetemcomitans* isolated from orodental infections. *Eur. J. Pharmaceut. Med. Res.*, 3(7), 577–581.

Karner, L., Drechsler, S., Metzger, M., Hacobian, A., Schadi, B., Slezak, P., et al. 2020. Antimicrobial photodynamic therapy fighting polymicrobial infections—A journey from *in vitro* to *in vivo. Photochem. Photobiol. Sci.*, 19, 1332–1343.

Kelly, D., McAuliffe, O., Ross, R.P., and Coffey, A. 2012. Prevention of *Staphylococcus aureus* biofilm formation and reduction in established biofilm density using a combination of phage K and modified derivatives. *Lett. Appl. Microbiol.*, 54(4), 286–291.

Kim, C., Hesek, D., Lee, M., and Mobashery, S. 2018. Potentiation of the activity of β-lactam antibiotics by farnesol and its derivatives. *Bioorg. Med. Chem. Lett.*, 28(4), 642–645.

Kim, H.S., Lee, S.H., Byun, Y., and Park, H.D. 2015a. 6-Gingerol reduces *Pseudomonas aeruginosa* biofilm formation and virulence via quorum sensing inhibition. *Sci. Rep.*, 5, 8656.

Kim, M.M., and Darafsheh, A. 2020. Light sources and dosimetry techniques for photodynamic therapy. *Photochem. Photobiol.*, 96(2), 280–294.

Kim, Y.G., Lee, J.H., Kim, S.I., Baek, K.H., and Lee, J. 2015b. Cinnamon bark oil and its components inhibit biofilm formation and toxin production. *Int. J. Food Microbiol.*, 195, 30–39.

Kjelleberg, S., and Molin, S. 2002. Is there a role for quorum sensing signals in bacterial biofilms? *Curr. Opin. Microbiol.*, 5, 254–258.

Klein, E.Y., Van Boeckel, T.P., Martínez, E.M., Pant, S., Gandra, S., Levin, S.A., et al. 2018. Global increase and geographic convergence in antibiotic consumption between 2000 and 2015. *Proc. Natl. Acad. Sci. U. S. A.*, 115(15), E3463–E3470.

Klingler, F.M., Wichelhaus, T.A., Frank, D., Bernal, J.C., El-Delik, J., Muller, H.F., et al. 2015. Approved drugs containing thiols as inhibitors of Metallo-β-lactamases: Strategy to combat multidrug-resistant bacteria. *J. Med. Chem.*, 58(8), 3626–3630.

Koch, G., Nadal-Jimenez, P., Reis, C.R., Muntendam, R., Bokhove, M., Mellilo, E., et al. 2014. Reducing virulence of the human pathogen *Burkholderia* by altering the substrate specificity of the quorum-quenching acylase PvdQ. *Proc. Natl. Acad. Sci. U. S. A.*, 111(4), 1568–1573.

Koh, C.L., Sam, C.K., Yin, W.F., Tan, L.Y., Krishnan, T., Chong, Y.M., and Chan, K.G. 2013. Plant-derived natural products as sources of anti-quorum sensing compounds. *Sensors*, 13(5), 6217–6228.

Konreddy, A.K., Rani, G.U., Lee, K., and Choi, Y. 2019. Recent drug-repurposing-driven advances in the discovery of novel antibiotics. *Curr. Med. Chem.*, 26(28), 5363–5388.

Kordbacheh, H., Eftekhar, F., and Ebrahimi, S.N. 2017. Anti-quorum sensing activity of *Pistacia atlantica* against *Pseudomonas aeruginosa* PAO1 and identification of its bioactive compounds. *Microb. Pathog.*, 110, 390–398.

Korkorian, N., and Mohammadi-Sichani, M. 2017. Anti-quorum sensing and antibacterial activity of *Rumex alveolatus. Zahedan J. Res. Med. Sci.*, 19(11), e56009.

Krishna, H.C. and Kondapalli, K. 2018. Isolation and purification of antibiofilm peptides from *Tridax procumbens*, a medicinal plant. *Int. J. Pharm. Biol. Sci.*, 8(3), 1130–1135.

Kumar, A., Khan, I.A., Koul, S., Koul, J.L., Taneja, S.C., Ali, I., et al. 2008. Novel structural analogues of piperine as inhibitors of the NorA efflux pump of *Staphylococcus aureus. J. Antimicrob. Chemother.*, 61, 1270–1276.

Kumar, L., Chhibber, S., Kumar, R., Kumar, M., and Harjai, K. 2015. Zingerone silences quorum sensing and attenuates virulence of *Pseudomonas aeruginosa. Fitoterapia*, 102, 84–95.

Kumar, M., Sarma, D.K., Shubham, S., Kumawat, M., Verma, V., Nina, P.B., et al. 2021. Futuristic non-antibiotic therapies to combat antibiotic resistance: A review. *Front. Microbiol.*, 12, 609459.

Lahiri, D., Nag, M., Dutta, B., Dey, S., Mukherjee, D., Joshi, S.J., and Ray, R.R. 2021a. Antibiofilm and anti-quorum sensing activities of eugenol and linalool from *Ocimum tenuiflorum* against *Pseudomonas aeruginosa* biofilm. *J. Appl. Microbiol.*, 131(6), 2821–2837.

Lahiri, D., Nag, M., Dutta, B., Mukherjee, I., Ghosh, S., Dey, A., et al. 2021b. Catechin as the most efficient bioactive compound from *Azadirachta indica* with antibiofilm and anti-quorum sensing activities against dental biofilm: An *in vitro* and *in silico* study. *Appl. Biochem. Biotechnol.*, 193(6), 1617–1630.

Larsson, D.G.J., Flach, C.F. 2022. Antibiotic resistance in the environment. *Nat. Rev. Microbiol.*, 20, 257–269.

Lazdunski, A.M., Ventre, I., Sturgis, J.N. 2004. Regulatory circuits and communication in Gram-negative bacteria. *Nat. Rev. Microbiol.*, 2, 581–592.

Lee, J.H., Kim, Y.G., Ryu, S.Y., Cho, M.H., and Lee, J. 2014. Ginkgolic acids and *Ginkgo biloba* extract inhibit *Escherichia coli* O157: H7 and *Staphylococcus aureus* biofilm formation. *Int. J. Food Microbiol.*, 174, 47–55.

Lee, J.H., Regmi, S.C., Kim, J.A., Cho, M.H., Yun, H., Lee, C.S., and Lee, J. 2011. Apple flavonoid phloretin inhibits *Escherichia coli* O157: H7 biofilm formation and ameliorates colon inflammation in rats. *Infect. Immun.*, 79(12), 4819–4827.

Lee, K., Kim, D.W., Lee, D.H., Kim, Y.S., Bu, J.H., Cha, J.H., et al. 2020. Mobile resistome of human gut and pathogen drives anthropogenic bloom of antibiotic resistance. *Microbiome*, 8, 2.

Lewis, K. 2020. The science of antibiotic discovery. *Cell*, 181, 29–45.

Li, G., Yan, C., Xu, Y., Feng, Y., Wu, Q., Lv, X., et al. 2014. Punicalagin inhibits *Salmonella* virulence factors and has anti-quorum-sensing potential. *Appl. Environ. Microbiol.*, 80(19), 6204–6211.

Li, W.R., Ma, Y.K., Xie, X.B., Shi, Q.S., Wen, X., Sun, T.L., et al. 2019. Diallyl disulfide from garlic oil inhibits *Pseudomonas aeruginosa* quorum sensing systems and corresponding virulence factors. *Front. Microbiol.*, 10, 3222.

Li, X.H., and Lee, J.H. 2017. Antibiofilm agents: A new perspective for antimicrobial strategy. *J. Microbiol.*, 55(10), 753–766.

Lian, C., Piksa, M., Yoshida, K., Persheyev, S., Pawlik, K.J., Matczyszyn, K., and Samuel, I.D.W. 2019. Flexible organic light-emitting diodes for antimicrobial photodynamic therapy. *NPJ Flexible Electron.*, 3, 18.

Liu, B., Trout, R.E.L., Chu, G.H., McGarry, D., Jackson, R.W., Hamrick, J.C., et al. 2020. Discovery of Taniborbactam (VNRX-5133): A broad-spectrum Serine- and Metallo-β-lactamase inhibitor for carbapenem-resistant bacterial infections. *J. Med. Chem.*, 63(6), 2789–2801.

Liu, Y., Qin, R., Zaat, S.A.J., Breukink, E., and Heger, M. 2015. Antibacterial photodynamic therapy: Overview of a promising approach to fight antibiotic-resistant bacterial infections. *J. Clin. Transl. Res.*, 1(3), 140–167.

Liu, Y., Tong, Z., Shi, J., Li, R., Upton, M., and Wang, Z. 2021. Drug repurposing for next-generation combination therapies against multidrug-resistant bacteria. *Theranostics*, 11(10), 4910–4928.

Liu, Y., Xu, Y., Song, Q., Wang, F., Sun, L., Liu, L., et al. 2017. Antibiofilm activities from *Bergenia crassifolia* leaves against *Streptococcus mutans*. *Front. Microbiol.*, 8, 1738.

Livermore, D.M., Mushtaq, S., Warner, M., Vickers, A., and Woodford, N. 2017. *In vitro* activity of cefepime/zidebactam (WCK 5222) against Gram-negative bacteria. *J. Antimicrob. Chemother.*, 72(5), 1373–1385.

Lopes, B.S., Hanafiah, A., Nachimuthu, R., Muthupandian, S., Md Nesran, Z.N., and Patil, S. 2022. The Role of antimicrobial peptides as antimicrobial and antibiofilm agents in tackling the silent pandemic of antimicrobial resistance. *Molecules*, 27(9), 2995.

López-Jácome, L.E., Garza-Ramos, G., Hernandez-Duran, M., Franco-Cendejas, R., Loarca, D., Romero-Martínez, D., et al. 2019. AiiM Lactonase strongly reduces quorum sensing controlled virulence factors in clinical strains of *Pseudomonas aeruginosa* isolated from burned patients. *Front. Microbiol.*, 10, 2657.

Luciardi, M.C., Blázquez, M.A., Alberto, M.R., Cartagena, E., and Arena, M.E. 2021. Lemon oils attenuate the pathogenicity of *Pseudomonas aeruginosa* by quorum sensing inhibition. *Molecules*, 26(10), 2863.

Luciardi, M.C., Blázquez, M.A., Cartagena, E., Bardon, A., and Arena, M.E. 2016. Mandarin essential oils inhibit quorum sensing and virulence factors of *Pseudomonas aeruginosa*. *LWT–Food Sci. Technol.*, 68, 373–380.

Ludacka, P., Kubat, P., Bosakova, Z., and Mosinger, J. 2022. Antibacterial nanoparticles with natural photosensitizers extracted from spinach leaves. *ACS Omega.*, 7(1), 1505–1513.

Luo, A., Wang, F., Sun, D., Liu, X., and Xin, B. 2022. Formation, development, and cross-species interactions in biofilms. *Front. Microbiol.*, 12, 757327.

Luo, J., Dong, B., Wang, K., Cai, S., Liu, T., Cheng, X., et al. 2017. Baicalin inhibits biofilm formation, attenuates the quorum sensing-controlled virulence and enhances *Pseudomonas aeruginosa* clearance in a mouse peritoneal implant infection model. *PloS One*, 12 (4), e0176883.

Luo, J., Kong, J.L., Dong, B.Y., Huang, H., Wang, K., Wu, L.H., et al. 2016. Baicalein attenuates the quorum sensing-controlled virulence factors of *Pseudomonas aeruginosa* and relieves the inflammatory response in *P. aeruginosa*-infected macrophages by downregulating the MAPK and NFκB signal-transduction pathways. *Drug. Des. Dev. Ther.*, 10, 183–203.

Lyles, J.T., Kim, A., Nelson, K., Bullard-Roberts, A.L., Hajdari, A., Mustafa, B., and Quave, C.L. 2017. The chemical and antibacterial evaluation of St. John's Wort oil macerates used in Kosovar traditional medicine. *Front. Microbiol.*, 8, 1639.

Maisuria, V.B., De Los Santos, Y.L., Tufenkji, N., and Déziel, E. 2016. Cranberry-derived proanthocyanidins impair virulence and inhibit quorum sensing of *Pseudomonas aeruginosa. Sci. Rep.*, 6, 30169.

Makhfian, M., Hassanzadeh, N., Mahmoudi, E., and Zandyavari, N. 2015. Anti-quorum sensing effects of ethanolic crude extract of *Anethum graveolens* L. *J. Essent. Oil Bearing. Plants.*, 18(3), 687–696.

Malik, B., Bhattacharyya, S. 2019. Antibiotic drug-resistance as a complex system driven by socio-economic growth and antibiotic misuse. *Sci. Rep.*, 9, 9788.

Manohar, P., Royam, M.M., Loh, B., Bozdogan, B., Nachimuthu, R., and Leptihn, S. 2022. Synergistic effects of phage-antibiotic combinations against *Citrobacter amalonaticus. ACS Infect. Dis.*, 8, 59–65.

Mansuri, A., Lokhande, K., Kore, S., Gaikwad, S., Nawani, N., Swamy, K.V., et al. 2022. Antioxidant, anti-quorum sensing, biofilm inhibitory activities and chemical composition of Patchouli essential oil: *In vitro* and *in silico* approach. *J. Biomol. Struct. Dyn.*, 40(1), 154–165.

Martínez, O.F., Rigueiras, P.O., da Silva Pires, A., Porto, W.F., Silva, O.N., de la Fuente-Nunez, C., and Franco, O.L. 2019. Interference with quorum-sensing signal biosynthesis as a promising therapeutic strategy against multidrug-resistant pathogens. *Front. Cell. Infect. Microbiol.*, 8, 444.

Maryam, L., Usmani, S.S., Raghava, G.P.S. 2021. Computational resources in the management of antibiotic resistance: Speeding up drug discovery. *Drug Discov. Today*, 26(9), 2138–2151.

Mattila, S., Ruotsalainen, P., and Jalasvuori, M. 2015. On-demand isolation of bacteriophages against drug-resistant bacteria for personalized phage therapy. *Front. Microbiol.*, 6, 1271.

Memariani, H., Memariani, M., and Ghasemian, A. 2019. An overview on antibiofilm properties of quercetin against bacterial pathogens. *World J. Microbiol. Biotechnol.* 35(9), 1–16.

Méndez, D.A.C., Cuellar, M.R.C., Pedrinha, V.F., Espedilla, E.G.V., de Andrade, F.B., de Almeida Rodrigues, P., and Cruvinel, T. 2021. Effect of curcumin-mediated antimicrobial photodynamic therapy associated to different chelators against *Enterococcus faecalis* biofilms. *Photodiagnosis Photodyn. Ther.*, 35, 102464.

Michael, C.A., Dominey-Howes, D., and Labbate, M. 2014. The antimicrobial resistance crisis: Causes, consequences, and management. *Front. Public Health*, 2, 145.

Miethke, M., Pieroni, M., Weber, T., Bronstrup, M., Hammann, P., Halby, L., et al. 2021. Towards the sustainable discovery and development of new antibiotics. *Nature. Rev. Chem.*, 5, 726–749.

Mishra, B., Lushnikova, T., Golla, R.M., Wang, X., and Wang, G. 2017. Design and surface immobilization of short antibiofilm peptides. *Acta Biomater.*, 49, 316–328.

Mishra, R., Panda, A.K., Mandal, S.D., Shakeel, M., Bisht, S.S., and Khan, J. 2020. Natural antibiofilm agents: Strategies to control biofilm-forming pathogens. *Front. Microbiol.*, 11, 566325.

Moghadam, M.T., Khoshbayan, A., Chegini, Z., Farahani, I., and Shariati, A. 2020. Bacteriophages, a new therapeutic solution for inhibiting multidrug-resistant bacteria causing wound infection: Lesson from animal models and clinical trials. *Drug Des. Devel. Ther.*, 14, 1867–1883.

Mojica, M.F., Rossi, M.A., Vila, A.J., and Bonomo, R.A. 2022. The urgent need for metallo-β-lactamase inhibitors: An unattended global threat. *Lancet Infect. Dis.*, 22, e28–34.

Moradi, F., Hadi, N., and Bazargani, A. 2020. Evaluation of quorum-sensing inhibitory effects of extracts of three traditional medicine plants with known antibacterial properties. *New Microbes. New Infect.*, 38, 100769.

Mu, Y., Zeng, H., and Chen, W. 2021. Quercetin inhibits biofilm formation by decreasing the production of EPS and altering the composition of EPS in *Staphylococcus epidermidis. Front. Microbiol.*, 12, 631058.

Mu, Y., Zeng, H., and Chen, W. 2020. Okanin in *Coreopsis tinctoria* Nutt is a major quorum-sensing inhibitor against *Chromobacterium violaceum. J. Ethnopharmacol.*, 260, 113017.

Muhammad, M.H., Idris, A.L., Fan, X., Guo, Y., Yu, Y., Jin, X., et al. 2020. Beyond risk: Bacterial biofilms and their regulating approaches. *Front. Microbiol.*, 11, 928.

Mukherjee, S., and Bassler, B.L. 2019. Bacterial quorum sensing in complex and dynamically changing environments. *Nat. Rev. Microbiol.*, 17, 371–382.

Munita, J.M., and Arias, C.A. 2016. Mechanisms of antibiotic resistance. *Microbiol. Spectrum.*, 4(2), doi: 10.1128/microbiolspec.VMBF-0016-2015

Murata, R.M., Branco-de-Almeida, L.S., Franco, E.M., Yatsuda, R., dos Santos, M.H., de Alencar, S.M., et al. 2010. Inhibition of *Streptococcus mutans* biofilm accumulation and development of dental caries *in vivo* by 7-epiclusianone and fluoride. *Biofouling*, 26(7), 865–872.

Murray, C.J.L., Ikuta, K.S., Sharara, F., Swetschinski, L., Aguilar, G.R., Gray, A., et al. 2022. Global burden of bacterial antimicrobial resistance in 2019: A systematic analysis. *Lancet*, 399, 629–655.

Murugaiyan, J., Kumar, P.A., Rao, G.S., Iskandar, K., Hawser, S., Hays, J.P., et al. 2022. Progress in alternative strategies to combat antimicrobial resistance: Focus on antibiotics. *Antibiotics*, 11(2), 200.

Murugan, K., Sekar, K., Sangeetha, S., Ranjitha, S., and Sohaibani, S.A. 2013. Antibiofilm and quorum sensing inhibitory activity of *Achyranthes aspera* on cariogenic *Streptococcus mutans*: An *in vitro* and *in silico* study. *Pharm. Biol.*, 51(6), 728–736.

Muzondiwa, D., Mutshembele, A., Pierneef, R.E., and Reva, O.N., 2020. Resistance Sniffer: An online tool for prediction of drug resistance patterns of *Mycobacterium tuberculosis* isolates using next-generation sequencing data. *Int. J. Med. Microbiol.*, 310(2), 151399.

Myszka, K., Schmidt, M.T., Majcher, M., Juzwa, W., Olkowicz, M., and Czaczyk, K. 2016. Inhibition of quorum sensing-related biofilm of *Pseudomonas fluorescens* KM121 by *Thymus vulgare* essential oil and its major bioactive compounds. *Int. Biodeter. Biodegr.*, 114, 252–259.

Nagar, N., Aswathanarayan, J.B., and Vittal, R.R. 2020. Anti-quorum sensing and biofilm inhibitory activity of *Apium graveolens* L. oleoresin. *J. Food. Sci. Technol.*, 57(7), 2414–2422.

Namasivayam, S.K.R., and Roy, E.A. 2013. Antibiofilm effect of medicinal plant extracts against clinical isolate of biofilm of *Escherichia coli*. *Int. J. Pharm. Pharmaceut. Sci*, 5(2), 486–489.

Ngenge, T.A., Kucukaydin, S., Ceylan, O., and Duru, M.E. 2021. Evaluation of enzyme inhibition and anti-quorum sensing potentials of *Melaleuca alternifolia* and *Citrus sinensis* essential oils. *Nat. Prod. Commun.*, 16(9), 1–8.

Niu, C., and Gilbert, E.S. 2004. Colorimetric method for identifying plant essential oil components that affect biofilm formation and structure. *Appl. Environ. Microbiol.*, 70(12), 6951–6956.

Noumi, E., Snoussi, M., Alreshidi, M.M., Rekha, P.D., Saptami, K., Caputo, L., et al. 2018. Chemical and biological evaluation of essential oils from *Cardamom* species. *Molecules*, 23(11), 2818.

Noumi, E., Snoussi, M., Merghni, A., Nazzaro, F., Quindos, G., Akdamar, G., et al. 2017. Phytochemical composition, antibiofilm and anti-quorum sensing potential of fruit, stem and leaves of *Salvadora persica* L. methanolic extracts. *Microb. Pathog.*, 109, 169–176.

Okshevsky, M., and Meyer, R.L. 2015. The role of extracellular DNA in the establishment, maintenance and perpetuation of bacterial biofilms. *Crit. Rev. Microbiol.*, 41, 341–352.

Oliveira, B.D., Rodrigues, A.C., Cardoso, B.M.I., Ramos, A.L.C.C., Bertoldi, M.C., Taylor, J.G., et al. 2016. Antioxidant, antimicrobial and anti-quorum sensing activities of *Rubus rosaefolius* phenolic extract. *Ind. Crop Prod.*, 84, 59–66.

Oluwatuyi, M., Kaatz, G.W., and Gibbons, S. 2004. Antibacterial and resistance modifying activity of *Rosmarinus officinalis*. *Phytochemistry*, 65, 3249–3254.

Packiavathy, I.A.S.V., Agilandeswari, P., Musthafa, K.S., Pandian, S.K., and Ravi, A.V. 2012. Antibiofilm and quorum sensing inhibitory potential of *Cuminum cyminum* and its secondary metabolite methyl eugenol against Gram-negative bacterial pathogens. *Food Res. Int.*, 45, 85–92.

Paczkowski, J.E., Mukherjee, S., McCready, A.R., Cong, J.P., Aquino, C.J., Kim, H., et al. 2017. Flavonoids suppress *Pseudomonas aeruginosa* virulence through allosteric inhibition of quorum-sensing receptors. *J. Biol. Chem.*, 292(10), 4064–4076.

Pang, Z., Raudonis, R., Glick, B.R., Lin, T.J., Cheng, Z. 2019. Antibiotic resistance in *Pseudomonas aeruginosa*: Mechanisms and alternative therapeutic strategies. *Biotechnol. Adv.*, 37(1), 177–192.

Papenfort, K., and Bassler, B.L. 2016. Quorum-sensing signal-response systems in Gram-negative bacteria. *Nat. Rev. Microbiol.*, 14(9), 576–588.

Papp-Wallace, K.M., Bethel, C.R., Caillon, J., Barnes, M.D., Potel, G., Bajaksouzian, S., et al. 2019. Beyond piperacillin-tazobactam: Cefepime and AAI101 as a potent β-Lactam-β-Lactamase inhibitor combination. *Antimicrob. Agents Chemother.*, 63(5), e00105–19.

Parasuraman, P., Antony, A.P., Sruthil Lal, S.B., Sharan, A., Siddhardha, B., Kasinathan, K., et al. 2019. Antimicrobial photodynamic activity of toluidine blue encapsulated in mesoporous silica nanoparticles against *Pseudomonas aeruginosa* and *Staphylococcus aureus*. *Biofouling*, 35(1), 89–103.

Parsek, M.R., Val, D.L., Hanzelka, B.L., Cronan Jr, J.E., and Greenberg, E.P. 1999. Acyl homoserine-lactone quorum-sensing signal generation. *Proc. Natl. Acad. Sci. U. S. A.*, 96(8), 4360–4365.

Patel, B., Kumari, S., Banerjee, R., Samanta, M., and Das, S. 2017. Disruption of the quorum sensing regulated pathogenic traits of the biofilm-forming fish pathogen *Aeromonas hydrophila* by tannic acid, a potent quorum quencher. *Biofouling*, 33(7), 580–590.

Pattnaik, S., Barik, S., Muralitharan, G., and Busi, S. 2018. Ferulic acid encapsulated chitosan-tripolyphosphate nanoparticles attenuate quorum sensing regulated virulence and biofilm formation in *Pseudomonas aeruginosa* PAO1. *IET Nanobiotechnol.*. 12(8), 1056–1061.

Peerzada, Z., Kanhed, A.M., and Desai, K.B. 2022. Effects of active compounds from *Cassia fistula* on quorum sensing mediated virulence and biofilm formation in *Pseudomonas aeruginosa. RSC Adv.*, 12, 15196–15214.

Peterson, E., and Kaur, P. 2018. Antibiotic resistance mechanisms in bacteria: Relationships between resistance determinants of antibiotic producers, environmental bacteria, and clinical pathogens. *Front. Microbiol.*, 9, 2928.

Pires, D.P., Melo, L.D.R., Boas, D.V., Sillankorva, S., and Azeredo, J. 2017. Phage therapy as an alternative or complementary strategy to prevent and control biofilm-related infections. *Curr. Opin. Microbiol.*, 39, 48–56.

Poli, J.P., Guinoiseau, E., de Rocca Serra, D., Sutour, S., Paoli, M., Tomi, F., et al. 2018. Anti-quorum sensing activity of 12 Essential Oils on *Chromobacterium violaceum* and specific action of cis-cis-p-Menthenolide from Corsican *Mentha suaveolens* ssp. Insularis. *Molecules*, 23(9), 2125.

Porfírio, E.M., Melo, H.M., Pereira, A.M.G., Cavalcante, T.T.A., Gomes, G.A., Carvalho, M.G.D., et al. 2017. *In vitro* antibacterial and antibiofilm activity of *Lippia alba* essential oil, citral, and carvone against *Staphylococcus aureus. Sci. World J.*, 2017, 4962707.

Prateeksha, Singh B.R., Shoeb, M., Sharma, S., Gupta, V.K., and Singh, B.N. 2017. Scaffold of selenium nanovectors and honey phytochemicals for inhibition of *Pseudomonas aeruginosa* quorum sensing and biofilm formation. *Front. Cell. Infect. Microbiol.* 7, 93.

Pustelny, C., Albers, A., Büldt-Karentzopoulos, K., Parschat, K., Chhabra, S.R., Cámara, M., et al. 2009. Dioxygenase-mediated quenching of quinolone-dependent quorum sensing in *Pseudomonas aeruginosa. Chem. Biol.*, 16(12), 1259–1267.

Radchenko, M., Symersky, J., Nie, R., Lu, M. 2015. Structural basis for the blockade of MATE multidrug efflux pumps. *Nat. Commun.*, 6, 7995.

Rahman, M., Kim, S., Kim, S.M., Seol, S.Y., and Kim, J. 2011. Characterization of induced *Staphylococcus aureus* bacteriophage SAP-26 and its antibiofilm activity with rifampicin. *Biofouling*, 27(10), 1087–1093.

Rahman, T., Yarnall, B., and Doyle, D.A. 2017. Efflux drug transporters at the forefront of antimicrobial resistance. *Eur. Biophys. J.*, 46(7), 647–653.

Rajesh, S., Koshi, E., Philip, K., and Mohan, A. 2011. Antimicrobial photodynamic therapy: An overview. *J. Indian. Soc. Periodontol.*, 15(4), 323–327.

Rajkumari, J., Borkotoky, S., Murali, A., and Busi, S. 2018c. Anti-quorum sensing activity of *Syzygium jambos* (L.) Alston against *Pseudomonas aeruginosa* PAO1 and identification of its bioactive components. *South African. J. Bot.*, 118, 151–157.

Rajkumari, J., Borkotoky, S., Murali, A., Suchiang, K., Mohanty, S.K., and Busi, S. 2018b. Cinnamic acid attenuates quorum sensing associated virulence factors and biofilm formation in *Pseudomonas aeruginosa* PAO1. *Biotechnol. Lett.* 40(7), 1087–1100.

Rajkumari, J., Borkotoky, S., Murali, A., Suchiang, K., Mohanty, S.K., and Busi, S. 2018a. Attenuation of quorum sensing controlled virulence factors and biofilm formation in *Pseudomonas aeruginosa* by pentacyclic triterpenes, betulin and betulinic acid. *Microb. Pathog.*, 118, 48–60.

Rajkumari, J., Borkotoky, S., Reddy, D., Mohanty, S.K., Kumavath, R., Murali, A., Suchiang, K., and Busi, S. 2019. Anti-quorum sensing and antibiofilm activity of 5-hydroxymethylfurfural against *Pseudomonas aeruginosa* PAO1: Insights from *in vitro*, *in vivo* and *in silico* studies. *Microbiol. Res.*, 226, 19–26.

Randhawa, H.K., Hundal, K.K., Ahirrao, P.N., Jachak, S.M., and Nandanwar, H.S. 2016. Efflux pump inhibitory activity of flavonoids isolated from *Alpinia calcarata* against methicillin-resistant *Staphylococcus aureus. Biologia*, 71(5), 484–493.

Raorane, C.J., Lee, J.H., Kim, Y.G., Rajasekharan, S.K., García-Contreras, R., and Lee, J. 2019. Antibiofilm and antivirulence efficacies of flavonoids and curcumin against *Acinetobacter baumannii. Front. Microbiol.*, 10, 990.

Reffuveille, F., de la Fuente-Nunez, C., Mansour, S., Hancock, R.E.W. 2014. A broad-spectrum antibiofilm peptide enhances antibiotic action against bacterial biofilms. *Antimicrob. Agents Chemother.* 58, 5363–5371.

Reygaert, W.C. 2018. An overview of the antimicrobial resistance mechanisms of bacteria. *AIMS Microbiol.*, 4(3), 482–501.

Rezaie, P., Pourhajibagher, M., Chiniforush, N., Hosseini, N., and Bahador, A. 2018. The effect of quorum-sensing and efflux pumps interactions in *Pseudomonas aeruginosa* against photooxidative stress. *J. Lasers. Med. Sci.*, 9(3), 161–167.

Rutherford, S.T., and Bassler, B.L. 2012. Bacterial quorum sensing: Its role in virulence and possibilities for its control. *Cold Spring Harb. Perspect. Med.*, 2(11), a012427.

Ryan, E.M., Alkawareek, M.Y., Donnelly, R.F., and Gilmore, B.F. 2012. Synergistic phage-antibiotic combinations for the control of *Escherichia coli* biofilms in vitro. *FEMS Immunol. Med. Microbiol.*, 65(2), 395–398.

Sagar, P.K., and Singh, R. 2021. Anti-quorum sensing activity of clove (*Syzygium aromaticum*) bud extract and its combined efficacy with ceftazidime against multidrug-resistant *Pseudomonas aeruginosa*. *Res. J. Biotechnol.*, 16(3), 75–82.

Salini, R., Sindhulakshmi, M., Poongothai, T., and Pandian, S.K. 2015. Inhibition of quorum sensing mediated biofilm development and virulence in uropathogens by *Hyptis suaveolens*. *Antonie Van Leeuwenhoek*, 107, 1095–1106.

Santajit, S., Indrawattana, N. 2016. Mechanisms of antimicrobial resistance in ESKAPE pathogens. *Biomed. Res. Int.*, 2016, 2475067.

Sarabhai, S., Sharma, P., and Capalash, N. 2013. Ellagic acid derivatives from *Terminalia chebula* Retz. downregulate the expression of quorum sensing genes to attenuate *Pseudomonas aeruginosa* PAO1 virulence. *PloS One*, 8(1), e53441.

Sarkar, R., Chaudhary, S.K., Sharma, A., Yadav, K.K., Nema, N.K., Sekhoacha, M., et al. 2014. Antibiofilm activity of Marula– a study with the standardized bark extract. *J. Ethnopharmacol.*, 154(1), 170–175.

Saxena, P., Joshi, Y., Rawat, K., and Bisht, R. 2019. Biofilms: Architecture, resistance, quorum sensing and control mechanisms. *Ind. J. Microbiol.*, 59(1), 3–12.

Schittek, B., Paulmann, M., Senyurek, I., and Steffen, H. 2008. The role of antimicrobial peptides in human skin and in skin infectious diseases. *Infect. Disord. Drug Targets*, 8(3), 135–143.

Schultz, F., Anywar, G., Tang, H., Chassagne, F., Lyles, J.T., Garbe, L.A., and Quave, C.L. 2020. Targeting ESKAPE pathogens with anti-infective medicinal plants from the Greater Mpigi region in Uganda. *Sci. Rep.*, 10(1), 11935.

Scott, L.C., Wilson, M.J., Esser, S.M., Lee, N.L., Wheeler, M.E., Aubee, A., and Aw, T.G. 2021. Assessing visitor use impact on antibiotic resistant bacteria and antibiotic resistance genes in soil and water environments of Rocky Mountain National Park. *Sci. Total Environ.*, 785, 147122.

Seukep, A.J., Kuete, V., Nahar, L., Sarker, S.D., and Guo, M. 2020. Plant-derived secondary metabolites as the main source of efflux pump inhibitors and methods for identification. *J. Pharmaceut. Anal.*, 10, 277–290.

Sharchi, R., Shayegh, J., and Hosseinzadeh, S. 2020. Anti-quorum sensing and antibacterial activities of *Satureja sahendica* hydroalcoholic extract against avian isolate of *Salmonella typhimurium*. *Iranian J. Vet. Sci. Technol.*, 12(1), 71–79.

Sharifi, A., Ahmadi, A., and Mohammadzadeh, A. 2018a. *Streptococcus pneumoniae* quorum sensing and biofilm formation are affected by *Thymus daenensis*, *Satureja hortensis*, and *Origanum vulgare* essential oils. *Acta Microbiol. Immunol. Hung.*, 65(3), 345–359.

Sharifi, A., Mohammadzadeh, A., Salehi, T.Z., and Mahmoodi, P. 2018b. Antibacterial, antibiofilm and antiquorum sensing effects of *Thymus daenensis* and *Satureja hortensis* essential oils against *Staphylococcus aureus* isolates. *J. Appl. Microbiol.*, 124(2), 379–388.

Sharma, A., Gupta, V.K., and Pathannia, R. 2019. Efflux pump inhibitors for bacterial pathogens: From bench to bedside. *Indian J. Med. Res.* 149(2), 129–145.

Sharma, S.K., Dai, T., Kharkwal, G.B., Huang, Y.Y., Huang, L., De Arce, V.J.B., et al. 2011. Drug discovery of antimicrobial photosensitizers using animal models. *Curr. Pharmaceut. Des.*, 17, 1303–1319.

Sikdar, R., and Elias, M. 2020. Quorum quenching enzymes and their effects on virulence, biofilm and microbiomes: A review of recent advances. *Expert Rev. Anti Infect. Ther.*, 18(12), 1221–1233.

Silveira, G.G.O.S., Torres, M.D.T., Ribeiro, C.F.A., Meneguetti, B.T., Carvalho, C.M.E., de la Fuente-Nunez, C., et al. 2021. Antibiofilm peptides: Relevant preclinical animal infection models and translational potential. *ACS Pharmacol. Transl. Sci.*, 4(1), 55–73.

Singh, B.R., Singh, B.N., Singh, A., Khan, W., Naqvi, A.H., and Singh, H.B. 2015. Mycofabricated biosilver nanoparticles interrupt *Pseudomonas aeruginosa* quorum sensing systems. *Sci. Rep.*, 5, 13719.

Singh, M., Jadaun, G.P., Ramdas, Srivastava K., Chauhan, V., Mishra, R., et al. 2011. Effect of efflux pump inhibitors on drug susceptibility of ofloxacin resistant *Mycobacterium tuberculosis* isolates. *Indian J. Med. Res.*, 133, 535–540.

Siriyong, T., Srimanote, P., Chusri, S., Yingyongnarongkul, B.E., Suaisom, C., Tipmanee, V., and Voravuthikunchai, S.P. 2017. Conessine as a novel inhibitor of multidrug efflux pump systems in *Pseudomonas aeruginosa*. *BMC Complement. Altern. Med.*, 17(1), 405.

Smoum, R., Rubinstein, A., Dembitsky, V.M., and Srebnik, M. 2012. Boron containing compounds as protease inhibitors. *Chem. Rev.*, 112(7), 4156–4220.

Snoussi, M., Noumi, E., Punchappady-Devasya, R., Trabelsi, N., Kanekar, S., Nazzaro, F., et al. 2018. Antioxidant properties and anti-quorum sensing potential of *Carum copticum* essential oil and phenolics against *Chromobacterium violaceum*. *J. Food. Sci. Technol.*, 55(8), 2824–2832.

Somma, A.D., Moretta, A., Canè, C., Cirillo, A., and Duilio, A. 2020. Antimicrobial and antibiofilm peptides. *Biomolecules*, 10(4), 652.

Song, Z., Kong, K.F., Wu, H., Maricic, N., Ramalingam, B., Priestap, H., et al. 2010. *Panax ginseng* has anti-infective activity against opportunistic pathogen *Pseudomonas aeruginosa* by inhibiting quorum sensing, a bacterial communication process critical for establishing infection. *Phytomedicine*, 17(13), 1040–1046.

Songca, S.P. and Adjei, Y. 2022. Applications of antimicrobial photodynamic therapy against bacterial biofilms. *Int. J. Mol. Sci.*, 23, 3209.

Songsantiphap, Vanichanan J., Chatsuwan, T., Asawonanda, P., and Boontaveeyuwat, E. 2022. Methylene blue–mediated antimicrobial photodynamic therapy against clinical isolates of extensively drug resistant gram-negative bacteria causing nosocomial infections in Thailand, An *in vitro* study. *Front. Cell. Infect. Microbiol.*, 12, 929242.

Srinivasan, R., Devi, K.R., Kannappan, A., Pandian, S.K., and Ravi, A.V. 2016. *Piper betle* and its bioactive metabolite phytol mitigates quorum sensing mediated virulence factors and biofilm of nosocomial pathogen *Serratia marcescens in vitro*. *J. Ethnopharmacol.*, 193, 592–603.

Stavri, M., Piddock, L.J., and Gibbons, S. 2007. Bacterial efflux pump inhibitors from natural sources. *J. Antimicrob. Chemother.*, 59, 1247–1260.

Subhaswaraj, P., Barik, S., Macha, C., Chiranjeevi, P.V., and Siddhardha, B. 2018. Antiquorum sensing and antibiofilm efficacy of cinnamaldehyde encapsulated chitosan nanoparticles against *Pseudomonas aeruginosa* PAO1. *LWT–Food Sci. Technol.*, 97, 752–759.

Subhaswaraj, P., Syed, A., and Siddhardha, B. 2020. Novel nanotherapeutics as next-generation anti-infective agents: Current trends and future prospectives. *Curr. Drug. Discov. Target.*, 17, 455–467.

Swift, B.M.C., Bennett, M., Waller, K., Dodd, C., Murray, A., Gomes, R.L., et al. 2019. Anthropogenic environmental drivers of antimicrobial resistance in wildlife. *Sci. Total Environ.*, 649, 12–20.

Taconelli, E., Carrara, E., Savoldi, A., Harbarth, S., Mendelson, M., Monnet, D.L., et al. 2018. Discovery, research, and development of new antibiotics: The WHO priority list of antibiotic-resistant bacteria and tuberculosis. *Lancet Infect. Dis.*, 18(3), 318–327.

Taganna, J.C., Quanico, J.P., Perono, R.M.G., Amor, E.C., and Rivera, W.L. 2011. Tannin-rich fraction from *Terminalia catappa* inhibits quorum sensing (QS) in *Chromobacterium violaceum* and the QS-controlled biofilm maturation and LasA Staphylolytic activity in *Pseudomonas aeruginosa*. *J. Ethnopharmacol.*, 134(3), 865–871.

Tambat, R., Jangra, M., Mahey, N., Chandal, N., Kaur, M., Chaudhary, S., et al. 2019. Microbe-derived indole metabolite demonstrates potent multidrug efflux pump inhibition in *Staphylococcus aureus*. *Front. Microbiol.*, 10, 2153.

Tan, S.Y., Chew, S.C., Tan, S.Y.Y., Givskov, M., and Yang, L. 2014. Emerging frontiers in detection and control of bacterial biofilms. *Curr. Opin. Biotechnol.*, 26, 1–6.

Tapia-Rodriguez, M.R., Hernandez-Mendoza, A., Gonzalez-Aguilar, G.A., Martínez-Tellez, M.A., Martins, C.M., and Ayala-Zavala, J.F. 2017. Carvacrol as potential quorum sensing inhibitor of *Pseudomonas aeruginosa* and biofilm production on stainless steel surfaces. *Food. Control.*, 75, 255–261.

Tateda, K., Comte, R., Pechere, J.C., Kohler, T., Yamaguchi, K., and Delden, C.V. 2001. Azithromycin inhibits quorum sensing in *Pseudomonas aeruginosa*. *Antimicrob. Agents Chemother.*, 45(6), 1930–1933.

Thakur, V., Uniyal, A., and Tiwari, V. 2021. A comprehensive review on pharmacology of efflux pumps and their inhibitors in antibiotic resistance. *Eur. J. Pharmacol.*, 903, 174151.

Thomson, K.S., AbdelGhani, S., Snyder, J.W., and Thomson, G.K. 2019. Activity of cefepime-zidebactam against multidrug-resistant (MDR) gram-negative pathogens. *Antibiotics*, 8(1), 32.

Tintino, S.R., Morais-Tintino, C.D., Campina, F.F., Costa, M.D.S., Menezes, I.R.A., de Matos, Y.M.L.S., et al. 2017. Tannic acid affects the phenotype of *Staphylococcus aureus* resistant to tetracycline and erythromycin by inhibition of efflux pumps. *Bioorg. Chem.*, 74, 197–200.

Tintino, S.R., Oliveira-Tintino, C.D., Campina, F.F., Silva, R.L., Costa Mdo, S., Menezes, I.R., et al. 2016. Evaluation of the tannic acid inhibitory effect against the NorA efflux pump of *Staphylococcus aureus*. *Microb. Pathog*. 97, 9–13.

Tran, H.T., Solnier, J., Pferschy-Wenzig, E.M., Kunert, O., Martin, L., Bhakta, S., et al. 2020. Antimicrobial and efflux pump inhibitory activity of Carvotacetones from *Sphaeranthus africanus* against mycobacteria. *Antibiotics.*, 9, 390.

Truong, T., Bui, H.D., Pham, T.T.V., Tran, L.T., Nguyen, D.H., Ng, C., and Le, T.H. 2022. Occurrences of antibiotic resistant bacteria in a tropical river impacted by anthropogenic activities in Ho Chi Minh city. *Int. J. Environ. Sci. Technol.*, 19, 7049–7058.

Tüfekci, E.F., Alkateeb, A., Akar, S., Çorum, O., Altunoğlu, Y.Ç., Baloğlu, M.C., et al. 2020. Antimicrobial and anti-quorum sensing activities of giant fennel (*Ferula elaeochytris* Korovin) from the Hatay region. *Eurasian. J. Vet. Sci.*, 36(3), 214–220.

Uroz, S., Chhabra, S.R., Cámara, M., Williams, P., Oger, P., and Dessaux, Y. 2005. N-Acylhomoserine lactone quorum-sensing molecules are modified and degraded by *Rhodococcus erythropolis* W2 by both amidolytic and novel oxidoreductase activities. *Microbiology*, 151(10), 3313–3322.

Uruen, C., Chopo-Escuin, G., Tommassen, J., Mainar-Jaime, R.C., and Arenas, J. 2021. Biofilms as promoters of bacterial antibiotic resistance and tolerance. *Antibiotics*, 10(1), 3.

Utari, P.D., Setroikromo, R., Melgert, B.N., and Quax, W.J. 2018. PvdQ quorum quenching acylase attenuates *Pseudomonas aeruginosa* virulence in a mouse model of pulmonary infection. *Front. Cell. Infect. Microbiol.*, 8, 119.

Utari, P.D., Vogel, J., and Quax, W.J. 2017. Deciphering Physiological Functions of AHL quorum quenching acylases. *Front. Microbiol.*, 8, 1123.

Vadakkan, K., Hemapriya, J., and Selvaraj, V. 2019a. Quorum quenching intervened in vivo attenuation and immunological clearance enhancement by *Solanum torvum* root extract against *Pseudomonas aeruginosa* instigated pneumonia in Sprague Dawley rats. *Clin. Phytosci.*, 5, 24.

Vadakkan, K., Vijayanand, S., Hemapriya, J., and Gunasekaran, R. 2019b. Quorum sensing inimical activity of *Tribulus terrestris* against Gram-negative bacterial pathogens by signalling interference. *3 Biotech*, 9, 163.

Vandeputte, O.M., Kiendrebeogo, M., Rasamiravaka, T., Stevigny, C., Duez, P., Rajaonson, S., et al. 2011. The flavanone naringenin reduces the production of quorum sensing-controlled virulence factors in *Pseudomonas aeruginosa* PAO1. *Microbiology*, 157(7), 2120–2132.

Venkatramanan, M., Ganesh, P.S., Senthil, R., Akshay, J., Ravi, A.V., Langeswaran, K., et al. 2020. Inhibition of quorum sensing and biofilm formation in *Chromobacterium violaceum* by fruit extracts of *Passiflora edulis*. *ACS Omega.*, 5(40), 25605–25616.

Venter, H., Mowla, R., Ohene-Agyei, T., and Ma, S. 2015. RND-type drug efflux pumps from Gram-negative bacteria: Molecular mechanism and inhibition. *Front. Microbiol.*, 6, 377.

Ventola, C.L. 2015. The antibiotic resistance crisis. *Pharm. Therpeut.*, 40(4), 277–283.

Vera, C., Tulli, F., and Borsarelli, C.D. 2021. Photosensitization with supramolecular arrays for enhanced antimicrobial photodynamic treatments. *Front. Bioeng. Biotechnol.*, 9, 655370.

Vijayakumar, K., and Ramanathan, T. 2020. *Musa acuminata* and its bioactive metabolite 5-Hydroxymethylfurfural mitigates quorum sensing (las and rhl) mediated biofilm and virulence production of nosocomial pathogen *Pseudomonas aeruginosa in vitro*. *J. Ethnopharmacol.*, 246, 112242.

Viju, N., Satheesh, S., and Vincent, S.G.P. 2013. Antibiofilm activity of coconut (*Cocos nucifera* Linn.) husk fibre extract. *Saudi. J. Biol. Sci.*, 20(1), 85–91.

Vikram, A., Jesudhasan, P.R., Jayaprakasha, G.K., Pillai, S.D., and Patil, B.S. 2011. *Citrus limonoids* interfere with *Vibrio harveyi* cell–cell signalling and biofilm formation by modulating the response regulator LuxO. *Microbiology*, 157(1), 99–110.

Vrenna, G., Artini, M., Ragno, R., Relucenti, M., Fiscarelli, E.V., Tuccio Guarna Assanti, V., et al. 2021. Anti-virulence properties of *Coridothymus capitatus* Essential Oil against *Pseudomonas aeruginosa* clinical isolates from cystic fibrosis patients. *Microorganisms*, 9, 2257.

Wang, G., Li, X., and Wang, Z. 2016. APD3: The antimicrobial peptide database as a tool for research and education. *Nucleic Acids Res.*, 44(D1), D1087–D1093.

Williams, N.L.R., Siboni, N., McLellan, S.L., Potts, J., Scanes, P., Johnson, C., et al. 2022. Rainfall leads to elevated levels of antibiotic resistance genes within seawater at an Australian beach. *Environ. Pollut.*, 307, 119456.

Wolloscheck, D., Krishnamoorthy, G., Nguyen, J., and Zgurskaya, H.I. 2018. Kinetic control of quorum sensing in *Pseudomonas aeruginosa* by multidrug efflux pumps. *ACS Infect. Dis.*, 4(2), 185–195.

Wu, H., Moser, C., Wang, H.Z., Hoiby, N., and Song, Z.J. 2015. Strategies for combating bacterial biofilm infections. *Int. J. Oral Sci.*, 7, 1–7.

Wu, L., and Luo, Y. 2021. Bacterial quorum-sensing systems and their role in intestinal bacteria-host crosstalk. *Front. Microbiol.*, 12, 611413.

Xiang, H., Cao, F., Ming, D., Zheng, Y., Dong, X., and Zhong, X., et al. 2017. Aloe-emodin inhibits *Staphylococcus aureus* biofilms and extracellular protein production at the initial adhesion stage of biofilm development. *Appl. Microbiol. Biotechnol.*, 101(17), 6671–6681.

Xu, G.M. 2016. Relationships between the regulatory systems of quorum sensing and multidrug resistance. *Front. Microbiol.*, 7, 958.

Xu, Q., Hu, X., and Wang, Y. 2021. Alternatives to conventional antibiotic therapy: Potential therapeutic strategies of combating antimicrobial-resistance and biofilm-related infections. *Mol. Biotechnol.*, 63, 1103–1124.

Xu, Z., Zhang, H., Yu, H., Dai, Q., Xiong, J., Sheng, H., et al. 2019. Allicin inhibits *Pseudomonas aeruginosa* virulence by suppressing the rhl and pqs quorum-sensing systems. *Can. J. Microbiol.*, 65(8), 563–574.

Yada, S., Kamalesh, B., Sonwane, S., Guptha, I., and Swetha, R.K. 2015. Quorum sensing inhibition, relevance to periodontics. *J. Int. Oral. Health.*, 7(1), 67–69.

Yadav, M.K., Park, S.W., Chae, S.W., and Song, J.J. 2014. Sinefungin, a natural nucleoside analogue of S-adenosylmethionine, inhibits *Streptococcus pneumoniae* biofilm growth. *Biomed. Res. Int.*, 2014, 156987.

Yamaguchi, S., Huster, D., Waring, A., Lehrer, R.I., Kearney, W., Tack, B.F., and Hong, M. 2001. Orientation and dynamics of an antimicrobial peptide in the lipid bilayer by solid-state NMR spectroscopy. *Biophys. J.*, 81(4), 2203–2214.

Yamanaka, A., Kimizuka, R., Kato, T., and Okuda, K. 2004. Inhibitory effects of cranberry juice on attachment of oral *Streptococci* and biofilm formation. *Oral Microbiol. Immunol.*, 19(3), 150–154.

Yan, X., Gu, S., Shi, Y., Cui, X., Wen, S., and Ge, J. 2017. The effect of emodin on *Staphylococcus aureus* strains in planktonic form and biofilm formation *in vitro*. *Arch. Microbiol.*, 199(9), 1267–1275.

Yang, Y.X., Xu, Z.H., Zhang, Y.Q., Tian, J., Weng, L.X., and Wang, L.H. 2012. A new quorum-sensing inhibitor attenuates virulence and decreases antibiotic resistance in *Pseudomonas aeruginosa*. *J. Microbiol.*, 50(6), 987–993.

Yap, P.S.X., Krishnan, T., Yiap, B.C., Hu, C.P., Chan, K.G., and Lim, S.H. 2014. Membrane disruption and anti-quorum sensing effects of synergistic interaction between *Lavandula angustifolia* (lavender oil) in combination with antibiotic against plasmid-conferred multidrug-resistant *Escherichia coli*. *J. Appl. Microbiol.*, 116(5), 1119–1128.

Yoneda, T., Sakata, H., Yamasaki, S., Hayashi-Nishino, M., and Nishino, K. 2022. Analysis of multidrug efflux transporters in resistance to fatty acid salts reveals a TolC-independent function of EmrAB in *Salmonella enterica*. *PloS One*, 17(4), e0266806.

Zaidi, S., Misba, L., and Khan, A.U. 2017. Nano-therapeutics: A revolution in infection control in post-antibiotic era. *Nanomedicine*, 13(7), 2281–2301.

Zhang, Y., Sass, A., Van Acker, H., Wille, J., Verhasselt, B., Van Nieuwerburgh, F., et al. 2018. Coumarin reduces virulence and biofilm formation in *Pseudomonas aeruginosa* by affecting quorum sensing, type III secretion and c-di-GMP levels. *Front. Microbiol.*, 9, 1952.

Zhang, Z., Zhang, Q., Wang, T., Xu, N., Lu, T., Hong, W., et al. 2022. Assessment of global health risk of antibiotic resistance genes. *Nat. Commun.*, 13, 1553.

Zhao, X., Yu, Z., and Ding, T. 2020. Quorum-sensing regulation of antimicrobial resistance in bacteria. *Microorganisms.*, 8, 425.

Zhao, Z., Ma, J., Wang, Y., Xu, Z., Zhao, L., Zhao, J., et al. 2021. Antimicrobial photodynamic therapy combined with antibiotic in the treatment of rats with third-degree burns. *Front. Microbiol.*, 12, 622410.

Zhen, X., Lundborg, C.S., Sun, X., Hu, X., and Dong, H. 2019. Economic burden of antibiotic resistance in ESKAPE organisms: A systematic review. *Antimicrob. Resist. Infect. Control*, 8, 137.

Zheng, D., Yin, G., Liu, M., Chen, C., Jiang, Y., Hou, L., and Zheng, Y. 2021. A systematic review of antibiotics and antibiotic resistance genes in estuarine and coastal environments. *Sci. Total Environ.*, 777, 146009.

Zhong, S., and He, S. 2021. Quorum sensing inhibition or quenching in *Acinetobacter baumannii*: The novel therapeutic strategies for new drug development. *Front. Microbiol.*, 12, 558003.

Zhou, J.W., Hou, B., Liu, G.Y., Jiang, H., Sun, B., Wang, Z.N., et al. 2018a. Attenuation of *Pseudomonas aeruginosa* biofilm by hordenine: A combinatorial study with aminoglycoside antibiotics. *Appl. Microbiol. Biotechnol.*, 102(22), 9745–9758.

Zhou, J.W., Luo, H.Z., Jiang, H., Jian, T.K., Chen, Z.Q., and Jia, A.Q. 2018b. Hordenine: A novel quorum sensing inhibitor and antibiofilm agent against *Pseudomonas aeruginosa*. *J. Agric. Food Chem.*, 66(7), 1620–1628.

Zhou, J.W., Muhammad, J., Sun, B., Yang, R., Wadood, A., Wang, J.S., and Jia, A.Q. 2019. Metabolomic analysis of quorum sensing inhibitor hordenine on *Pseudomonas aeruginosa*. *Appl. Microbiol. Biotechnol.*, 103(15), 6271–6285.

Zhou, L., Zhang, Y., Ge, Y., Zhu, X., and Pan, J. 2020. Regulatory mechanisms and promising applications of quorum sensing-inhibiting agents in control of bacterial biofilm formation. *Front. Microbiol.*, 11, 589640.

Zhu, X., Radovic-Moreno, A.F., Wu, J., Langer, R., and Shi, J. 2014. Nanomedicine in the management of microbial infection-overview and perspectives. *Nano Today.*, 9(4), 478–498.

Zhu, Y.G., Zhao, Y., Li, B., Huang, C.L., Zhang, S.Y., Yu, S., et al. 2017. Continental-scale pollution of estuaries with antibiotic resistance genes. *Nat. Microbiol.*, 2, 16270.

6 Antimicrobial Photodynamic Therapy (aPDT) and Mechanism

V.T. Anju, Siddhardha Busi, Simi Asma Salim, and Madhu Dyavaiah
Pondicherry University, Pondicherry, India

CONTENTS

6.1 INTRODUCTION

The emerging global crisis of antibiotic resistance is associated with enhanced mortality rate (Geralde et al., 2017). Current statistics says around 1 million deaths are attributed to resistant pathogens (Jamrozik and Heriot, 2022). The inappropriate intake of antibiotics for bacterial infections has led to the antibiotic resistance crisis. Though antibiotics have saved millions of lives, presently, most antibiotics, including the last line of antibiotic classes, exhibited ineffectiveness against many bacterial infections (Lushniak, 2014). The ineffectiveness in the bacterial infection treatment led to the evolution of antimicrobial resistance (AMR). Scientists reported that AMR may cause 10 million deaths/year by the year 2050. Thus, an effective global plan is needed to reduce the burden of AMR. It is predicted that if AMR is neglected, it may cause more deaths in the coming years, and also enhance economic burdens (Murray et al., 2022).

Among all AMR pathogens, novel and alternative therapeutics are on urgent demand for the priority pathogens called as ESKAPE. The ESKAPE pathogens are *Enterococcus faecium*, *Staphylococcus aureus*, *Klebsiella pneumoniae*, *Acinetobacter baumannii*, *Pseudomonas aeruginosa*, and *Enterobacter* sp. (Chang et al., 2022). ESKAPE pathogens contribute to several chronic infections in immune deficient patients owing to their AMR (Rice, 2010). The ESKAPE pathogens carry antibiotic resistance genes (ARGs) in their chromosome, plasmid or transposons and can transfer to other organisms (Giedraitiene et al., 2011). There are several mechanisms through which

DOI: 10.1201/9781003345299-6

microorganisms gain antibiotic resistance. They include alteration or inactivation of drugs/drug targets, altered cell membrane permeability causing reduced drug accumulation, and formation of drug resistant biofilms (Santajit and Indrawattana, 2016). Bacteria produces certain enzymes (e.g., β lactamases) that destroy or neutralise drugs in irrevocable manner. Target sites can be modified using enzymes, ribosomal sites, or cell wall precursor modifications. The modified target by any of these methods causes decreased binding of drugs to the target site. The presence of efflux pumps causes decreased penetration of drugs. In addition, loss of porin protein in the outer membrane also leads to decreased accumulation of drugs inside the bacteria (De Oliveira et al., 2020). Furthermore, bacterial biofilms can favour the emergence of drug resistance. The improved tolerance of biofilms towards antibiotics and other stresses is mainly due to their immobile structure. Biofilms are populations of microorganisms enclosed in a self-produced extracellular matrix (ECM) comprising exopolysaccharides, extracellular DNA, proteins etc. Treatment of biofilms demand alternative therapeutic methods, as drugs cannot penetrate the deeper layers of ECM (Sharma et al., 2019).

Biofilms are associated with persistent and chronic infections (Hall-Stoodley et al., 2004). The available antibiotics fail to control novel antibiotic resistant strains. There is a declining trend in the antibiotic discovery pipeline and effectiveness to control and manage AMR. Thus, there is a requirement for new and effective strategies to eradicate and control the infections caused by antibiotic resistant strains and biofilms (Chang et al., 2022). Antimicrobial photodynamic therapy (aPDT) is evolved as an excellent substitute to traditional therapeutic strategy for the eradication of planktonic and biofilms cells which confer AMR. aPDT is a non-invasive method employing light to illuminate a photosensitiser and to synthesise reactive oxygen species (ROS). The ROS produced through aPDT mediates the microbial cell damage and death (Hu et al., 2018). In this chapter, the aPDT and its mechanism of action on microorganisms are well discussed.

6.2 OVERVIEW OF PHOTODYNAMIC THERAPY

The origin of photodynamic therapy was referred to as heliotherapy in the ancient times, especially the regions of India, China, Rome, Greece, and Egypt. In primeval times, people used to worship the sun, and thought it had the power to heal many diseases. In that concept, sunlight was exploited for the management and cure of skin cancer, rickets, psoriasis, vitiligo and psychosis (Daniell and Hill, 1991).

The first permitted PDT drug by Food and Drug Administration was in 1999 with Arbold Rikli (Abdel-Kader, 2016). It was Oscar Raab who first introduced the concept of dyes as photosensitiser in 1898. In one of his experiments, the toxic effects of acridine orange on Paramecium was observed, from which he considered light can activate acridine orange to kill paramecia. Later, his observations lead to the foundation for the development of PDT (Abdel-Kader, 2016). PDT has established its roots in the field of cancer and non-cancer disease treatments in its journey. Though PDT has been known and studied for many years, it has only attained wide attention and interest recently.

Photodynamic therapy is the method used in the management of diseases using a photosensitiser which upon exposure to a visible non-thermal light of appropriate wavelength is excited to a triplet state and generates cytotoxic molecules such as singlet oxygen. The uptake or targeted delivery of photosensitising agent and subsequent light exposure, leads to the production of highly reactive ROS. The produced ROS cause damage and death of cells (Ackroyd et al., 2001). In the treatment of diseases using PDT, three major steps are involved. Initial step is the administration of PS which allows the distribution in the target sites in body followed by the illumination of disease site or body part. Once, the PS get excited appropriate ROS are generated causing the destruction of tumors or infectious parts (Fitzgerald, 2017).

Along with the oncologic applications, PDT is extended towards antimicrobial treatment where it serves as an alternative to antibiotics or antiviral drugs. In past years, antimicrobial photodynamic therapy (aPDT) was used extensively to inactivate bacteria, yeasts, viruses, and fungi. The procedure of aPDT involves the photoactivation of non-toxic dyes or PS in presence of visible light to generate ROS which act on microorganisms. The main advantage of using aPDT over traditional antimicrobial therapeutics (antibiotics) is the elimination of antibiotic resistant strains as ROS is the killing agent (Correia et al., 2021). In aPDT, PS are taken by microorganisms and accumulate in the cells, especially in cytoplasm and near to the cell membranes. The photoactivated PS inside the bacteria generates ROS which targets and damages the cell membranes and other cellular targets (Darabpour et al., 2016). Recently, aPDT has been used against the treatment of highly contagious COVID-19 infections (Pourhajibagher et al., 2021). Although *in vivo* and clinical approaches in the treatment of COVID-19 are scant, *in vitro* studies represent promising results (Teixeira et al., 2021).

6.3 IMPORTANCE OF ANTIMICROBIAL PHOTODYNAMIC THERAPY (aPDT)

Photodynamic antimicrobial chemotherapy (PACT) or aPDT is a non-invasive treatment modality to eradicate the infectious microorganism and its biofilms (Kornman et al., 1997). aPDT has acquired great attention in the field of medicine as it possess broad spectrum activity against microbes, involving Gram-negative and -positive bacteria and multidrug resistant strains (Sobotta et al., 2019). World health organisation (WHO) has reported antimicrobial resistance (AMR) as severe worldwide threat which can also affect economy. Even still, it is a dangerous situation where general antibiotics are compromised in their efficacy, and AMR can contribute to enhanced mortality rates. The British Society for Antimicrobial Chemotherapy (BSAC) has predicted approximately 10 million AMR deaths per year by the year 2050; more than that of cancer and diabetics (WHO, 2014; Smith and Coast, 2013; BSAC, 2018).

Biofilms, a resistant microbial phenotype, serve as a significant factor in the development of antibiotic resistance and failure of treatment. Biofilms are highly resistant groups of cells, self-protected from antibiotics and other extreme conditions through the production of exopolymer substances (EPS matrix). Biofilms causes more than 80% of diseases linked with microbes, and contribute to accelerated morbidity and mortality rates (López et al., 2010). Biofilms can even survive up to 1,000 times higher concentration of antimicrobial agents (Donlan, 2002). The different parameters associated with biofilm recalcitrance and decreased susceptibility towards antimicrobial agents are the production of EPS matrix, differential physiological activity, and formation of non-dividing persister cells (Ciofu et al., 2017; Lewis, 2007). Thus, aPDT has been introduced to eliminate the challenges related to the AMR and associated chronic infections caused by biofilms. At a glimpse, aPDT can be applied to pathogens which already have gained antibiotic resistance (Inada et al., 2021).

To date, several PS are reported for their efficacy towards photoinactivation of bacteria, viruses, and fungi. The most widely discussed PS in aPDT are phenothiaziniums, porphyrins, phthalocyanines, fullerenes, naturally occurring photosensitisers such as hypericin and curcumin (Liu et al., 2015).

The supreme characteristics for a PS are maximum phototoxicity, less dark toxicity, optimum pharmacokinetics, low or negligible toxicity to host cells, maximum yield of ROS, ability to entrap or bind to the cell membrane, and more PS uptake and accumulation in bacteria (Tim, 2015; Carrera et al., 2016). There is always a difference in the binding and intake of charged PS by Gram-negative and -positive bacteria, owing to the uniqueness in their cell walls. The maximum absorption of anionic and neutral PS by Gram-negative bacteria is owing to the presence of thick, pore structures of peptidoglycan. Cationic PS are preferred by Gram-negative and positive bacteria, due to higher phototoxicity (Merchat et al., 1996; Liu et al., 2015).

Few examples of aPDT are discussed here where different types of photosensitisers are employed for the eradication of pathogenic microbes. In a study, a cationic PS, porphyrin was conjugated to methyl ethylenediamine derivative to eliminate clinical pathogens, methicillin resistant *Staphylococcus aureus*, *Escherichia coli*, and *Pseudomonas aeruginosa* in laboratory and *in vivo* conditions. Also, the PS conjugate showed accelerated wound healing in rat wounds infected with these bacteria. The study observed that the aPDT of bacterial pathogens were contributed by the synthesis of ROS (singlet oxygen) (Zhao et al., 2021). Grizante Barião and colleagues discussed antifungal photodynamic inactivation of a MDR fungal pathogen, *Candida auris* using phenothiazinium PSs namely, methylene blue (MB), toluidine blue, and derivatives of MB. Along with the antibiofilm activity of phenothiazinium PSs, an enhanced survival of infected *Galleria mellonella* insect model was also observed in antifungal photodynamic therapy (Grizante Barião et al., 2022). Similarly, antiviral activity mediated by PDT was also studied. aPDT was successfully studied *in vitro* for the inactivation of SARS-CoV-2. Two photosensitisers, MB and radahlorin inhibited SARS-CoV-2 in Vero cells (Svyatchenko et al., 2021).

In addition to the translational medicine, aPDT mediated strategies are well-studied in several *in vivo* models. Similar kind of studies for the management of wound, body cavity and topical skin and corneal infections recommends the clinical applications of aPDT in future (Oyim et al., 2021).

6.4 PHOTOCHEMICAL MECHANISM OF aPDT

PDT comprises PS, visible or infrared light, and molecular oxygen. The two photochemical mechanisms occurring in PDT through which biomolecules of target system is damaged are type I and II pathways (Moreira et al., 2012) (Figure 6.1). During photodynamic therapy, non-toxic PS is administrated appropriately to the target site and then exposed to light of suitable wavelength. PS illuminated with light gets excited and interact with molecular oxygen present surrounding to the target site and ends with the generation of cytotoxic radicals. The generated radicals cause damage and finally death of tissue or target system (Robertson et al., 2009).

PS in their singlet state having ground state energy absorb the photons generated by light, and become excited to the more energetic state with shorter life. The singlet state is explained as containing two electrons in lower energy orbital with opposite spin. During the development of excited PS, one electron in the ground state is transferred to a higher energy orbit. Sometimes PS may return to the normal state by losing the absorbed photon through fluorescence or heat energy loss mechanisms. Another possibility is that excited PS can transform to triplet state with lower energy and more life span. Formation of triplet state occurs through intersystem crossing mechanisms along

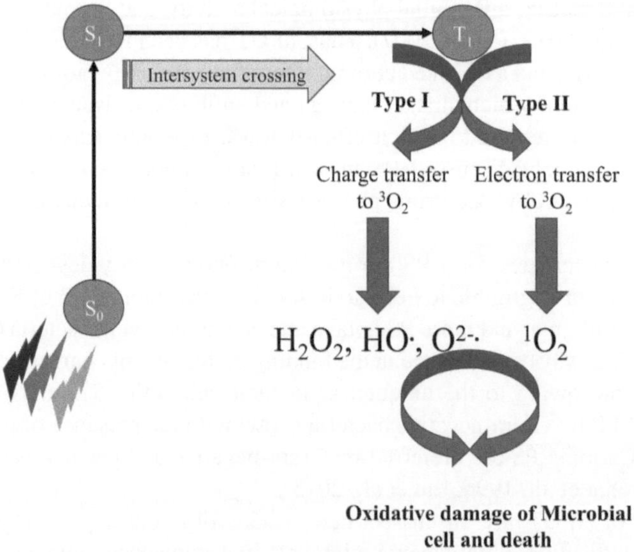

FIGURE 6.1 The Jablonski figure, illustrating the pathways involved in the antimicrobial photodynamic therapy leading to the production of free radicals. S_0 is denoted as PS in its ground singlet state, S_1 is PS in the excited state after receiving appropriate photons from light and T_1 is the triplet excited state of photosensitiser formed through intersystem crossing. The ground state of oxygen is presented as $^3O_2$. Free radicals are displayed in $^1O_2$ (singlet oxygen), H_2O_2 (hydrogen peroxide), HO. (hydroxyl radical), and O^{2-}. (superoxide anion).

with an alteration in electron spin. The electron spin is observed in parallel form in the triplet state of PS. Triplet PS state allocate more energy to other molecules than the singlet state (Li et al., 2022; Juzeniene and Moan, 2007; Robertson et al., 2009). The triplet state of PS undergoes the competing reactions i.e., I or II.

6.4.1 Type I and II PDT Mechanism

In type I, electrons or protons from triplet state are transferred to oxygen or surrounding molecules forming a charged radical (anion or cation respectively). Reactive oxygen species are produced when these charged radicals react with molecular oxygen. The major product of type I pathways are superoxide radicals that cause less damage to the biological entities. However, superoxide radicals interact to produce hydrogen peroxide through dismutation. The produced H_2O_2 causes cellular destruction by passing through biological membranes. Through Fenton reaction, superoxide contribute electron to metal ions and catalysis the reduction of hydrogen peroxide to hydroxyl radical (HO•) and a hydroxide ion (HO). Likewise, superoxide also involved in the release of other ROS such as singlet oxygen ($^1O_2$) or peroxynitrite by reacting with HO• and nitric oxide, respectively (Castano et al., 2004; Plaetzer et al., 2009; Allison and Moghissi, 2013).

In type II pathways, triplet state PS interacts with molecular oxygen yielding $^1O_2$. In contrast to type I pathway, energy transfer occurs in II pathway where the produced singlet oxygen serves as the major cytotoxic species in PDT. The energy transfer occurs between oxygen having a ground triplet state and PS in triplet state to yield transient and extremely reactive $^1O_2$. During this reaction, electron spin is not reversed, as the reaction is in between two triplets (Sharma et al., 2012). Though two pathways can happen at a similar time during PDT, different factors determine the ratio between the two pathways. This ratio can vary depending on the oxygen availability, PS concentrations, type of PS used, and PDT parameters (light) (Sai et al., 2021). The ROS produced through type I reaction and singlet oxygen by type II together exert damage to various important biological molecules such as DNA, proteins etc (Castano et al., 2004) (Figure 6.2).

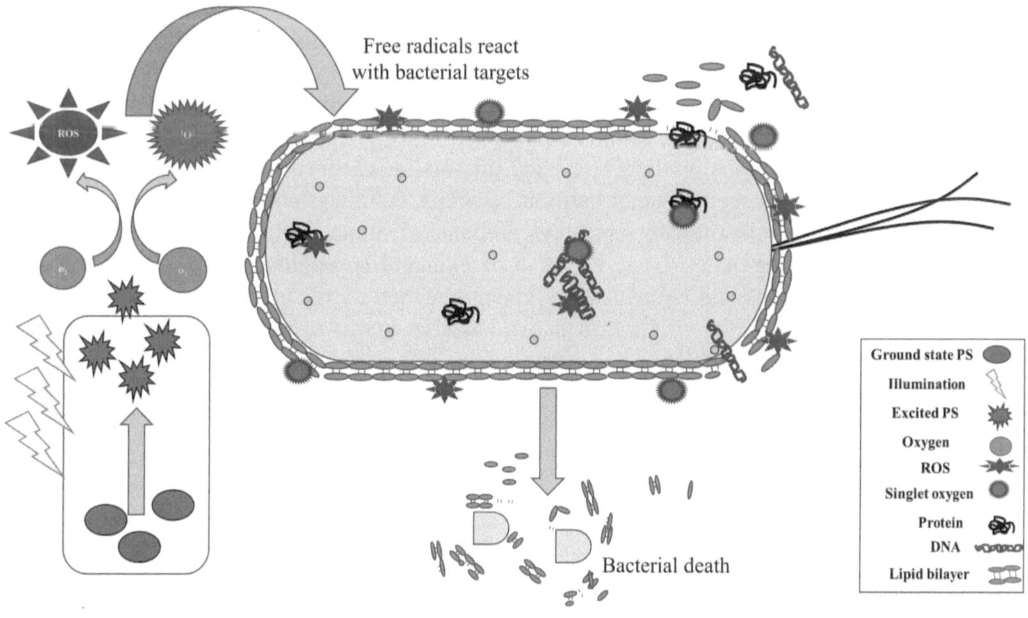

FIGURE 6.2 The putative cellular targets of antimicrobial photodynamic therapy in microorganisms. The major cytotoxic species of PDT such as ROS and singlet oxygen oxidatively damages the vital biomolecules of bacteria causing the death.

The mechanisms of photodynamic inactivation on microorganisms and various cellular targets of microbes are discussed below. The ROS oxidises the biomolecules depending on the distance between the accumulated PS and biomolecule, abundance of target molecules, and specific rate constant of reactions. Singlet oxygen exhibits the highest cytotoxic behaviour compared to the free radicals generated by type I pathway. However, it has been reported that free radicals released through type I reactions can also be the major oxidising species in microorganisms. The ultimate damage and death of microorganisms is contributed by the synergistic effect of cytotoxic species produced by I and II pathway but not by any one of the pathways (Benov, 2015).

6.5 INTRACELLULAR TARGETS OF aPDT

6.5.1 Cell Wall and Membrane

Cell membrane of microorganisms is the primary target of aPDT. The PSs first attach to the cell membrane and penetrate through the semipermeable region to accumulate near the target site. There is a difference in the binding mechanism of PS to Gram-positive and negative bacteria as there is a variation in the structure of cell membrane and cell wall. Most of the amphipathic PS are widely accepted as it can easily interact and pass through the hydrophobic cell membrane (Anas et al., 2021). The negative charge of cell membrane and cell wall demands a positively charged PS for effective phototoxicity. Anionic and neutral PSs can easily bind and penetrate through Gram-positive bacterial plasma membrane whereas it is not possible with that of Gram-negative bacteria. Gram-negative bacteria are mostly prone to cationic PSs. In contrast, a pre-treatment for Gram-positive bacteria with polycations or EDTA is preferred to disturb the LPS of outer membrane and make it susceptible to PSs (Bertolini et al., 2006; Sperandio et al., 2013). Similarly, EDTA act as a cell-damaging compound on Gram-positive bacteria where it chelates with the Mg^{2+} and Ca^{2+} ions. The treatment with EDTA not only enhance the penetration of PS but also the antibiotics. It is suggested that care should be taken while selecting EDTA treatment for some bacteria and the type of PS in aPDT. There was evidence of decreased uptake of methylene blue by *A. actinomycetemcomitans* upon higher the concentrations of EDTA. This is contributed by the reduction in the appropriate binding sites for methylene blue in outer membrane (George et al., 2009).

The disruption of cell membrane integrity is an indicator of cytoplasmic contents loss (proteins and nucleic acids). Cieplik and co-authors discussed the effect of phenalen-1-one on photodynamic inactivation of periodontal biofilms comprising *Actinomyces naeslundii*, *Fusobacterium nucleatum*, and *Porphyromonas gingivalis*. In the above study, authors found that the strong PS, phenalen-1-one did not target cytoplasmic membrane of biofilms. Thereby, there was no reported release of nucleic acid contents. Still, the probable targets of high amounts of singlet oxygen produced by the aPDT mechanism employing phenalen-1-one is yet to be explored (Cieplik et al., 2018b). In contrast, metallacycle 1 (PS) self-assembled with viral protein having a transacting activator of transduction (TAT) peptide enhanced aPDT of Gram-negative bacteria. It was observed that the PS aggregate exhibited cell membrane intercalation and enhanced the aPDT efficacy against *E. coli*. The excellent antibacterial effect on *E. coli* was substantiated by the higher amount of ROS production and leakage of cellular contents through ruptured cell membrane (Gao et al., 2019). Two PSs, methylene blue and TMPyP exhibited more cell membrane damaged cells of *E. coli* and *S. aureus* owing to the efficient aPDT mechanism with a reduction of 5 $\log_{10}$. However, authors observed that cell membrane is not only the major cellular target of aPDT, and further studies are recommended (Muehler et al., 2022).

6.5.2 Proteins

The major molecular target of aPDT are proteins owing to their high-rate constants and abundance in cell. The mechanism of oxidation varies with each amino acid residues as they possess diverse rate constants. The most affected residues in a peptide chain are histidine, tyrosine, tryptophan, methionine, and cysteine. The end products of protein oxidation by ROS and singlet oxygen are numerous and follows a complex mechanism of action (Davies, 2004; Plowman et al., 2013). A cationic and broad spectrum antimicrobial PS, methylene blue exhibited photodynamic oxidation of outer membrane proteins than other biological targets. It has shown cytotoxic effects on core proteins of HIV-1 virus. The free radicals have the ability to disintegrate the viral protein structure and inactivate enzymes by pulling out the hydrogen from their proteins (Wainwright, 2010; Dias and Bagnato, 2020). Phthalocyanines and its derivatives are excellent photosensitisers with antimicrobial properties. The zinc-phthalocyanines revealed its antimicrobial photoinactivation property against the pathogen, *Streptococcus mitis*, by disrupting the stability of its membrane proteins (Spesia and Durantini, 2013).

A natural PS, sorbicillinoids with water soluble property exhibited photodynamic inactivation of *S. aureus*. The higher amounts of singlet oxygen generated by the UV irradiated novel PS could exert phototoxicity on Gram-positive bacteria by targeting the cellular contents such as proteins, DNA etc (Yang et al., 2020). It was also observed that during aPDT some enzymes (catalytic proteins) are inactivated. The inactivated intracellular enzymes important in vital biological activities can lead to microbial death. Likewise, the activity of ATPase and antioxidant enzymes (SOD, CAT, POD) were reduced drastically after aPDT of *Staphylococcus saprophyticus* using curcumin under blue LED light. The decreased enzyme activity is correlated to the ion leakage, alteration in cell structure, functional disorder, weak defence mechanisms and ultimately death (Wang et al., 2021; Ning et al., 2019).

6.5.3 Lipid Peroxides

Another vital biomolecule attacked by free radicals produced through aPDT are lipids. The ROS mediated damage of cellular targets of bacteria including lipids can cause the microbial cell injury and cell death (Xuan et al., 2018). The major proportion of cell membrane and other membranes around cell organelle are lipids and proteins. The light mediated oxidation of lipid layers cause membrane permeabilisation and cell death (Bacellar et al., 2018; Tasso et al., 2019). Photooxidation of lipids occurs in two ways such as contact dependent and independent pathways. In contact dependent pathway, excited PS interact with the target directly such as the unsaturated lipids leading to the production of lipid peroxides by removing hydrogens. In the independent pathway, excited PS react with lipid bilayers and other lipids through the synthesis of a mediator i.e., singlet oxygen (Bacellar et al., 2018).

Also, studies reported that lipid peroxides produced through antimicrobial photodynamic inactivation process can regenerate singlet oxygen or free radicals. This can be efficiently utilised to kill biofilms of pathogenic microorganisms (Tonon et al., 2022). A cationic photosensitiser, tricationic porphyrin irreversibly converted lipids to lipid hydroperoxides in photoactivated *E. coli* cells. The Gram-negative bacteria are inactivated by the oxidation of phospholipids and formation of high lipid peroxides. Majorly, the phosphatidylethanolamines having a carbon chain of 16 and 18 were converted to peroxides (Alves et al., 2013). Several studies showed that high amount of lipid peroxide formation in light toxicity compared to the dark incubation occurred as a result of more ROS production in Gram-negative and positive bacteria. More production of lipid peroxides and oxidation of other cellular targets led to the reduction of sessile bacteria and their biofilms (Parasuraman et al., 2019; Anju et al., 2018).

6.5.4 DNA

Another potential target responsible for the reduced microbial viability during antimicrobial photodynamic therapy is DNA. The damage to the DNA occurs when the appropriate PS accumulate in the vicinity of respective target. The damage to DNA occurs through various ways. For example, photodynamic inactivation results in the breakage of DNA in their single (ss) and double (ds) strand form. Even, supercoiled plasmid converts into nicked DNA in Gram-positive and -negative bacteria. Further, PS may intercalate more into ds DNA leading to the damage. Among different nucleotide bases, guanine is more prone to ROS oxidative damage than other bases (Hamblin and Hasan, 2004;

Cieplik et al., 2018a). The reaction of singlet oxygen with guanine bases ends up with the release of oxidised product of base, 8-oxo-7,8-dihydro-20-deoxyguanosine (Cadet et al., 2007). In general, aPDT mediates the nucleic acid damage through three pathway methods. In the first method, electron oxidation occurs where electrons are abstracted from the bases of DNA by the triplet state of PS. In the second, free radicals of type I cause oxidation of DNA. The third method involves the damage induced by singlet oxygen. As PSs which can intercalate to DNA by accumulating in the vicinity of target can cause more DNA damage (Cadet et al., 2006; Cadet et al., 2012; Epe, 2012; Benov, 2015).

In another method, reduced cell survival is parallelly associated with less DNA. Authors have observed that damage of cell membrane may lead to the leakage of cellular contents including DNA. Reported aPDT of *Yersinia pseudotuberculosis* disrupted cell membrane and release of DNA (Walther et al., 2009). In *S. aureus* excellent photodynamic inactivation using porphyrin molecules, zinc phthalocyanine, methylene blue, and toluidine blue was achieved when DNA repair mechanism was inhibited. In the above study, an enhanced DNA damage was observed along with the upregulated expression of DNA repair proteins. It was reported that treatment with a DNA repair inhibitor along with the photodynamic treatment displays reduced cell viability (Grinholc et al., 2015).

Still, there are some reports stating that DNA is not the prime target of antimicrobial photodynamic inactivation. Some bacteria exhibit good DNA repairing mechanisms, but efficiently inactivated or inhibited the growth by PS mediated photodynamic therapy (Hadi et al., 2021; Schäfer et al., 1998). For instance, Zn porphyrin PS illuminated with visible light mediated efficient photodynamic inactivation of *E. coli*. The authors observed that the major target of PDI was not DNA, but other cellular components such as proteins. Thus, it is not necessary that damage occurs at the primary targets of PDT (Awad et al., 2016). Some of the photosensitisers and its mechanism of action of different cellular contents of microorganisms are provided in table (Table 6.1).

TABLE 6.1
Examples of Photosensitisers and Their Antimicrobial Photoinactivation Mechanism on Different Microbes

Sl. No	PS	Microorganisms	PDT Parameters	Targets	References
1	Cationic PS, meso-tetra-4-N-methyl pyridyl porphine	*S. aureus*	Blue light	Proteins found on cytoplasmic membrane	Dosselli et al. (2012)
2	Tetracationic metalloporphyrin Zn(II) meso-tetrakis (N-n-hexylpyridinium-2-yl) porphyrin	*C. albicans*	Blue LED	Hyphae formation and biofilms	Souza et al. (2022)
3	Metalloporphyrin	MRSA and *E. coli*	Blue light	Bacterial cell membranes	Skwor et al. (2016)
4	Porphyrin derivatives	*E. coli*	White light	Membrane phospholipids	Lopes et al. (2014)
5	Curcumin, Eosin Y and MB	*Vibrio* sp.	Blue LED and red laser	DNA	Notaro et al. (2021)
6	Hypericin	*P. aeruginosa*	Orange light	Cell wall	Alam et al. (2019)
7	Curcumin	Amastigotes of *Leishmania braziliensis* and *L. major*	LED	Mitochondria	Pereira et al. (2021)

The light used for phototoxicity and the cellular target of aPDT is also provided.

6.5.5 Lysosomes and Mitochondria

Photosensitisers are also found to be localised in lysosomes leading to the enhanced photodynamic inactivation of bacteria (Mete et al., 2021). A number of studies have indicated that some PS possess more negative charge which cannot easily cross plasma membranes. Thus, PSs are entrapped inside through endocytosis and localise in lysosomes (Boyle and Dolphin, 1996). In addition, the disruption of lysosome membrane through aPDT cause the release of hydrolytic enzymes which may improve the outcome of aPDT (Benov, 2015). Zhuang and co-workers reported lysosome targeted photodynamic inactivation of *S. aureus* using novel PS (Polat and Kang, 2021; Zhuang et al., 2020).

Some lipophilic PSs tend to accumulate in mitochondria owing to the presence of lipophilic membranes. In a study, some PS were tagged with molecules which target mitochondria and led to the photodynamic inactivation of pathogenic bacteria *E. faecalis* and *S. aureus* under blue light (Kirakci et al., 2019). It has been observed that metalloporphyrins with more aliphatic chain carbons localise more in the mitochondria than free distribution in cytoplasm (Ezzeddine et al., 2013). Mitochondrial targeting is more exhibited during PDT of cancer. In cancer therapy, structural changes occurs in mitochondria which causes interruption of electron transport chain, degeneration of mitochondrial membrane potential and bulge of the organelle (Benov, 2015).

6.6 CONCLUSION

Pathogenic microorganisms and its biofilms remain as a global threat involving AMR which urges the development of effective antimicrobial therapies. The introduction of antimicrobial photodynamic therapy came into the light paving new opportunities in antimicrobial drug discovery and elimination of AMR. aPDT is an antimicrobial modality involving the combination of PS, light, and O_2 yielding free radicals. The ROS (hydrogen peroxide, superoxide radicals, singlet oxygen) released through the type I and II of PDT reduce the survival of microorganisms without producing resistant strains. The major cellular target of aPDT is cytoplasmic membrane. The disruption of membrane integrity leads to the release of cellular contents such as proteins, DNA, and lipid out of cells. Finally, oxidation of cellular contents causes photodynamic inactivation of bacteria. The oxidation of targets and further irreversible damage is associated with the localisation of PS in the vicinity of targets. Lysosome and mitochondria mediated aPDT is also studied, but evidence is scarce. Therefore, more studies are required on other cellular targets of photodynamic therapy of bacteria to fully understand the mechanism of action.

REFERENCES

Abdel-Kader, M.H. 2016. The Journey of PDT throughout History: PDT from Pharos to Present. In *Photodynamic Medicine: From Bench to Clinic*, ed. H. Kostron and T. Hasan, 1–21. Cambridge, UK: RSC Publishing.

Ackroyd, R., C. Kelty, N. Brown, M. Reed. 2001. The history of photodetection and photodynamic therapy. *Photochem. Photobiol.*, 74(5), 656–669. https://doi.org/10.1562/0031-8655(2001)074<0656:thopap>2.0.co;2

Alam, S.T., T.A.N. Le, J.S. Park, H.C. Kwon, K. Kang. 2019. Antimicrobial biophotonic treatment of ampicillin-resistant *Pseudomonas aeruginosa* with hypericin and ampicillin cotreatment followed by orange light. *Pharmaceutics.*, 11, 641. https://doi.org/10.3390/pharmaceutics11120641

Allison, R.R., K. Moghissi. 2013. Photodynamic therapy (PDT): PDT mechanisms. *Clin. Endosc.*, 46, 24. https://doi.org/10.5946/ce.2013.46.1.24

Alves, E., N. Santos, T. Melo, E. Maciel, M.L. Dõria, M.A.F. Faustino, J.P.C. Tomé, M.G.P.M.S. Neves, J.A.S. Cavaleiro, Â. Cunha, L.A. Helguero, P. Domingues, A. Almeida, M.R.M. Domingues. 2013. Photodynamic oxidation of *Escherichia coli* membrane phospholipids: New insights based on lipidomics. *Rapid Commun. Mass Spectrom.*, 27(23), 2717–2728. https://doi.org/10.1002/rcm.6739

Anas, A., J. Sobhanan, K.M. Sulfiya, C. Jasmin, P.K. Sreelakshmi, V. Biju. 2021. Advances in photodynamic antimicrobial chemotherapy. *J. Photochem. Photobiol. C Photochem. Rev.*, 49, 100452. https://doi.org/10.1016/j.jphotochemrev.2021.100452

Anju, V.T., P. Paramanantham, S.L. Sb, A. Sharan, M.H. Alsaedi, T.M.S. Dawoud, S. Asad, S. Busi. 2018. Antimicrobial photodynamic activity of rose bengal conjugated multi-walled carbon nanotubes against planktonic cells and biofilm of *Escherichia coli*. *Photodiagnosis Photodyn. Ther.*, 24, 300–310. https://doi.org/10.1016/j.pdpdt.2018.10.013

Awad, M.M., A. Tovmasyan, J.D. Craik, I. Batinic-Haberle, L.T. Benov. 2016. Important cellular targets for antimicrobial photodynamic therapy. *Appl. Microbiol. Biotechnol.*, 100, 7679–7688. https://doi.org/10.1007/s00253-016-7632-3

Bacellar, I.O.L., M.C. Oliveira, L.S. Dantas, E.B. Costa, H.C. Junqueira, W.K. Martins, A.M. Durantini, G. Cosa, P. Di Mascio, M. Wainwright, R. Miotto, R.M. Cordeiro, S. Miyamoto, M.S. Baptista. 2018. Photosensitized membrane permeabilization requires contact-dependent reactions between photosensitizer and lipids. *J. Am. Chem. Soc.*, 140(30) 9606–9615. https://doi.org/10.1021/jacs.8b05014

Benov, L. 2015. Photodynamic therapy: Current status and future directions, *Med. Princ. Pract.*, 24, 14–28. https://doi.org/10.1159/000362416

Bertolini, G., F. Rossi, G. Valduga, G. Jori, J. Lier. 2006. Photosensitizing activity of water- and lipid-soluble phthalocyanines on *Escherichia coli*. *FEMS Microbiol. Lett.*, 71, 149–155. https://doi.org/10.1111/j.1574-6968.1990.tb03814.x

Boyle, R.W., D. Dolphin. 1996. Structure and biodistribution relationships of photodynamic sensitizers. *Photochem. Photobiol.*, 64, 469–485. https://doi.org/10.1111/j.1751-1097.1996.tb03093.x

British Society for Antimicrobial Chemotherapy. 2018. *Antimicrobial Stewardship from Principles to Practice*. Birmingham, UK: British Society for Antimicrobial Chemotherapy, BSAC. Available from: http://www.bsac.org.uk/antimicrobialstewardshipebook/BSAC-AntimicrobialStewardship-FromPrinciplestoPractice-eBook.pdf Accessed August 9, 2022.

Cadet, J., T. Douki, C. Badouard, A. Favier, J-L. Ravanat. 2007. Oxidatively generated damage to cellular DNA: Mechanistic aspects. *Acc. Chem. Res.*, 41(8), 1075–1083. https://doi.org/10.1021/ar700245e

Cadet, J., T. Douki, J.-L. Ravanat. 2006. One-electron oxidation of DNA and inflammation processes. *Nat. Chem. Biol.*, 2, 348–349. https://doi.org/10.1038/nchembio0706-348

Cadet, J., J.L. Ravanat, M. TavernaPorro, H. Menoni, D. Angelov. 2012. Oxidatively generated complex DNA damage: Tandem and clustered lesions. *Cancer Lett.*, 327, 5–15. https://doi.org/10.1016/j.canlet.2012.04.005

Carrera, E.T., H.B. Dias, S.C.T. Corbi, R.A.C. Marcantonio, A.C.A. Bernardi, V.S. Bagnato, M.R. Hamblin, A.N.S. Rastelli. 2016. The application of antimicrobial photodynamic therapy (aPDT) in dentistry: A critical review. *Laser Phys.*, 26, 123001. https://doi.org/10.1088/1054-660X/26/12/123001

Castano, A.P., T.N. Demidova, M.R. Hamblin. 2004. Mechanisms in photodynamic therapy: part one—photosensitizers, photochemistry and cellular localization. *Photodiagnosis Photodyn. Ther.*, 1, 279–293. https://doi.org/10.1016/S1572-1000(05)00007-4

Chang, R.Y.K., V Nang, H.-K. Chan, J. Li. 2022. Novel antimicrobial agents for combating antibiotic-resistant bacteria. *Adv. Drug Deliv. Rev.*, 187, 114378. https://doi.org/10.1016/j.addr.2022.114378

Cieplik, F., D. Deng, W. Crielaard, W. Buchalla, E. Hellwig, A. Al-Ahmad, T. Maisch. 2018a. Antimicrobial photodynamic therapy–what we know and what we don't. *Crit. Rev. Microbiol.*, 44, 571–589. https://doi.org/10.1080/1040841X.2018.1467876

Cieplik, F., V.S. Steinwachs, D. Muehler, K.A. Hiller, T. Thurnheer, G.N. Belibasakis, W. Buchalla, T. Maisch. 2018b. Phenalen-1-one-mediated antimicrobial photodynamic therapy: Antimicrobial efficacy in a periodontal biofilm model and flow cytometric evaluation of cytoplasmic membrane damage. *Front. Microbiol.*, 9, 688. https://doi.org/10.3389/fmicb.2018.00688

Ciofu, O., E. Rojo-Molinero, M.D. Macià, A. Oliver. 2017. Antibiotic treatment of biofilm infections. *APMIS*, 125, 304–319. https://doi.org/10.1111/apm.12673

Correia, J.H., J.A. Rodrigues, S. Pimenta, T. Dong, Z. Yang. 2021. Photodynamic therapy review: Principles, photosensitizers, applications, and future directions. *Pharmaceutics*, 13(9), 1332. https://doi.org/10.3390/pharmaceutics13091332

Daniell, M.D., J.S. Hill. 1991. A history of photodynamic therapy. *Aust. N. Z. J. Surg.*, 61, 340–348. https://doi.org/10.1111/j.1445-2197.1991.tb00230.x

Darabpour, E., N. Kashef, S. Mashayekhan. 2016. Chitosan nanoparticles enhance the efficiency of methylene blue-mediated antimicrobial photodynamic inactivation of bacterial biofilms: An in vitro study. *Photodiagnosis Photodyn. Ther.*, 14, 211–217. https://doi.org/10.1016/j.pdpdt.2016.04.009

Davies, M.J. 2004. Reactive species formed on proteins exposed to singlet oxygen. *Photochem. Photobiol. Sci.*, 3, 17–25. https://doi.org/10.1039/b307576c

De Oliveira, D.M.P., B.M. Forde, T.J. Kidd, P.N.A. Harris, M.A. Schembri, S.A. Beatson, D.L. Paterson, M.J. Walker. 2020. Antimicrobial resistance in ESKAPE pathogens. *Clin. Microbiol. Rev.*, 33, e00181-19. https://doi.org/10.1128/CMR.00181-19

Dias, L.D., V.S. Bagnato. 2020. An update on clinical photodynamic therapy for fighting respiratory tract infections: a promising tool against COVID-19 and its co-infections. *Laser Phys. Lett.*, 17, 083001. https://doi.org/10.1088/1612-202X/ab95a9

Donlan, R.M. 2002. Biofilms: Microbial Life on Surfaces. *Emerg. Infect. Dis.*, 8, 881–890. https://doi.org/10.3201/eid0809.020063

Dosselli, R., R. Millioni, L. Puricelli, P. Tessari, G. Arrigoni, C. Franchin, A. Segalla, E. Teardo, E. Reddi. 2012. Molecular targets of antimicrobial photodynamic therapy identified by a proteomic approach. *J. Proteomics*, 77, 329–343. https://doi.org/10.1016/j.jprot.2012.09.007

Epe, B. 2012. DNA damage spectra induced by photosensitization. *Photochem. Photobiol. Sci.*, 11, 98–106. https://doi.org/10.1039/c1pp05190c

Ezzeddine, R., A. Al-Banaw, A. Tovmasyan, J.D. Craik, I. Batinic-Haberle, L.T. Benov. 2013. Effect of molecular characteristics on cellular uptake, subcellular localization, and phototoxicity of Zn(II) N-Alkylpyridylporphyrins. *J. Biol. Chem.*, 288, 36579–36588. https://doi.org/10.1074/jbc.M113.511642

Fitzgerald, F. 2017. *Photodynamic Therapy (PDT): Principles, Mechanisms and Applications*. New York: Nova Biomedical.

Gao, S., X. Yan, G. Xie, M. Zhu, X. Ju, P.J. Stang, Y. Tian, Z. Niu. 2019. Membrane intercalation-enhanced photodynamic inactivation of bacteria by a metallacycle and TAT-decorated virus coat protein. *Proc. Natl. Acad. Sci.*, 116, 23437–23443. https://doi.org/10.1073/pnas.1911869116

George, S., M.R. Hamblin, A. Kishen. 2009. Uptake pathways of anionic and cationic photosensitizers into bacteria. *Photochem. Photobiol. Sci.*, 8, 788. https://doi.org/10.1039/b809624d

Geralde, M.C., I.S. Leite, N.M. Inada, A.C.G. Salina, A.I. Medeiros, W.M. Kuebler, C. Kurachi, V.S. Bagnato. 2017. Pneumonia treatment by photodynamic therapy with extracorporeal illumination—an experimental model. *Physiol. Rep.*, 5, e13190. https://doi.org/10.14814/phy2.13190

Giedraitiene, A., V Vitkauskiene, R. Naginiene, A. Pavilonis. 2011. Antibiotic resistance mechanisms of clinically important bacteria. *Medicina*, 47(3), 19. https://doi.org/10.3390/medicina47030019

Grinholc, M., A. Rodziewicz, K. Forys, A. Rapacka-Zdonczyk, A. Kawiak, A. Domachowska, G. Golunski, C. Wolz, L. Mesak, K. Becker, K.P. Bielawski. 2015. Fine-tuning recA expression in *Staphylococcus aureus* for antimicrobial photoinactivation: importance of photo-induced DNA damage in the photoinactivation mechanism. *Appl. Microbiol. Biotechnol.*, 99, 9161–9176. https://doi.org/10.1007/s00253-015-6863-z

Grizante Barião, P.H., L. Tonani, G.T.P. Brancini, E. Nascimento, G.Ú.L. Braga, M. Wainwright, M.R. von Zeska Kress. 2022. *In vitro* and *in vivo* photodynamic efficacies of novel and conventional phenothiazinium photosensitizers against multidrug-resistant *Candida auris*. *Photochem. Photobiol. Sci.*, https://doi.org/10.1007/s43630-022-00258-4

Hadi, J., S. Wu, A. Soni, A. Gardner, G. Brightwell. 2021. Genetic factors affect the survival and behaviors of selected bacteria during antimicrobial blue light treatment. *Int. J. Mol. Sci.*, 22(19), 10452. https://doi.org/10.3390/ijms221910452

Hall-Stoodley, L., J.W. Costerton, P. Stoodley. 2004. Bacterial biofilms: From the natural environment to infectious diseases. *Nat. Rev. Microbiol.*, 2, 95–108. https://doi.org/10.1038/nrmicro821

Hamblin, M.R., T. Hasan. 2004. Photodynamic therapy: A new antimicrobial approach to infectious disease? *Photochem. Photobiol. Sci.*, 3, 436–450. https://doi.org/10.1039/b311900a

Hu, X., Y.Y. Huang, Y. Wang, X. Wang, M.R. Hamblin. 2018. Antimicrobial photodynamic therapy to control clinically relevant biofilm infections. *Front. Microbiol.*, 9, 1299. https://doi.org/10.3389/fmicb.2018.01299

Jamrozik, E., G.S. Heriot. 2022. Ethics and antibiotic resistance. *Br. Med. Bull.*, 141, 4–14. https://doi.org/10.1093/bmb/ldab030

Juzeniene, A., J. Moan. 2007. The history of PDT in Norway. Part one: Identification of basic mechanisms of general PDT. *Photodiagnosis Photodyn. Ther.*, 4, 3–11. https://doi.org/10.1016/j.pdpdt.2006.11.002

Kirakci, K., J. Zelenka, M. Rumlová, J. Cvačka, T. Ruml, K. Lang. 2019. Cationic octahedral molybdenum cluster complexes functionalized with mitochondria-targeting ligands: Photodynamic anticancer and antibacterial activities. *Biomater. Sci.*, 7, 1386–1392. https://doi.org/10.1039/c8bm01564c

Kornman, K.S., R.C. Page, M.S. Tonetti. 1997. The host response to the microbial challenge in periodontitis: Assembling the players. *Periodontol. 2000*, 14, 33–53. https://doi.org/10.1111/j.1600-0757.1997.tb00191.x

Lewis, K. 2007. Persister cells, dormancy and infectious disease. *Nat. Rev. Microbiol.*, 5, 48–56. https://doi.org/10.1038/nrmicro1557

Li, J., Z. Zhuang, Z. Zhao, B.Z. Tang. 2022. Type I AIE photosensitizers: Mechanism and application. *View*, 3, 20200121. https://doi.org/10.1002/viw.20200121

Liu, Y., R. Qin, S.A.J. Zaat, E. Breukink, M. Heger. 2015. Antibacterial photodynamic therapy: overview of a promising approach to fight antibiotic-resistant bacterial infections. *J. Clin. Transl. Res.*, 1, 140–167. https://doi.org/10.18053/jctres.201503.002

Lopes, D., T. Melo, N. Santos, L. Rosa, E. Alves, M.C. Gomes, Â. Cunha, M.G.P.M.S. Neves, M.A.F. Faustino, M.R.M. Domingues, A. Almeida. 2014. Evaluation of the interplay among the charge of porphyrinic photosensitizers, lipid oxidation and photoinactivation efficiency in *Escherichia coli*. *J. Photochem. Photobiol. B Biol.*, 141, 145–153. https://doi.org/10.1016/j.jphotobiol.2014.08.024

López, D., H. Vlamakis, R. Kolter. 2010. Biofilms. *Cold Spring Harb. Perspect. Biol.*, 2, a000398. https://doi.org/10.1101/cshperspect.a000398

Lushniak, B.D. 2014. Antibiotic resistance: A public health crisis. *Public Health Rep.*, 129, 314–316. https://doi.org/10.1177/003335491412900402

Inada, N.M., L.D. Dias, K.C. Blanco, G. Kassab, H. Buzzá, V.S. Bagnato. 2021. Antimicrobial Photodynamic Therapy of the Respiratory Tract: From the Proof of Principles to Clinical Application. In *Photodynamic Therapy- From Basic Science to Clinical Research*, ed. N.M. Inada, H.H. Buzzá, K.C. Blanco, and L.D. Dias. London: IntechOpen. https://doi/10.5772/intechopen.95602

Moreira, L.M., J.P. Lyon, A.P. Romani, D. Severino, M.R. Rodrigues, H.P.M. de Oliveira. 2012. Phenotiazinium Dyes as Photosensitizers (PS) in Photodynamic Therapy (PDT): Spectroscopic Properties and Photochemical Mechanisms. In *Advanced Aspects of Spectroscopy*, ed. M.A. Farrukh. London: IntechOpen, https://doi/10.5772/48087

Merchat, M., G. Bertolini, P. Giacomini, A. Villanueva, G. Jori. 1996. Meso-substituted cationic porphyrins as efficient photosensitizers of Gram-positive and gram-negative bacteria. *J. Photochem. Photobiol. B Biol.*, 32, 153–157. https://doi.org/10.1016/1011-1344(95)07147-4

Mete, E., N. Kabay, F. Dumoulin, V. Ahsen, S. Tuncel Kostakoğlu, Ç. Ergin. 2021. Photodynamic inactivation of *Staphylococcus aureus* using tetraethylene glycol-substituted Zn(II) phthalocyanine. *Biotech. Histochem.*, 311–314. https://doi.org/10.1080/10520295.2020.1854855

Muehler, D., E. Brandl, K.A. Hiller, F. Cieplik, T. Maisch. 2022. Membrane damage as mechanism of photodynamic inactivation using Methylene blue and TMPyP in *Escherichia coli* and *Staphylococcus aureus*. *Photochem. Photobiol. Sci.*, 21, 209–220. https://doi.org/10.1007/s43630-021-00158-z

Murray, C.J., K.S. Ikuta, F. Sharara, L. Swetschinski, G. Robles Aguilar, A. Gray, C. Han, C. Bisignano, P. Rao, E. Wool, S.C. Johnson, A.J. Browne, M.G. Chipeta, F. Fell, S. Hackett, G. Haines-Woodhouse, B.H. Kashef Hamadani, E.A.P. Kumaran, B. McManigal, R. Agarwal, S. Akech, S. Albertson, J. Amuasi, J. Andrews, A. Aravkin, E. Ashley, F. Bailey, S. Baker, B. Basnyat, A. Bekker, R. Bender, A. Bethou, J. Bielicki, S. Boonkasidecha, J. Bukosia, C. Carvalheiro, C. Castañeda-Orjuela, V. Chansamouth, S. Chaurasia, S. Chiurchiù, F. Chowdhury, A.J. Cook, B. Cooper, T.R. Cressey, E. Criollo-Mora, M. Cunningham, S. Darboe, N.P.J. Day, M. De Luca, K. Dokova, A. Dramowski, S.J. Dunachie, T. Eckmanns, D. Eibach, A. Emami, N. Feasey, N. Fisher-Pearson, K. Forrest, D. Garrett, P. Gastmeier, A.Z. Giref, R.C. Greer, V. Gupta, S. Haller, A. Haselbeck, S.I. Hay, M. Holm, S. Hopkins, K.C. Iregbu, J. Jacobs, D. Jarovsky, M. Javanmardi, N. Khorana, E. Kissoon, T. Kobeissi, F. Kostyanev, F. Krapp, R. Krumkamp, A. Kumar, H.H. Kyu, C. Lim, D. Limmathurotsakul, M.J. Loftus, M. Lunn, J. Ma, N. Mturi, T. Munera-Huertas, P. Musicha, M.M. Mussi-Pinhata, T. Nakamura, R. Nanavati, S. Nangia, P. Newton, C. Ngoun, A. Novotney, D. Nwakanma, C.W. Obiero, A. Olivas-Martinez, P. Olliaro, E. Ooko, E. Ortiz-Brizuela, A.Y. Peleg, C. Perrone, N. Plakkal, A. Ponce-de-Leon, M. Raad, T. Ramdin, A. Riddell, T. Roberts, J.V. Robotham, A. Roca, A. Rudd, N. Russell, J. Schnall, J.A.G. Scott, M. Shivamallappa, J. Sifuentes-Osornio, J. Steenkeste, A.J. Stewardson, T. Stoeva, N. Tasak, A. Thaiprakong, G. Thwaites, C. Turner, P. Turner, H.R. van Doorn, S. Velaphi, A. Vongpradith, H. Vu, T. Walsh, S. Waner, T. Wangrangsimakul, T. Wozniak, P. Zheng, B. Sartorius, A.D. Lopez, A. Stergachis, A. Moore, C. Dolecek, M. Naghavi. 2022. Global burden of bacterial antimicrobial resistance in 2019: a systematic analysis. *Lancet*, 399, 629–655. https://doi.org/10.1016/S0140-6736(21)02724-0

Ning, H.Q., Y.Q. Li, Z.S. Wang, H.Z. Mo. 2019. The synergistic antibacterial properties of glycinin basic peptide against bacteria via membrane damage and inactivation of enzymes. *Food Biophys.*, 14, 132–141. https://doi.org/10.1007/s11483-018-09564-w

Notaro, D.A., S.C. Culloty, S.A. Lynch. 2021. A pilot study investigating the potential of antimicrobial photodynamic therapy (aPDT) to control *Vibrio* spp. development in microalgae and seawater. *Aquac. Int.*, 29, 355–372. https://doi.org/10.1007/s10499-020-00631-z

Oyim, J., C.A. Omolo, E.K. Amuhaya. 2021. Photodynamic antimicrobial chemotherapy: Advancements in porphyrin-based photosensitize development. *Front. Chem.*, 9, 635344. https://doi.org/10.3389/fchem.2021.635344

Pereira, A.H.C., L.M.C. Marcolino, J.G. Pinto, J. Ferreira-Strixino. 2021. Evaluation of the photodynamic therapy with curcumin on *L. braziliensis* and *L. major* amastigotes. *Antibiotics*, 10, 634. https://doi.org/10.3390/antibiotics10060634

Parasuraman, P., V.T. Anju, L.S.B. Sruthil, A. Sharan, S. Busi, K. Kaviyarasu, M. Arshad, T.M.S. Dawoud, A. Syed. 2019. Synthesis and antimicrobial photodynamic effect of methylene blue conjugated carbon nanotubes on *E. coli* and *S. aureus*. *Photochem. Photobiol. Sci.*, 18, 563–576. https://doi.org/10.1039/C8PP00369F

Plaetzer, K., B. Krammer, J. Berlanda, F. Berr, T. Kiesslich. 2009. Photophysics and photochemistry of photodynamic therapy: fundamental aspects. *Lasers Med. Sci.*, 24, 259–268. https://doi.org/10.1007/s10103-008-0539-1

Plowman, J.E., S. Deb-Choudhury, A.J. Grosvenor, J.M. Dyer. 2013. Protein oxidation: Identification and utilisation of molecular markers to differentiate singlet oxygen and hydroxyl radical-mediated oxidative pathways. *Photochem. Photobiol. Sci.*, 12, 1960–1967. https://doi.org/10.1039/c3pp50182e

Polat, E., K. Kang. 2021. Natural photosensitizers in antimicrobial photodynamic therapy. *Biomedicine*, 9(6), 584. https://doi.org/10.3390/biomedicines9060584

Pourhajibagher, M., M. Azimi, V. Haddadi-Asl, H. Ahmadi, M. Gholamzad, S. Ghorbanpour, A. Bahador. 2021. Robust antimicrobial photodynamic therapy with curcumin-poly (lactic-co-glycolic acid) nanoparticles against COVID-19: A preliminary in vitro study in Vero cell line as a model. *Photodiagnosis Photodyn. Ther.*, 34, 102286. https://doi.org/10.1016/j.pdpdt.2021.102286

Rice, L.B. 2010. Progress and challenges in implementing the research on ESKAPE pathogens. *Infect. Control Hosp. Epidemiol.*, 31(S1), S7–S10. https://doi.org/10.1086/655995

Robertson, C.A., D.H. Evans, H. Abrahamse. 2009. Photodynamic therapy (PDT): A short review on cellular mechanisms and cancer research applications for PDT. *J. Photochem. Photobiol. B Biol.*, 96, 1–8. https://doi.org/10.1016/j.jphotobiol.2009.04.001

Sai, D.L., J. Lee, D.L. Nguyen, Y.P. Kim. 2021. Tailoring photosensitive ROS for advanced photodynamic therapy. *Exp. Mol. Med.*, 53, 495–504. https://doi.org/10.1038/s12276-021-00599-7

Santajit, S., N. Indrawattana. 2016. Mechanisms of antimicrobial resistance in ESKAPE pathogens. *Biomed. Res. Int.*, 2016, 2475067. https://doi.org/10.1155/2016/2475067

Schäfer, M., C. Schmitz, G. Horneck. 1998. High sensitivity of *Deinococcus radiodurans* to photodynamically produced singlet oxygen. *Int. J. Radiat. Biol.*, 74, 249–253. https://doi.org/10.1080/095530098141636

Sharma, D., L. Misba, A.U. Khan. 2019. Antibiotics versus biofilm: An emerging battleground in microbial communities. *Antimicrob. Resist. Infect. Control*, 8, 76. https://doi.org/10.1186/s13756-019-0533-3

Sharma, S.K., P. Mroz, T. Dai, Y.Y. Huang, T.G.S. Denis, M.R. Hamblin. 2012. Photodynamic therapy for cancer and for infections: What is the difference? *Isr. J. Chem.*, 52, 691–705. https://doi.org/10.1002/ijch.201100062

Skwor, T.A., S. Klemm, H. Zhang, B. Schardt, S. Blaszczyk, M.A. Bork. 2016. Photodynamic inactivation of methicillin-resistant *Staphylococcus aureus* and *Escherichia coli*: A metalloporphyrin comparison. *J. Photochem. Photobiol. B Biol.*, 165, 51–57. https://doi.org/10.1016/j.jphotobiol.2016.10.016

Smith, R., J. Coast. 2013. The true cost of antimicrobial resistance. *BMJ*, 346, f1493. https://doi.org/10.1136/bmj.f1493

Sobotta, L., P. Skupin-Mrugalska, J. Piskorz, J. Mielcarek. 2019. Porphyrinoid photosensitizers mediated photodynamic inactivation against bacteria. *Eur. J. Med. Chem.*, 175, 72–106. https://doi.org/10.1016/j.ejmech.2019.04.057

Souza, S.O., B.L. Raposo, J.F. Sarmento-Neto, J.S. Rebouças, D.P.C. Macêdo, R.C.B.Q. Figueiredo, B.S. Santos, A.Z. Freitas, P.E. Cabral Filho, M.S. Ribeiro, A. Fontes. 2022. Photoinactivation of Yeast and Biofilm Communities of *Candida albicans* Mediated by ZnTnHex-2-PyP4+ Porphyrin. *J. Fungi.*, 8, 556. https://doi.org/10.3390/jof8060556

Sperandio, F., Y.Y. Huang, M. Hamblin. 2013. Antimicrobial photodynamic therapy to kill gram-negative bacteria. *Recent Pat. Antiinfect. Drug Discov.*, 8, 108–120. https://doi.org/10.2174/1574891x113089990012

Spesia, M.B., E.N. Durantini. 2013. Photodynamic inactivation mechanism of *Streptococcus mitis* sensitized by zinc (II) 2,9,16,23-tetrakis[2-(N,N,N-trimethylamino) ethoxy]phthalocyanine. *J. Photochem. Photobiol. B Biol.*, 125, 179–187. https://doi.org/10.1016/j.jphotobiol.2013.06.007

Svyatchenko, V.A., S.D. Nikonov, A.P. Mayorov, M.L. Gelfond, V.B. Loktev. 2021. Antiviral photodynamic therapy: Inactivation and inhibition of SARS-CoV-2 in vitro using methylene blue and Radachlorin. *Photodiagnosis Photodyn. Ther.*, 33, 102112. https://doi.org/10.1016/j.pdpdt.2020.102112

Tasso, T.T., J.C. Schlothauer, H.C. Junqueira, T.A. Matias, K. Araki, É. Liandra-Salvador, F.C.T. Antonio, P. Homem-De-Mello, M.S. Baptista. 2019. Photobleaching efficiency parallels the enhancement of membrane damage for porphyrazine photosensitizers. *J. Am. Chem. Soc.*, 141, 15547–15556. https://doi.org/10.1021/jacs.9b05991

Teixeira, I.S., F.S. Leal, R.Y. Tateno, L.F. Palma, L. Campos. 2021. Photobiomodulation therapy and antimicrobial photodynamic therapy for orofacial lesions in patients with COVID-19: A case series. *Photodiagnosis Photodyn. Ther.*, 34, 102281. https://doi.org/10.1016/j.pdpdt.2021.102281

Tim, M. 2015. Strategies to optimize photosensitizers for photodynamic inactivation of bacteria. *J. Photochem. Photobiol. B Biol.*, 150, 2–10. https://doi.org/10.1016/j.jphotobiol.2015.05.010

Tonon, C.C., S. Ashraf, A.N. de Souza Rastelli, G. Ghosh, T. Hasan, Q. Xu, A. Greer, A.M. Lyons. 2022. Evaluation of photosensitizer-containing superhydrophobic surfaces for the antibacterial treatment of periodontal biofilms. *J. Photochem. Photobiol. B Biol.*, 233, 112458. https://doi.org/10.1016/j.jphotobiol.2022.112458

Wainwright, M. 2010. "Safe" photoantimicrobials for skin and soft-tissue infections. *Int. J. Antimicrob. Agents*, 36, 14–18. https://doi.org/10.1016/j.ijantimicag.2010.03.002

Walther, J., M.J. Bröcker, D. Wätzlich, M. Nimtz, M. Rohde, D. Jahn, J. Moser. 2009. Protochlorophyllide: A new photosensitizer for the photodynamic inactivation of Gram-positive and Gram-negative bacteria. *FEMS Microbiol. Lett.*, 290, 156–163. https://doi.org/10.1111/j.1574-6968.2008.01413.x

Wang, Z., Y. Jia, W. Li, M. Zhang. 2021. Antimicrobial photodynamic inactivation with curcumin against *Staphylococcus saprophyticus*, in vitro and on fresh dough sheet. *LWT*, 147, 111567. https://doi.org/10.1016/j.lwt.2021.111567

WHO. 2014. *Antimicrobial Resistance. Global Report on Surveillance*. World Health Organization, https://doi.org/10.1007/s13312-014-0374-3

Xuan, W., Y. He, L. Huang, Y.Y. Huang, B. Bhayana, L. Xi, J.A. Gelfand, M.R. Hamblin. 2018. Antimicrobial photodynamic inactivation mediated by tetracyclines in vitro and in vivo: photochemical mechanisms and potentiation by potassium iodide. *Sci. Rep.*, 8, 17130. https://doi.org/10.1038/s41598-018-35594-y

Yang, Z., Y. Qiao, J. Li, F.G. Wu, F. Lin. 2020. Novel type of water-soluble photosensitizer from *Trichoderma reesei* for photodynamic inactivation of Gram-positive bacteria. *Langmuir*, 36, 13227–13235. https://doi.org/10.1021/acs.langmuir.0c02109

Zhao, Z., J. Ma, Y. Wang, Z. Xu, L. Zhao, J. Zhao, G. Hong, T. Liu. 2021. Antimicrobial photodynamic therapy combined with antibiotic in the treatment of rats with third degree burns. *Front. Microbiol.*, 12, 622410. https://doi.org/10.3389/fmicb.2021.622410

Zhuang, J., H. Yang, Y. Li, B. Wang, N. Li, N. Zhao. 2020. Efficient photosensitizers with aggregation-induced emission characteristics for lysosome- and Gram-positive bacteria-targeted photodynamic therapy. *Chem. Commun.*, 56, 2630–2633. https://doi.org/10.1039/d0cc00394h

7 Applications of Antimicrobial Photodynamic Therapy (aPDT)

Madangchanok Imchen and Siddhardha Busi
Pondicherry University, Puducherry, India

Jamseel Moopantakath
Central University of Kerela, Kasaragod, India

Ranjith Kumavath
Pondicherry University, Puducherry, India

CONTENTS

7.1 INTRODUCTION

Penicillin have been used in large quantities since its discovery in 1928 by Alexander Fleming. During his Nobel lecture in 1945, he warned of antibiotic resistance when used in suboptimal concentrations. Today, antibiotic resistance has spread throughout the globe, including the pristine Artic cold deserts (Van Goethem et al., 2018). Resistance to antibiotic by pathogens has accelerated due to increase social mobility, inadequate hygiene, ease in availability of antibiotics, excessive prescription of antibiotics, over usage of antibiotics in agriculture and farming, and the overall socioeconomic status. Furthermore, inadequate wastewater treatment and pollution of environmental settings have been strongly linked to the prevalence of antibiotic resistance genes (ARGs) (Singh et al., 2022). Infections due to antibiotic resistance have spread worldwide, burdening the country's economy and the family. In the USA alone, it is estimated that antibiotic resistance causes over 2 million cases each year costing over US$ 55 billion (Youf et al., 2021). Globally, it is estimated that by 2050, tens of millions of lives will be lost annually due to antibiotic resistance.

DOI: 10.1201/9781003345299-7

TABLE 7.1
aPDT Combinatorial Therapy with Antibiotics Reduces the Antibiotic Dosage and Increases Efficiency in Antimicrobial Activity Including Biofilm Destruction

Photosensitiser or Nanoconjugate System Used	Antibiotic Drug Used	Organism	Impact/Target	Reference
ALA-PDT	Amikacin and clarithromycin	*M. fortuitum*	Wound healing	Gong et al. (2016)
2-aminothiazolo[4,5-c]-2,7,12,17-tetrakis(methoxyethyl) porphycene (ataztmpo)-PDT	Gentamicin conjugate	*S. aureus* and *E. coli*	Enhance the therapeutic index of gentamicin	Nieves et al. (2020)
Meso-tetrakis(N-methyl-4-pyridyl) porphine tetra-tosylate (TMP)-PDT	Vancomycin	*S. aureus* biofilms.	Combination blocks cell wall synthesis, and the damaged biofilms may be more susceptible to host defences	Di Poto et al. (2009)
Meso-tetra (4-aminophenyl) porphine-PDT	Chloramphenicol	*S. epidermidis* ATCC 35984	Reduction of growth	Perez-Laguna et al. (2021)
Methylene blue-PDT	Amikacin	*M. fortuitum*	(≥2 Log_{10} reduction)	Shih and Huang (2011)
ALA-PDT	Multiple antibiotics	*M. fortuitum*	Wound healing	Sun et al. (2017)
Tetracationic porphyrin 5,10,15,20-tetrakis(1-methylpyridinium-4-yl)porphyrin tetra-iodide (Tetra-Py+-Me)	Ampicillin	*S. aureus*	Inactivating the bacterium	Branco et al. (2018)
Rose bengal -pdt and methylene blue -apdt	Mupirocin or linezolid	*S. aureus*	Inactivated	Perez-Laguna et al. (2017)
Chlorin e6	Ciprofloxacin	*Escherichia coli*	Inhibiting bacterial growth	Tichaczek-Goska et al. (2019)
Deuteroporphyrin	Oxacillin	*Staphylococcus aureus*	Damage in the cell wall and cell membrane	Iluz et al. (2018)
Toluidine blue (TB)-PDT	Gentamicin	*Staphylococcus aureus*	Decreased bacterial virulence and infectivity	Liu et al. (2022)
Temoporfin-PDT	Ampicillin	*Staphylococcus aureus*	Enhanced the bactericidal effect	Shih et al. (2022)
Temoporfin-PDT	Chlorhexidine	*Staphylococcus aureus*	Enhanced the bactericidal effect	Shih et al. (2022)

Antimicrobial resistance has become a severe setback for public health. Antibiotics for treating pathogens reduce efficacy due to the ever-increasing prevalence of antibiotic resistance genes (ARGs). Furthermore, resistance to multiple antibiotics has become a common phenome among clinically relevant pathogens. Hence, alternative approaches have been considered a necessity for treating antibiotic-resistant infections. In this context, antimicrobial photodynamic therapy (aPDT) has gained tremendous attention for high its efficacy, specificity, and minimal side effects. Recent developments in antimicrobial photodynamic therapy have made it feasible to implement it in treating localised infections. aPDT relies on the photosensitizing agents activated by light, followed by the generation of reactive oxygen species (ROS). aPDT is a chemical reaction of a photosensitiser through light activation, generating reactive oxygen species (Rajesh et al., 2011). Its main advantage is its localised treatment efficiency and ability to deter antibiotic resistance genes from spreading. Initially, the cell walls and membranes are disrupted through hydrophobic and electrostatic interaction, followed by targeting external components (Hu et al., 2018). aPDT also weakens the biofilm

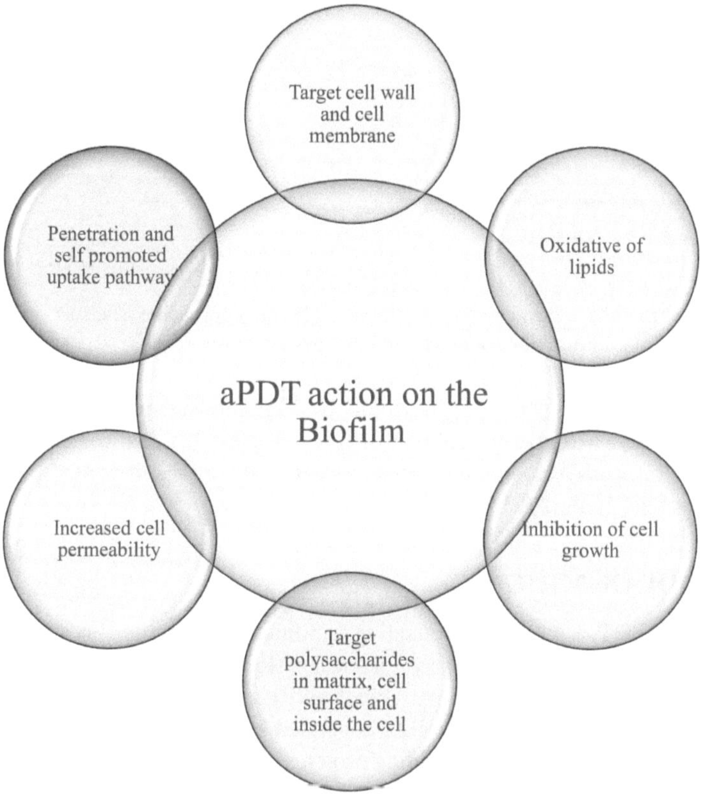

FIGURE 7.1 The basic principle in aPDT is the absorption of light by the PS that eventually leads to the formation of ROS, which can disrupt the biomolecules in the bacterial cell.

EPS matrix, thereby increasing its effectiveness in reaching bacterial cells within the biofilm. The weakened biofilm matrix also makes the microbial cells accessible to conventional antibiotics for combinatorial therapy (Figure 7.1).

aPDT has many applications, such as antibacterial, antifungal, and anticancer therapeutics. It can also be used with other agents to increase its efficiency. Its inhibition mechanism, ROS generation, weakening of microbial cell wall and membrane, and EPS matrix disruption make it suitable for biofilm-forming pathogens. Advantages of aPTD include the inactivation of virulence factors and pathogenic cells (Perni et al., 2021). In cancer treatment, aPDT can be used to kill abnormal tissues through necrosis or apoptosis with minimum side effects (Carrera et al., 2016). Therefore, the strength of aPDT is multifaceted: it has synergistic action among the photosensitiser, and it can be in conjunction with other agents, combined with chemotherapy, anticancer, antibacterial, and antifungal applications. Such versatilities of aPDT and the wide range of photosensitisers have put forth aPDT as a promising treatment for multidrug-resistant (MDR) infections (Youf et al., 2021). aPDT is an effective antimicrobial and cytotoxic, targeting rapidly dividing cells. Hence, it has specificity against pathogens and cancer cells. Treatment of peri-implantitis through aPTD has shown aPTD has a significantly higher efficiency than traditional treatment (Zhao et al., 2022). Hence, aPDT has been considered a novel promising strategy to treat infections due to MDR pathogens (Figure 7.2).

The light source is a critical factor in the activation of aPDT. Several types of light sources, such as LEDs, laser beams, xenon lamps, etc., can be used for aPDT. LEDs, in particular, have been considered an alternative option since its safe, cost-effective, and easier to handle. Overall, longer wavelength light sources are preferred for better tissue penetration (Polat and Kang, 2021). PS has a stable ground state (^1PS) which, upon activation by a suitable light source, is promoted to an unstable excited state ($^1PS^*$) or longer-living excited triplet state PS ($^3PS^*$) (the * symbol indicates

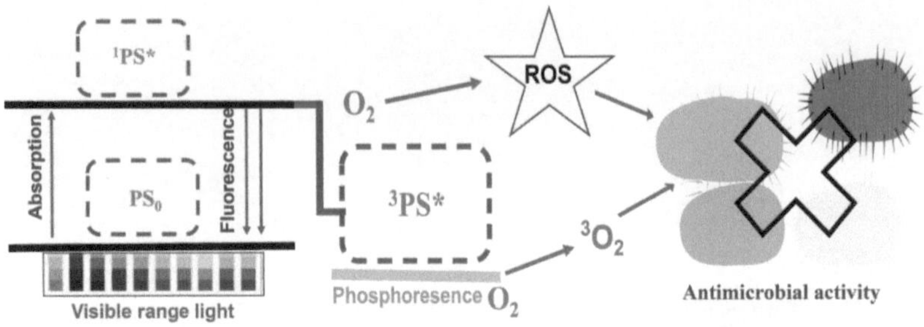

FIGURE 7.2 aPDT inhibits microbial growth through several approaches such as disruption of the cell membrane, lipid oxidation, EPS disruption, increased cell permeability, etc.

excited state) that leads to conversions of molecular oxygen into ROS and short-lived singlet oxygen ($^1O_2$) which causes lethal oxidation of cellular organelles. Unstable free radicals, to attain stability, obtain electrons from nearby compounds, which leads to the attacked compounds being unstable (Celiešiūtė-Germanienė et al., 2021).

7.2 aPDT MODE OF ACTION

The mechanism of aPDT is based on activating a nontoxic molecule, known as photosensitiser (PS), by visible light leading to reactive oxygen species (ROS) generation, which can unselectively kill microbial cells through oxidative burst (Polat and Kang 2021; Pourhajibagher et al., 2018). The primary mechanism of aPDT is that ROS generated by the excited photosensitiser attack cell membrane and intracellular materials, including proteins and DNA, that causes microbial cell death. Three components are required to establish a photodynamic reaction: photosensitiser (nontoxic material), visible light, and molecular oxygen (Dai et al., 2012). Due to the absorption of energy (photon) from visible light, the ground singlet PS forms an excited singlet state (Calzavara-Pinton et al., 2007). The excited PS loses energy in the form of heat or fluorescence through internal conversion, or it can form a triplet state PS with a longer lifespan. The triplet state PS returns to the ground state PS through phosphorescence emission or by ROS generation via two mechanisms: (i) electrons transfer (charge) from the triplet state to the surrounding biomolecules for the generation of ROS such as superoxide radicals ($O_2^{\bullet-}$) that eventually forms hydrogen peroxide (H_2O_2) through dismutation, and free hydroxyl radicals ($HO^{\bullet}$); and (ii) the energy (not charge) reacting with the surrounding oxygen molecules to form singlet oxygen. After returning to the ground state, the PS can be affected by the incoming photons once again, leading to more ROS generation. The major advantage of photosensitisers is that they can produce more than 1,000 free radicals ($O2^{\bullet}$) before getting destroyed (Cieplik et al., 2018).

PDT is effective against biofilm by weakening the biofilm matrix and its adhesion towards the substratum (Kikuchi et al., 2015). ROS generated during aPDT leads to damage to the crucial biomolecules in the microbes. ROS also affects all adjacent cells, damaging the biomatrix and other cell components (Hu et al., 2018). aPDT affects microbes in several ways, such as lipid peroxidation in *Staphylococcus warneri* (Alves et al., 2013), Cell surface damage in *M. catarrhalis* using Photofrin-based aPDT (Luke-Marshall et al., 2014), alteration in outer membrane permeability of *C. albicans* treated with toluidine blue ortho (TBO) mediated PDT (Akhtar et al., 2021), damaged cellular materials in the *E. coli* (Wardlaw et al., 2012). Furthermore, aPDT acts against Gram-negative bacteria via hydrophobic and electrostatic interactions, leading to the cell wall and membrane damage.

7.3 aPDT PHOTOSYNTHESIZERS (PS)

Photosynthesizers are the most crucial component in aPDT for the generation of ROS. Photosynthesizers are broadly classified as porphyrins, chlorins, and dyes. They are also classified as 1st,

2nd, and 3rd generation PS (Polat and Kang, 2021). The 1st generation PSs are the porphyrins that are soluble in water. Some of the second-generation PSs are toluidine blue (TB), 5′-aminolevulinic acid (ALA), toluidine blue, and methylene blue. Third-generation PSs are an active area of research that aims to minimise side effects, better bioavailability, and have been commonly used through conjugation or combinatory therapy. The concept of an ideal PS has been put forth as a PS with an absorption maximum between 650 and 800 nm, efficient membrane permeation, good ROS production, high selectivity, no dark toxicity, no self-aggregation, long storage, cost-effective, and optimal ADME profile (absorption, distribution, metabolism, and excretion) (Polat and Kang, 2021).

Compounds of natural and synthetic sources have been reported as efficient photosynthesizers against clinically relevant pathogens. Plant metabolites such as nano Quercetin are also promising PS for PDT that can reduce virulence factor expression in the biofilm-forming *S. mutans* (Pourhajibagher et al., 2022). Photosensitiser aloe-emodin (AE) from *Aloe vera* is also efficient against *C. albicans* cells that induce cell damage in a dose-dependent manner (Ma et al., 2020). In another independent study, AE could inhibit *A. baumannii* isolates by rupturing the cell membrane, compromising genomic DNA integrity, and condensing ribosomes in a light dose-dependent manner (Li et al., 2020). A recent report of chlorin-based aPDT was reported against otopathogens such as *Moraxella catarrhalis*, *Streptococcus pneumoniae*, and non-typeable *Haemophilus influenzae*, which are common infectious agents of middle ear disease Otitis media (OM). As a photosensitiser, chlorin e6 (Ce6) elicits activity against such otopathogens in planktonic and biofilm states (Luke-Marshall et al., 2020). Chlorin e(6) PDT can also inhibit *S. aureus* within 24 hrs of treatment (Park et al., 2012).

Dye-based photosynthesizers have received special interest. Rose bengal, a red-coloured dye, has also been reported as a promising photosensitiser for aPDT against oral pathogens in planktonic and biofilm forms such as *S. mutans*, *S. sobrinus*, *S. sanguinis*, and *L. salivarius* (Hirose et al., 2021). Similarly, toluidine blue ortho (TBO), a blue-coloured dye, has also been reported as a potential photosynthesizer for aPDT, yet its effectiveness as an antifungal remains uncertain. Although it has excellent activity in reducing planktonic *C. albicans*, complete eradication is seldom reported. Similarly, TBO-aPDT partially reduced the biofilm in terms of weight and extracellular polymeric substance (EPS) (Wiench et al., 2021).

7.4 NANOPARTICLES AS PS FOR aPDT

Research has also focused on reducing light exposure duration in aPDT. This has been addressed to some extent through the combinatorial therapy of aPDT and nanotechnology. Nanoparticles (NPs) can enhance the effectiveness of antimicrobial photodynamic therapy by increasing the photosensitisers' performance (Awad et al., 2022). Methylene blue (MB) combined with silver nanoparticles increases antimicrobial activity and reduces the activation energy by ~50% (Maliszewska et al., 2021). Similarly, graphene quantum dots loaded with curcumin increased ROS production and reduced CFU of MRSA, *P. aureus*, *E. Coli*, and C. *albicans* (Mushtaq et al., 2022). Synergistic effects of photosensitiser were also observed when propolis nanoparticle (PNP) was combined with photosensitiser toluidine blue O (TBO) and photosensitiser chlorophyllin-phycocyanin mixture. The combinatory effect was significant against *S. mutans* and their biofilm formation (Afrasiabi et al., 2020). Likewise, rose bengal as photosensitiser conjugated with lignin nano complex had higher *Candida tropicalis* inhibition activity when the noncomplex were doped in hydrogels (Chandna et al., 2020). aPDT based on Bengal functionalised chitosan nanoparticles (CSRBnp) completely inhibited the growth of *Enterococcus faecalis* after 24 hrs of treatment (Shrestha and Kishen 2014). Interestingly, gold and silver NPs as PS for PDT exhibit high antimicrobial activity (Songca and Adjei 2022).

Although there are several elements for nanocomplex development, carbon-based nanomaterials have been shown to have higher mechanical strength, optical properties, biocompatibility, and minimum toxicity. In this context, nanocomplex such as fullerenes, nanotubes, nanohorns, and nanodots have been explored for aPDT (Albert and Hsu, 2016). Methylene blue conjugated carbon nanotubes

(MBCNTs) are also an efficient agent against *S. aureus* and *E. coli* by inhibiting biofilm, EPS production, ROS generation, and targeting cell membrane in particular (Parasuraman et al., 2019). Similarly, enhancement of aPDT through toluidine blue/multiwalled carbon nanotube conjugate (TBCNT) has also shown promising improvements in cell viability reduction and biofilm inhibition (Anju et al., 2019). Single-walled carbon nanotubes (SWCNTs) conjugated with porphyrin interact with the *S. aureus* cells that damage the membrane (Sah et al., 2018).

7.5 aPDT-BASED COMBINATORY THERAPY WITH ANTIBIOTICS

Antimicrobial photodynamic therapy has shown high synergetic activity when administered in combination with other therapy such as photothermal, magnetic, and antibiotic therapy (Songca and Adjei 2022). The antibiofilm activity of aPDT provides an excellent window of opportunity for other antimicrobials to treat pathogens with biofilm properties. aPDT combinatorial therapy with antibiotics can also reduce the antibiotic dosage (Willis et al., 2022). For instance, toluidine blue (TB) based aPDT combined with gentamicin significantly increases its antimicrobial activity against MDR *S. aureus* compared to aPDT or gentamicin (Liu et al., 2020). The authors report that aPDT increases the intracellular ROS and accumulation of gentamicin by increasing the membrane permeability. Thus, the threshold requirement for gentamicin is reduced significantly. Treatment of *E. coli* with MB-aPDT (100 μM) and ampicillin promotes bacterial inhibition by destroying envelope architecture, forming bulges, invagination, and developing vesicles (Garcez et al., 2020). Furthermore, *in vivo* investigation of *E. coli* infected *Galleria mellonella* (caterpillar) did not exhibit significant survival when treated with aPTD or MB-aPDT compared to PBS. However, the application of antibiotics after aPTD significantly reduced the mortality rate (Garcez et al., 2020). Similarly, when combined with ciprofloxacin, 100 μg/mL Ce6 at a dose of 120 J/cm^2 significantly inhibited urinary tract infections (UTI) agent *E. coli* rods (Tichaczek-Goska et al., 2019).

Combining aPDT with other standard routine procedures has a significant synergistic outcome in treating peri-implantitis (Zhao et al., 2022). Combinatorial treatment for cutaneous and mucosal infections using antibiotics such as terbinafine and itraconazole fortified with 5-aminolevulinic acid or phenothiazinium dye-based aPDT increases the antibiotic penetrations that significantly reduce the treatment duration, drug dosage, and toxicity (Perez-Laguna et al., 2019). The combination of different antibiotics with 5-aminolevulinic acid-mediated aPDT has been reported to have a high antimicrobial activity (Songca and Adjei 2022). Similarly, aPDT reduces post-treatment complications in scaling and root planning (SRP). For instance, in a clinical trial of chronic periodontitis treatment consisting of two groups with 15 subjects each, the subjects treated with scaling and root planning (SRP) and indocyanine green-based photodynamic therapy had a significantly lower anaerobic culture (Sethi and Raut, 2019). A combination of antibiotics with aPDT can be additive or syngenetic. For instance, meso-tetra (4-aminophenyl) porphine aPTD with chloramphenicol and tobramycin have an additive effective against *S. aureus* while it is syngenetic against MRSA (Dastgheyb et al., 2013). Thus, aPDT can be used alongside routine clinical practice to minimise the occurrences of infection.

7.6 HYDROGELS AND LIPOSOMES AS A DRUG DELIVERY SYSTEM

Hydrogels are polymers forming a 3D structure that can absorb a large volume of water. Hydrogels by volume are composed of 99% water, ranging from 1 to 1000 nm (Silvestre et al., 2021). Hydrogels are efficient in bioadhesion and permeability, making them suitable for drug delivery and release. The efficiency of a drug is highly dependent on the drug delivery system (DDS) for controlled and optimum drug release in the target site. For this purpose, hydrogels have been used as a DDS with promising results (Goergen et al., 2019). Hydrogels help to rehydrate dead tissues, absorb exudates and blood, and keep the site moist while providing antimicrobial activity. Yet, hydrogel biocompatibility is one of the main limitations. Hydrogels biocompatibility arises due to unreacted monomers, cross-linkers, or undesirable physical properties. Low molecular weight gelators, particularly lipids, have been of interest for DDS. A class of bipolar lipids, also known as bolalipids, is one of such low molecular weight gelators possessing. In symmetric bolalipids, the hydrophilic head groups connected by the hydrophobic ends may be identical, while the head groups in asymmetric bolalipids differ in size, charges, and other properties. Bolalipids occur naturally in members of the Archaeal domain, particularly among the acidophilic and thermophilic extremophiles. However, naturally occurring bolalipids usually occur as a combination of various head groups and alkyl chains. When incorporated with methylene blue, artificial bolalipids made up of C32 alkyl chain can inhibit *Staphylococcus aureus* and *Saccharomyces cerevisiae* (Goergen et al., 2019). Hydrogels made of chitosan are efficient antimicrobials due to the inherited antimicrobial properties of chitosan. Chitosan hydrogels are positively charged; hence, they bind to the negatively charged bacterial membrane, leading to permeability modulation (Frade et al., 2018). When chitosan hydrogel was formulated with methylene blue (MB) for aPDT, they have been demonstrated to be active against *Propionibacterium acnes* in planktonic (12.5 μg/mL) as well as biofilm phase (75 μg/mL) (Frade et al., 2018). Meanwhile, thermosensitive hydrogels loaded with methylene blue (MB) can reduce MRSA colony forming unit (CFU) (Leung et al., 2020). However, higher hydrogel concentration reduced antimicrobial activity efficiencies due to the increased viscosity, leading to lower MB diffusion into the bacterial cells.

Liposomes are tiny, closed vesicles made of bilayer phospholipids. Hence, they provide room for drug targets and delivery. They also exhibit attractive properties such as biocompatibility and non-toxicity on the host, and hold large quantities of hydrophilic and hydrophobic drugs, which can be released in a controlled manner (Silvestre et al., 2021). The ability to modify their size, lipid composition, and surface charge enables controlled customisation of their pharmacokinetics and biodistribution in the host. Liposomes also increase the half-life of the PS and drugs. Thus, several independent researchers have attempted delivery of aPDT through liposomes. Niosomes vesicles with zinc phthalocyanine promote PDT that can inhibit pathogenic microbes and cancers (de Siqueira et al., 2022). Cationic liposomes, loaded with aluminium-chloride-phthalocyanine (AlClPc), have a stronger affinity towards the microbial cells than host cells and reduced the cariogenic bacteria load by up to 82% in caries lesions (Longo et al., 2012). Similarly, positively charged liposomes loaded with zinc phthalocyanine have been shown to bind and inhibit *Porphyromonas gingivalis* bacteria cells with high efficiency (Ko et al., 2013). Thus, liposomes are a strong candidate for drug delivery and aPDT of cancer therapy and oral infections.

7.7 APPROACHES TO ENHANCE aPDT

The importance of aPDT as an alternative to antibiotics is limited by several factors such as extended exposure time, low affinity to the bacterial cell, and overall requirement for increased efficiency. Hence, PS can be combined with other compounds to reduce the exposure time and increase its affinity to the bacterial cell. The photosynthesizer toluidine blue O (TBO), in combination with poly-beta-amino esters (PBAEs), reduced the aPDT exposure time by up to 30 folds compared to TBO alone (Perni et al., 2021). The combinatorial therapy of PBAEs and TBO was found to have no toxic effect on *in vitro* models. Other interesting approaches to increase the efficiency of aPDT are by conjugating the photosensitiser with a compound having a high affinity towards bacterial membrane or by maintaining the photosynthesizers diffusion rate. In this context, rose bengal efficiency was enhanced by binding with Concanavalin A (ConA) (Cantelli et al., 2020). The conjugation significantly increased the efficiency due to the strong binding affinity of the ConA with lipopolysaccharides. ConA-RB conjugates increased the uptake of the conjugate, production of superoxide and peroxide, and increased *E. coli* inhibition activity by up to 117 times. The synergistic effect of aPDT between copper sulfide (CuS) and indocyanine green (ICG) has been reported to increase up to four times against MRSA, *Candida albicans*, *Staphylococcus aureus*, and *Pseudomonas aeruginosa* (Garin et al., 2021).

Adding Potassium Iodide (KI) also enhances the methylene blue (MB) based aPDT by producing long-lived bactericidal I_2/I_3^-. This approach eliminated uropathogenic *Escherichia coli* (UPEC) in an *in vivo* mouse model (Huang et al., 2018). The aPDT based on five different cationic porphyrins was potentiated by KI against MRSA, *E. coli*, and *Candida albicans* (Vieira et al., 2019). Furthermore, combination with the KI reduced the treatment duration and PS concentration while achieving similar bactericidal activity. Similarly, H_2O_2 can enhance PDT by detaching polymicrobial biofilm, i.e., *Candida albicans* and *Streptococcus mutans* (Li et al., 2022).

7.8 CONCLUSION

Pathogens gain antibiotic resistance within a few years of introduction into the market. Antibiotic resistance continues to spread worldwide. Hence, antibiotics discovered through traditional approaches are prone to antibiotic resistance. In this context, antimicrobial photodynamic therapy (aPDT) is a suitable approach to combat antibiotic resistance. aPDT is a promising candidate to treat multidrug-resistant microbes since they act synergistically with the existing antibiotic and are less prone to resistance. Furthermore, it has multiple applications such as antibacterial, antifungal, antibiofilm, and anticancer activities. The 3rd generation of PSs is particularly promising due to the ability to control diffusion, conjugation of target specificity, drug delivery, and synergy. Although research on aPTD has witnessed considerable progress, its application in the current medical industry remains limited. Hence, in-depth *in vivo* and clinical studies emphasising its specificity and safety are paramount to ascertain its suitability for its application in medical settings.

CONFLICTS OF INTEREST

The authors do not declare any conflict of interest.

ETHICAL APPROVAL

Not applicable.

AUTHOR CONTRIBUTIONS

MI and JM prepared the draft manuscript and figures. MI, JM, SB, and RK critically refined and revised the manuscript. All the authors read and approved the manuscript.

ACKNOWLEDGEMENT

Authors thank Pondicherry University, Central University of Kerela, for the support provided during the manuscript preparation. MI thanks to DBT-RA Program in Biotechnology & Life Sciences for the RA fellowship.

REFERENCES

Afrasiabi, S., Pourhajibagher, M., Chiniforush, N., & Bahador, A. 2020. Propolis nanoparticle enhances the potency of antimicrobial photodynamic therapy against Streptococcus mutans in a synergistic manner. *Scientific Reports*, 10(1), 1–6.

Akhtar, F., Khan, A.U., Qazi, B., Kulanthaivel, S., Mishra, P., Akhtar, K., & Ali, A. 2021. A nano phototheranostic approach of toluidine blue conjugated gold silver core shells mediated photodynamic therapy to treat diabetic foot ulcer. *Scientific Reports*, 11(1), 1–19.

Albert, K., & Hsu, H.Y. 2016. Carbon-based materials for photo-triggered theranostic applications. *Molecules*, 21(11), 1585.

Alves, E., Melo, T., Simões, C., Faustino, M.A., Tomé, J.P., Neves, M.G., … & Almeida, A. 2013. Photodynamic oxidation of Staphylococcus warneri membrane phospholipids: new insights based on lipidomics. *Rapid Communications in Mass Spectrometry*, 27(14), 1607–1618.

Anju, V.T., Paramanantham, P., Sb, S.L., Sharan, A., Syed, A., Bahkali, N.A., Alsaedi, M.H., Kaviyarasu, K., & Busi, S. 2019. Antimicrobial photodynamic activity of toluidine blue-carbon nanotube conjugate against Pseudomonas aeruginosa and Staphylococcus aureus-understanding the mechanism of action. *Photodiagnosis and Photodynamic Therapy*, 27, 305–316.

Awad, M., Thomas, N., Barnes, T.J., & Prestidge, C.A. 2022. Nanomaterials enabling clinical translation of antimicrobial photodynamic therapy. *Journal of Controlled Release*, 346, 300–316.

Branco, T.M., Valerio, N.C., Jesus, V.I.R., Dias, C.J., Neves, M.G., Faustino, M.A., & Almeida, A. 2018. Single and combined effects of photodynamic therapy and antibiotics to inactivate Staphylococcus aureus on skin. *Photodiagnosis and Photodynamic Therapy*, 21, 285–293.

Calzavara-Pinton, P.G., Venturini, M., & Sala, R. 2007. Photodynamic therapy: update 2006 Part 1: Photochemistry and photobiology. *Journal of the European Academy of Dermatology and Venereology*, 21(3), 293–302.

Cantelli, A., Piro, F., Pecchini, P., Di Giosia, M., Danielli, A., & Calvaresi, M. 2020. Concanavalin A-Rose Bengal bioconjugate for targeted Gram-negative antimicrobial photodynamic therapy. *Journal of Photochemistry and Photobiology B: Biology*, 206, 111852.

Carrera, E.T., Dias, H.B., Corbi, S.C., Marcantonio, R.A., Bernardi, A.C., Bagnato, V.S., Hamblin, M.R., & Rastelli, A.N. 2016. The application of antimicrobial photodynamic therapy (aPDT) in dentistry: a critical review. *Laser Physics*, 26(12), 123001.

Celiešiūtė-Germanienė, R., Šimonis, P., & Stirkė, A. 2021. Antimicrobial photodynamic therapy (aPDT) for biofilm treatments. Possible synergy between aPDT and pulsed electric fields. *Virulence*, 12(1):2247.

Chandna, S., Thakur, N.S., Kaur, R., & Bhaumik, J. 2020. Lignin–bimetallic nanoconjugate doped pH-responsive hydrogels for laser-assisted antimicrobial photodynamic therapy. *Biomacromolecules*, 21(8), 3216–3230.

Cieplik, F., Deng, D., Crielaard, W., Buchalla, W., Hellwig, E., Al-Ahmad, A., & Maisch, T. 2018. Antimicrobial photodynamic therapy–what we know and what we don't. *Critical Reviews in Microbiology*, 44(5), 571–589.

Dai, T., Fuchs, B.B., Coleman, J.J., Prates, R.A., Astrakas, C., St. Denis, T.G., … & Tegos, G.P. 2012. Concepts and principles of photodynamic therapy as an alternative antifungal discovery platform. *Frontiers in Microbiology*, 3, 120.

Dastgheyb, S.S., Eckmann, D.M., Composto, R.J., & Hickok, N.J. 2013. Photo-activated porphyrin in combination with antibiotics: therapies against Staphylococci. *Journal of Photochemistry and Photobiology B: Biology*, 129, 27–35.

de Siqueira, L.B.D.O., dos Santos Matos, A.P., Feuser, P.E., Machado-de-Ávila, R.A., Santos-Oliveira, R., & Ricci-Júnior, E. 2022. Encapsulation of photosensitiser in niosomes for promotion of antitumor and antimicrobial photodynamic therapy. *Journal of Drug Delivery Science and Technology*, 68, 103031.

Di Poto, A., Sbarra, M.S., Provenza, G., Visai, L., & Speziale, P. 2009. The effect of photodynamic treatment combined with antibiotic action or host defence mechanisms on Staphylococcus aureus biofilms. *Biomaterials*, 30(18), 3158–3166.

Frade, M.L., De Annunzio, S.R., Calixto, G.M., Victorelli, F.D., Chorilli, M., & Fontana, C.R. 2018. Assessment of chitosan-based hydrogel and photodynamic inactivation against Propionibacterium acnes. *Molecules*, 23(2), 473.

Garcez, A.S., Kaplan, M., Jensen, G.J., Scheidt, F.R., Oliveira, E.M., & Suzuki, S.S. 2020. Effects of antimicrobial photodynamic therapy on antibiotic-resistant Escherichia coli. *Photodiagnosis and Photodynamic Therapy*, 32, 102029.

Garin, C., Alejo, T., Perez-Laguna, V., Prieto, M., Mendoza, G., Arruebo, M., Sebastian, V., & Rezusta, A. 2021. Chalcogenide nanoparticles and organic photosensitisers for synergetic antimicrobial photodynamic therapy. *Journal of Materials Chemistry B*, 9(31), 6246–6259.

Goergen, N., Wojcik, M., Drescher, S., Pinnapireddy, S.R., Brüßler, J., Bakowsky, U., & Jedelská, J. 2019. The use of artificial gel forming bolalipids as novel formulations in antimicrobial and antifungal therapy. *Pharmaceutics*, 11(7), 307.

Gong, N., Tan, Y., Li, M., Lu, W., & Lei, X. 2016. ALA-PDT combined with antibiotics for the treatment of multiple skin abscesses caused by Mycobacterium fortuitum. *Photodiagnosis and Photodynamic Therapy*, 15, 70–72.

Hirose, M., Yoshida, Y., Horii, K., Hasegawa, Y., & Shibuya, Y. 2021. Efficacy of antimicrobial photodynamic therapy with Rose Bengal and blue light against cariogenic bacteria. *Archives of Oral Biology*, 122, 105024.

Hu, X., Huang, Y.Y., Wang, Y., Wang, X., & Hamblin, M.R. 2018. Antimicrobial photodynamic therapy to control clinically relevant biofilm infections. *Frontiers in Microbiology*, 9, 1299.

Huang, Y.Y., Wintner, A., Seed, P.C., Brauns, T., Gelfand, J.A., & Hamblin, M.R. 2018. Antimicrobial photodynamic therapy mediated by methylene blue and potassium iodide to treat urinary tract infection in a female rat model. *Scientific Reports*, 8(1), 1–9.

Iluz, N., Maor, Y., Keller, N., & Malik, Z. 2018. The synergistic effect of PDT and oxacillin on clinical isolates of Staphylococcus aureus. *Lasers in Surgery and Medicine*, 50(5), 535–551.

Kikuchi, T., Mogi, M., Okabe, I., Okada, K., Goto, H., Sasaki, Y., … & Mitani, A. 2015. Adjunctive application of antimicrobial photodynamic therapy in nonsurgical periodontal treatment: a review of literature. *International Journal of Molecular Sciences*, 16(10), 24111–24126.

Ko, A., Yee, M., Skupin-Mrugalska, P., & Düzgünes, N. 2013. Photodynamic therapy of Porphyromonas gingivalis via liposome-encapsulated sensitizers. *Journal of the California Dental Association*, 41(11), 827–830.

Leung, B., Dharmaratne, P., Yan, W., Chan, B.C., Lau, C.B., Fung, K.P., Ip, M., & Leung, S.S. 2020. Development of thermosensitive hydrogel containing methylene blue for topical antimicrobial photodynamic therapy. *Journal of Photochemistry and Photobiology B: Biology*, 203, 111776.

Li, J., Qin, M., Liu, C., Ma, W., Zeng, X., & Ji, Y. 2020. Antimicrobial photodynamic therapy against multidrug-resistant Acinetobacter baumannii clinical isolates mediated by aloe-emodin: An *in vitro* study. *Photodiagnosis and Photodynamic Therapy*, 29, 101632.

Li, Y., Du, J., Huang, S., Wang, S., Wang, Y., Cai, Z., Lei, L., & Huang, X. 2022. Hydrogen peroxide potentiates antimicrobial photodynamic therapy in eliminating Candida albicans and Streptococcus mutans dual-species biofilm from denture base. *Photodiagnosis and Photodynamic Therapy*, 37, 102691.

Liu, S., Mai, B., Jia, M., Lin, D., Zhang, J., Liu, Q., & Wang, P. 2020a. Synergistic antimicrobial effects of photodynamic antimicrobial chemotherapy and gentamicin on Staphylococcus aureus and multidrug-resistant Staphylococcus aureus. *Photodiagnosis and Photodynamic Therapy*, 30, 101703.

Liu, X., Chang, Q., Ferrer-Espada, R., Leanse, L.G., Goh, X.S., Wang, X., Gelfand, J.A., & Dai, T. 2020b. Photoinactivation of moraxella catarrhalis using 405-nm blue light: Implications for the treatment of otitis media. *Photochemistry and Photobiology*, 96(3), 611–617.

Liu, X., Liu, S., Mai, B., Su, X., Guo, X., Chang, Y., Dong, W., Wang, W., & Feng, X. 2022. Synergistic gentamicin-photodynamic therapy against resistant bacteria in burn wound infections. *Photodiagnosis and Photodynamic Therapy*, 39, 103034.

Longo, J.P.F., Leal, S.C., Simioni, A.R., de Fátima Menezes Almeida-Santos, M., Tedesco, A.C., & Azevedo, R.B. 2012. Photodynamic therapy disinfection of carious tissue mediated by aluminum-chloride-phthalocyanine entrapped in cationic liposomes: an *in vitro* and clinical study. *Lasers in Medical Science*, 27(3), 575–584.

Luke-Marshall, N.R., Hansen, L.A., Shafirstein, G., & Campagnari, A.A. 2020. Antimicrobial photodynamic therapy with chlorin e6 is bactericidal against biofilms of the primary human otopathogens. *Msphere*, 5(4), e00492–20.

Luke-Marshall, N.R., Mang, T.S., Hansen, L.A., & Campagnari, A.A. 2014. Moraxella catarrhalis is susceptible to antimicrobial photodynamic therapy with Photofrin. *Lasers in Surgery and Medicine*, 46(9), 712–717.

Ma, W., Liu, C., Li, J., Hao, M., Ji, Y., & Zeng, X. 2020. The effects of aloe emodin-mediated antimicrobial photodynamic therapy on drug-sensitive and resistant Candida albicans. *Photochemical & Photobiological Sciences*, 19(4), 485–494.

Maliszewska, I., Wanarska, E., Thompson, A.C., Samuel, I.D., & Matczyszyn, K. 2021. Biogenic gold nanoparticles decrease methylene blue photobleaching and enhance antimicrobial photodynamic therapy. *Molecules*, 26(3), 623.

Mushtaq, S., Yasin, T., Saleem, M., Dai, T., & Yameen, M.A. 2022. Potentiation of antimicrobial photodynamic therapy by curcumin-loaded graphene quantum dots. *Photochemistry and Photobiology*, 98(1), 202–210.

Nieves, I., Hally, C., Viappiani, C., Agut, M., & Nonell, S. 2020. A porphycene-gentamicin conjugate for enhanced photodynamic inactivation of bacteria. *Bioorganic Chemistry*, 97, 103661.

Parasuraman, P., Anju, V.T., Lal, S.B., Sharan, A., Busi, S., Kaviyarasu, K., Arshad, M., Dawoud, T., & Syed, A. 2019. Synthesis and antimicrobial photodynamic effect of methylene blue conjugated carbon nanotubes on *E. coli* and S. aureus. *Photochemical & Photobiological Sciences*, 18(2), 563–576.

Park, J.H., Ahn, M.Y., Kim, Y.C., Kim, S.A., Moon, Y.H., Ahn, S.G., & Yoon, J.H. 2012. *In vitro* and *in vivo* antimicrobial effect of photodynamic therapy using a highly pure chlorin e6 against Staphylococcus aureus Xen29. *Biological and Pharmaceutical Bulletin*, 35(4), 509–514.

Perez-Laguna, V., Gilaberte, Y., Millán-Lou, M.I., Agut, M., Nonell, S., Rezusta, A., & Hamblin, M.R. 2019. A combination of photodynamic therapy and antimicrobial compounds to treat skin and mucosal infections: a systematic review. *Photochemical & Photobiological Sciences*, 18(5), 1020–1029.

Perez-Laguna, V., García-Luque, I., Ballesta, S., Rezusta, A., & Gilaberte, Y. 2021. Photodynamic therapy combined with antibiotics or antifungals against microorganisms that cause skin and soft tissue infections: A planktonic and biofilm approach to overcome resistances. *Pharmaceuticals*, 14(7), 603.

Perez-Laguna, V., Pérez-Artiaga, L., Lampaya-Pérez, V., García-Luque, I., Ballesta, S., Nonell, S., ... & Rezusta, A. 2017. Bactericidal effect of photodynamic therapy, alone or in combination with mupirocin or linezolid, on Staphylococcus aureus. *Frontiers in Microbiology*, 8, 1002.

Perni, S., Preedy, E.C., & Prokopovich, P. 2021. Amplify antimicrobial photo dynamic therapy efficacy with poly-beta-amino esters (PBAEs). *Scientific Reports*, 11(1), 1–15.

Polat, E., & Kang, K. 2021. Natural photosensitisers in antimicrobial photodynamic therapy. *Biomedicines*, 9(6), 584.

Pourhajibagher, M., Alaeddini, M., Etemad-Moghadam, S., Rahimi Esboei, B., Bahrami, R., Mousavi, M., & Bahador, A. 2022. Quorum quenching of Streptococcus mutans via the nano-quercetin-based antimicrobial photodynamic therapy as a potential target for cariogenic biofilm. *BMC Microbiology*, 22(1), 1–18.

Pourhajibagher, M., Kazemian, H., Chiniforush, N., Hosseini, N., Pourakbari, B., Azizollahi, A., ... & Bahador, A. 2018. Exploring different photosensitisers to optimize elimination of planktonic and biofilm forms of Enterococcus faecalis from infected root canal during antimicrobial photodynamic therapy. *Photodiagnosis and Photodynamic Therapy*, 24, 206–211.

Rajesh, S., Koshi, E., Philip, K., & Mohan, A. 2011. Antimicrobial photodynamic therapy: An overview. *Journal of Indian Society of Periodontology*, 15(4), 323.

Sah, U., Sharma, K., Chaudhri, N., Sankar, M., & Gopinath, P. 2018. Antimicrobial photodynamic therapy: Single-walled carbon nanotube (SWCNT)-Porphyrin conjugate for visible light mediated inactivation of Staphylococcus aureus. *Colloids and Surfaces B: Biointerfaces*, 162, 108–117.

Sethi, K.S., & Raut, C.P. 2019. Antimicrobial photodynamic therapy using indocyanine green as a photosensitiser in treatment of chronic periodontitis: a clinico-microbial study. *Indian Journal of Dental Research*, 30(6), 870.

Shih, M.H., & Huang, F.C. 2011. Effects of photodynamic therapy on rapidly growing nontuberculous mycobacteria keratitis. *Investigative Ophthalmology & Visual Science*, 52(1), 223–229.

Shih, Y.H., Yu, C.C., Chang, K.C., Tseng, Y.H., Li, P.J., Hsia, S.M., ... & Shieh, T.M. 2022. Synergistic effect of combination of a temoporfin-based photodynamic therapy with potassium iodide or antibacterial agents on oral disease pathogens *in vitro*. *Pharmaceuticals*, 15(4), 488.

Shrestha, A., & Kishen, A. 2014. Antibacterial efficacy of photosensitiser functionalized biopolymeric nanoparticles in the presence of tissue inhibitors in root canal. *Journal of Endodontics*, 40(4), 566–570.

Silvestre, A.L.P., Di Filippo, L.D., Besegato, J.F., de Annunzio, S.R., de Camargo, B.A.F., de Melo, P.B.G., de Souza Rastelli, A.N., Fontana, C.R., & Chorilli, M. 2021. Current applications of drug delivery nanosystems associated with antimicrobial photodynamic therapy for oral infections. *International Journal of Pharmaceutics*, 592, 120078.

Singh, A.K., Kaur, R., Verma, S., & Singh, S. 2022. Antimicrobials and antibiotic resistance genes in water bodies: Pollution, risk, and control. *Frontiers in Environmental Science*, 10, 397.

Songca, S.P., & Adjei, Y. 2022. Applications of antimicrobial photodynamic therapy against bacterial biofilms. *International Journal of Molecular Sciences*, 23(6), 3209.

Sun, K., Yang, H., Huang, X., Gong, N., Qin, Q., Lu, W., & Lei, X. 2017. ALA-PDT combined with antibiotics for the treatment of atypical mycobacterial skin infections: Outcomes and safety. *Photodiagnosis and Photodynamic Therapy*, 19, 274–277.

Tichaczek-Goska, D., Wojnicz, D., Symonowicz, K., Ziółkowski, P., & Hendrich, A.B. 2019. Photodynamic enhancement of the activity of antibiotics used in urinary tract infections. *Lasers in Medical Science*, 34(8), 1547–1553.

Van Goethem, M.W., Pierneef, R., Bezuidt, O.K., Van De Peer, Y., Cowan, D.A., & Makhalanyane, T.P. 2018. A reservoir of 'historical' antibiotic resistance genes in remote pristine Antarctic soils. *Microbiome*, 6(1), 1–12.

Vieira, C., Santos, A., Mesquita, M.Q., Gomes, A.T., Neves, M.G.P., Faustino, M.A.F., & Almeida, A. 2019. Advances in aPDT based on the combination of a porphyrinic formulation with potassium iodide: Effectiveness on bacteria and fungi planktonic/biofilm forms and viruses. *Journal of Porphyrins and Phthalocyanines*, 23(4–5), 534–545.

Wardlaw, J.L., Sullivan, T.J., Lux, C.N., & Austin, F.W. 2012. Photodynamic therapy against common bacteria causing wound and skin infections. *The Veterinary Journal*, 192(3), 374–377.

Wiench, R., Skaba, D., Matys, J., & Grzech-Leśniak, K. 2021. Efficacy of toluidine blue—Mediated antimicrobial photodynamic therapy on Candida spp. A systematic review. *Antibiotics*, 10(4), 349.

Willis, J.A., Cheburkanov, V., Chen, S., Kassab, G., Soares, J.M., Blanco, K.C., Bagnato, V.S., de Figueiredo, P., & Yakovlev, V.V. 2022, March. Antimicrobial Photodynamic Therapy Combined with Antibiotics Reduces Resistance and Aids Elimination in Four Resistant Bacterial Strains. In *Photonic Diagnosis, Monitoring, Prevention, and Treatment of Infections and Inflammatory Diseases 2022* (Vol. 11939, pp. 5–14). SPIE.

Youf, R., Müller, M., Balasini, A., Thétiot, F., Müller, M., Hascoët, A., Jonas, U., Schönherr, H., Lemercier, G., Montier, T., & Le Gall, T. 2021. Antimicrobial photodynamic therapy: Latest developments with a focus on combinatory strategies. *Pharmaceutics*, 13(12), 1995.

Zhao, T., Song, J., Ping, Y., & Li, M. 2022. The application of antimicrobial photodynamic therapy (aPDT) in the treatment of peri-implantitis. *Computational and Mathematical Methods in Medicine*, 2022, 8.

8 Photosensitisers and their Role in Antimicrobial Photodynamic Therapy

Paramanantham Parasuraman and Siddhardha Busi
Pondicherry University, Puducherry, India

CONTENTS

8.1 INTRODUCTION

Infectious diseases can be understood simply as impairment of the normal physiological function of the living organism due to the attack of pathogenic microorganisms or their toxic substances. Humans struggled to combat infectious diseases, and in ancient times many lives were lost to contagious diseases like smallpox. Due to the advancement in science and technology in more recent times, we humans have gained a sound knowledge of infectious diseases and sufficient technology for the early determination of pathogen and treatment procedures (Halili et al., 2016). However, these infectious diseases are still an immense global burden that significantly impacts public health and economies worldwide. The recent encounter with COVID-19 is the best example of this (Shang et al., 2021). This situation alarmed the scientific community and healthcare sectors to pay significant attention to contagious diseases.

Even before the COVID-19 crisis, the scientific community was faced with a significant issue in the form of growing bacterial resistance to conventional treatment methods. Arguably, the discovery of antibiotics is the most outstanding human achievement, and it has led to control of numerous infectious diseases, and the saving of many lives. But in recent years, antibiotic treatment response

DOI: 10.1201/9781003345299-8

has decreased considerably due to the resistance mechanism of infectious pathogens (Aminov, 2010). In 2014, the United Kingdom government published a report on over 700,000 death cases yearly, reaching 10 million in 2050 due to antimicrobial resistance (de Kraker et al., 2016). These resistance mechanisms in the pathogen that cause acute and chronic diseases worsen the condition and transfer the resistance gene to other microbial communities by horizontal gene transfer. Due to the lack of new antibiotics in recent years to the standoff with pathogenic microorganisms leading to the pre-antibiotic era, this becomes an alarming situation where the antibiotics era may be approaching its end (Warnes et al., 2012). Even the World Health Organization (WHO) warned in April 2014 that we are approaching the "post-antibiotic era" (Viens and Littmann, 2015). This alarming situation brings an urgent necessity to develop new and alternative cures to control infectious pathogens and antimicrobial-resistant microorganisms. Moreover, these alternative cures should prevent the emergence of resistance by a multi-target mode of action, unlike the key-lock mechanism known from antibiotics.

Antimicrobial photodynamic therapy (aPDT) was suggested as an alternative approach to antibiotic treatment. Unlike conventional antibiotics, it encounters the pathogen in a multi-target mode of action, which creates less chance of possessing selective pressure on the pathogen (Liu et al., 2015). This concept dates back to the beginning of the last century; however, its potential has been underestimated or even neglected. Recently, aPDT has gained significant attention as an alternative therapeutic option for some localised infections and extensive research has been initiated (Mahmoudi et al., 2018). In dentistry, the application of aPDT has been commenced positively due to its significant results in periodontitis treatment and root canal disinfection; it was observed as a promising therapeutic method (Carrera et al., 2016).

The working principle of aPDT has based on the interaction between the photosensitiser (PS), specific light, and molecular oxygen. PS involves the transfer of energy to oxygen, subsequently leading to the development of reactive oxygen species (ROS) (Ghorbani et al., 2018). Since the reactive oxygen species are cytotoxic in nature, the treatment site is disinfected. Although these methods render several advantages, site specificity was found to be the most significant. Here, the nontoxic PS was first administered at the targeted area (localised infection), and the specific light source would be flashed at the desired location to activate the PS (Abrahamse and Hamblin, 2016). In the aPDT method, selecting the PS and the light source were crucial steps that could influence the treatment results. Several groups of compounds were investigated to determine the effective PS, and many compounds showed photodynamic effects in the presence of light (de Freitas et al., 2018; Perni et al., 2021). This chapter gives insight into the updated list of PS compounds that have been proven significant for antimicrobial photodynamic therapy.

8.2 A BRIEF HISTORY OF aPDT

The concept of aPDT was not new to humans because phototherapy has been employed in ancient times in several parts of the world, including Greece, Egypt, and India. In ancient Egypt, plant extracts and sunlight were used to treat skin diseases like leprosy lesions (Ackroyd et al., 2001; Mitton and Ackroyd, 2005). Soon after, the concept of phototherapy either disappeared, or historical pieces of evidence have not appeared yet. At the end of the 19th century, Oscar Raab first found that dye substance, like acridine, kills the paramecium in the presence of light. He found that incubating the paramecium with acridine in light inhibits the growth of paramecia. However, in a similar experiment of acridine in the dark, no change was made in the growth of paramecia. In the following year, Von Tappeiner was the first person to coin the term "photodynamic action", and explains the importance of oxygen in photodynamic therapy (Ghorbani et al., 2018).

In 1904, the first photodynamic therapy was performed on a human with skin cancer. It was executed by T. Appaeiner and H. Jesionek, and used Eosin and white light. Similarly, in the early 20th century, Niels Finsen, a Danish physician, successfully employed PDT with the help of an arc lamp to treat *lupus vulgaris*. During these periods, photodynamic therapy was extensively used to treat cancer. From 1960 onward, photodynamic therapy was applied to treat microbial infections, termed antimicrobial photodynamic therapy. Several groups of compounds were reported as PS, among a particular group of compounds like porphyrins, chlorins, bacteriochlorins, and phthalocyanines were found to be promising (Masiera et al., 2017). In the beginning, toluidine blue was used as PS against different microorganisms, including bacteria, algae, and yeast. Due to the advancement in science, some more dyes like methylene blue, rose bengal, eosine Y, neutral red, acridine orange, crystal violet, and rhodamine 6G have been reported as promising PS (Ghorbani et al., 2018; Carrera et al., 2016). A maximum of the dyes mentioned above are cationic compounds that hold a positive charge in their functional group and easily bind with a negatively charged surface like a bacterial cell wall (Qiu et al., 2020). However, entering penicillin and other antibiotics into the microbial infection treatment process impeded the development of aPDT technology. The crisis of emergence of multidrug-resistant bacteria has been attributed to the overuse and misuse of antibiotics that created a situation to look for alternative therapeutic methods such as aPDT.

8.3 ADVANTAGES OF aPDT

Compared to antibiotics, aPDT has a number of benefits. PDT is considered to be threefold site-specific due to the predominant binding of PS by targeted cells relative to other cells, less toxicity of non-irradiated PS, and the site-directed irradiation of the infected site being the first significant benefit. As a result, toxicity is mostly absent in non-irradiated areas. Another significant advantage is tha no resistance is created against the PS in the aPDT process. In this regard, repeated application of aPDT has is less likely to develop resistant strains. This is supported by four factors:

1. Insufficient time between the administration of PS and aPDT treatment for the pathogen to acquire resistance mechanism.
2. Photosensitiser does not show dark toxicity, and pathogens are not required to apply adaptive survival strategies. Additionally, bacteria find it challenging to detect that the usually nontoxic PS is the source of oxidative stress (Schastak et al., 2010).
3. aPDT caused significant damage to the cells, making it impossible for them to transfer cross-generational adaptivity.
4. Unlike traditional antibiotics, aPDT does not target a single site of the pathogen. The different bacterial cell structures and distinct metabolic processes are targeted by the ROS produced by aPDT.

These factors support the potential value of aPDT as a minimally intrusive, patient-friendly method of treating resistant strains (Y. Liu et al., 2015).

8.4 TYPES OF PHOTOSENSITISERS USED IN aPDT

As reported, aPDT utilises three major components: nontoxic and non-thermal photochemical, light, and molecular oxygen. Activation of PS in the presence of light, interacting with molecular oxygen, produces ROS that subsequently causes the microbicidal effect in the treatment site. The administered PS should be retained at length on the treatment site for photoactivation and production of phototoxic products. The delay time of the PS at the treatment site depends mainly upon the concentration of accumulated PS, which varies among different groups of PS. Hence, selecting the appropriate PS is a crucial step in aPDT that significantly determines the effect of aPDT.

The promising photosensitiser should have the following properties:

- Stable at room temperature.
- High phototoxicity and minimum dark toxicity.
- Effectively absorb light at a specific wavelength, preferably between ~600 and 800 nm, penetrating deeply into tissue.

- Have enough energy in the triplet state to transmit that energy to the ground state.
- Have the necessary quantum yield and extended lifespan in the triplet state.
- High photostability.
- Inexpensive and commercially available for effective utilisation.

Different groups of PS have been employed in aPDT based on their structure and origin. Photosensitisers are usually grouped as porphyrin or non-porphyrin-based compounds. Porphyrin-based compounds are further grouped as first, second, and third generation PS. Photosensitisers like hematoporphyrin and Photofrin® come under first-generation porphyrin derivatives. The second generation PS are the purified forms of 1st generation PS that absorb light at a longer wavelength. The third generation PS are formulated onto the carriers, including antibodies liposomes that increase site selectivity (O'Connor et al., 2009).

The optimal combination of PS and the typical light influences the desired effect of aPDT. Additionally, the types of microorganisms also potentially influence the photosensitiser accumulation and impact of aPDT. Gram-positive bacteria are significantly more sensitive to anionic and neutral PS because of their unique cell membrane structure with a thick peptidoglycan layer (Pang et al., 2020). Due to the additional outer membrane and permeability barrier provided by lipopolysaccharides, Gram-negative bacteria are less likely to absorb anionic or neutral PS substances (Sperandio et al., 2013).

The cationic PS are preferred in aPDT against both Gram-positive and Gram-negative bacteria. Cationic phenothiazinium, porphyrin and phthalocyanine derivatives have been reported, and they exhibited promising photodynamic effects on both the Gram species. Nevertheless, the higher concentration of the anionic or neutral dyes potentially enhanced the aPDT effect than the cationic photo-sensing dye. Although the accumulation of PS in the bacteria is lower, a higher concentration of PS near the cell membrane allows the generated ROS to affect the cell membrane. Moreover, the generated ROS manage to diffuse into the cell and damage the internal cellular components. Several PS were reported with greater inhibition efficacy of the growth of the Gram-positive bacteria in their native chemical state. However, in some instances, the functional group of the PS were modified to improve their aPDT efficacy against Gram-negative bacteria (Liu et al., 2015).

8.4.1 Phenothiaziniums

Photosensitisers like methylene blue, rose bengal, and toluidine blue O are phenothiazinium derivatives, and come under the first-generation group of PS. These PS are initially employed as PS in the tumour treatment process. Later these PS are also used in aPDT due to their greater binding affinity towards Gram species, including methicillin-susceptible *staphylococcus aureus* (MSSA), methicillin-resistant *S. aureus* (MRSA) and *Escherichia coli* (Halili et al., 2016; Tseng et al., 2017). In particular cases, phenothiazinium PS showed inherent toxicity by showing minimum growth reduction to *E. coli* without light (Phoenix et al., 2003). Under light conditions, phenothiazinium PS can even damage the cytoplasmic DNA of *E. coli* (Digby et al., 2021). Some of the phenothiazinium derivatives and their antimicrobial photodynamic properties are listed in Table 8.1. The chemical structure of some of the phenothiazinium derivatives are illustrated in Figure 8.1.

Methylene blue was the first phenothiazinium dye discovered by Heinrich Caro in 1870. In most cases, this dye follows the type I mechanism due to its singlet oxygen quantum yielding properties. They absorb the light with a wavelength range from ~600–680 nm, which is helpful for their use as PS since longer wavelength light penetrates tissues more effectively (Huang et al., 2019). Moreover, the chemical structure of the methylene blue was modified in its functional group with a cationic functional group and delivered two different compounds: dimethyl methylene blue and methylene green. These modified compounds have enhanced cationic strength, facilitating greater binding affinity towards bacterial cells and improving the efficacy of aPDT than native dye (Gollmer et al., 2015; Paz-Cristobal et al., 2014).

TABLE 8.1
Phenothiazinium PS and Their Antimicrobial Photodynamic Properties

S. No.	Name of the PS	Charge	Excitation Wavelength (nm)	Test Pathogen	aPDT Efficacy	Reference
1	Methylene blue	Cationic	632	*S. aureus* *E. coli*	~2 log ~1 log	Donnelly et al. (2009)
2	Methylene blue	Cationic	632	Dental plaque sample	32%	Fontana et al. (2009)
3	Methylene blue	Cationic	632	*C. albicans*	≥2 log	Rossoni et al. (2014)
4	Methylene green	Cationic	654	*E. faecalis*	Total elimination	Wainwright et al. (1997)
5	Toluidine blue	Cationic	635	*S. aureus* *E. coli*	>3 log >3 log	Kasimova et al. (2014)
6	Toluidine blue	Cationic	635	*S. mutans*	2–5 log	Zanin et al. (2005)
7	Toluidine blue	Cationic	635	*S. mutans* *S. sobrinus* *S. sanguinis*	~1 log ~1 log ≥3 log	Zanin et al. (2006)
8	Toluidine blue	Cationic	635	*S. mutans*	≥5 log	Teixeira et al. (2012)
9	Toluidine blue	Cationic	635	*E. faecalis*	Total elimination	Zand et al. (2014)
10	Toluidine blue	Cationic	635	*E. faecalis*	<1 log	Meire et al. (2012)
11	New methylene blue	Cationic	410	*A. baumannii*	6 log	Ragàs et al. (2010)
12	Dimethylmethylene blue	Cationic	410	*A. baumannii*	6 log	Ragàs et al. (2010)
13	Rose Bengal	Anionic	532	*S. aureus* *E. coli*	~4 log ~4 log	Turbay et al. (2014)

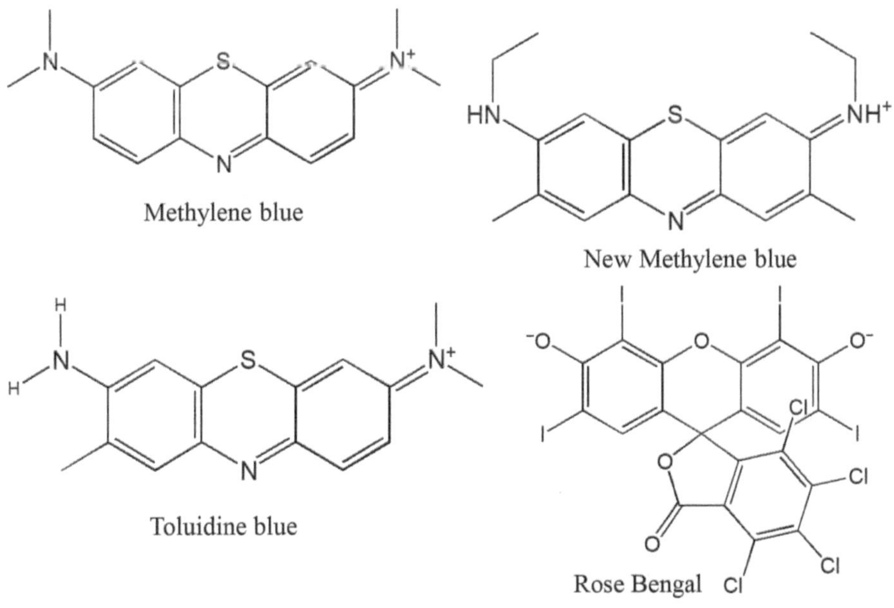

FIGURE 8.1 Chemical structure of some of the commonly used phenothiazinium derivatives in aPDT.

Similarly, other research was conducted to chemically alter the structure of the methylene blue to develop novel derivative compounds that potentially act as PS in aPDT better than its parental compound. For instance, methylene blue and its two derivatives were investigated for their toxicities and phototoxicity against vancomycin-resistant *Enterococcus* spp. The derivative compounds showed enhanced photobactericidal activity against vancomycin-resistant *Enterococcus* spp. due to the greater hydrophobicities nature than native methylene blue. The derivatives also showed more intrinsic photosensitising effects (Wainwright et al., 1999). Furthermore, several structural modifications were made to the native structure of methylene blue to reduce the drug-light interval and target only the pathogen without affecting the host tissue (Hamblin and Dai, 2010).

Toluidine blue O is another phenothiazinium salt that showed a promising aPDT effect against several pathogenic microorganisms (Nakano et al., 2011; Tandukar et al., 2013). Toluidine blue O and its derivatives were extensively investigated and reported as potential PS in aPDT process. Past two decades, extensive research has been conducted on toluidine blue O, and several novel derivatives were formulated with an exceptional aPDT activity compared to its native compound. Several derivatives of toluidine blue O have played a significant role as promising PS in dentistry treatment (Ichinose-Tsuno et al., 2014; Afkhami et al., 2020). For example, de Oliveira and co-workers reported a pilot study to determine the effect of toluidine blue-mediated aPDT against dental microcosms. The study results showed that the aPDT exhibited a significant reduction in bacteria with a dose dependent efficacy (de Oliveira et al., 2018). Efflux pumps are the one defence mechanism of multidrug-resistant bacteria that quickly eject the bacterial cell surface accumulated PS, giving rise to poor aPDT results.

In most cases, phenothiazinium group PS lose their efficacy due to the efflux pump mechanisms of the bacteria. This limitation could be overcome by combining the PS and efflux pump inhibitors. Toluidine blue O mediated aPDT showed significant bactericidal action on *S. aureus* and *Pseudomonas aeruginosa* when co-incubated with several efflux pump inhibitors (EPIs), targeting NorA and Mex-AB (Tegos et al., 2008). However, EPIs are often toxic, and this strategy must be researched further. Interestingly, a recent study has proved that efflux pump encoding genes of methicillin-resistant *S. aureus* were suppressed by using sub-lethal photodynamic therapy. In the study, it was observed that sub-lethal toluidine blue O mediated aPDT significantly reduced the expression of efflux pump genes and impede biofilm formation in MRSA cells (Yu et al., 2022). However, the exact mechanism of sub-lethal toluidine blue O mediated aPDT help to down-regulate the expression of the efflux pump gene requires detailed investigation.

8.4.2 Porphyrins

Porphyrins are one of the most widely utilised PS due to their high frequency, competitively high triplet state quantum yield, increased rate of ROS production, and simple chemical modification. Porphyrins required ~405–550 nm wavelength light for their excitation. Due to their aromatic character, porphyrins typically undergo electrophilic substitution processes at the meso locations, which are the most electron-dense and reactive (Amos-Tautua et al., 2019). The toxicity of porphyrins *in vitro* and *in vivo* is generally low, and can be modified to be water-soluble or insoluble (Mondal and Bera, 2014). Some of the porphyrin derivatives and their antimicrobial photodynamic properties are listed in Table 8.2. The chemical structure of some of the porphyrins derivatives are illustrated in Figure 8.2.

Numerous studies have demonstrated that porphyrins are effective PS for usage in aPDT. At very low concentrations (0.1–5 M), porphyrins have exhibited considerable antimicrobial activity against a broad spectrum of bacteria due to their strong propensity for attaching to biological constituents like membranes, protein and DNA (Kou et al., 2017; Nyman and Hynninen, 2004). In addition, porphyrins have a wider range of therapeutic windows in which they effectively inhibit the different bacterial groups, including multidrug resistance bacteria, without harming host cells. Furthermore, a vast array of mechanisms mediates the effects of porphyrins on microbial and viral pathogens, and the development of the resistance was significantly reduced (Stojiljkovic et al., 2001).

TABLE 8.2
Porphyrins PS and Their Antimicrobial Photodynamic Properties

S. No.	Name of the PS	Excitation Wavelength (nm)	Test Pathogen	aPDT Efficacy	Reference
1	5,10,15,20-tetrakis (1-methylpyridinium-4-yl) porphyrin tetra-iodide	546	*E. coli* *S. warneri*	0.2 log 2.8 log	Lambrechts et al. (2004)
2	5,10,15-tris(1-methylpyridinium-4-yl):20-(pentafluorophenyl) porphyrin tri-iodide	422	*E. coli* *S. warneri*	3.1 log 5.6 log	Lambrechts et al. (2004)
3	Hematoporphyrin monomethylether	405	*S. aureus*	90%	Sun et al. (2013)
4	Zn(II) N-alkylpyridylporphyrins	415	*E. coli*	~6 log	Alenezi et al. (2017)
5	5-(4-pentafluorophenyl)-10,15,20-tris[4-(3-*N,N*-dimethylaminopropoxy) phenyl]porphyrin	420	*S. aureus* *E. coli* *C. albicans*	~7 log ~7 log 5 log	Scanone et al. (2018)

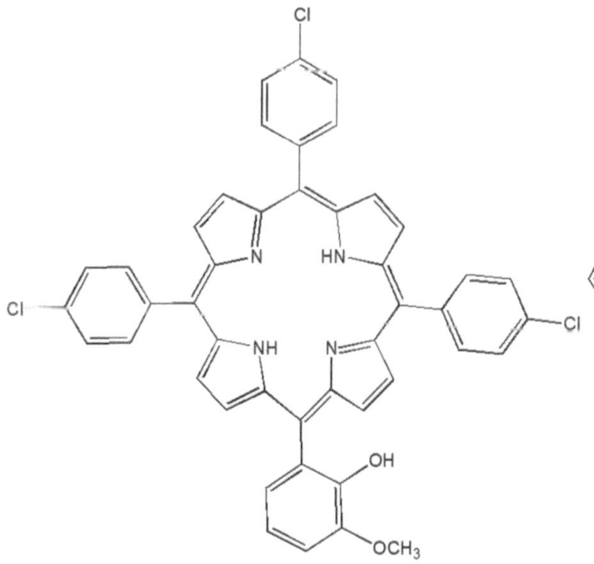

5,10,15-Tris(p-chlorophenyl)-20-(2-hydroxy-3-methoxyphenyl)-21H,23H-porphyrin

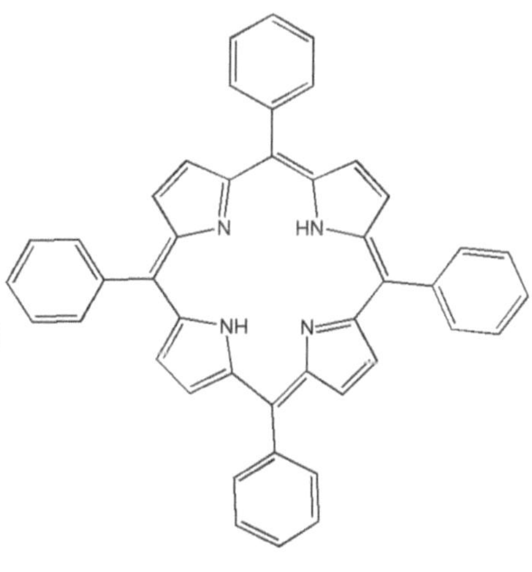

Tetraphenylporphyrin

FIGURE 8.2 Chemical structure of some of the commonly used porphyrins in aPDT.

Porphyrins are positively charged compounds that play a crucial role in binding to the bacterial cell surface and potentially influence the aPDT efficacy. Because of their cellular structure, some bacteria (e.g., *Propionibacterium acnes*, *Bacteroides* species, *Porphyromonas gingivalis*, *Prevotella* spp, and *Aggregatibacter actinomycetemcomtans*) allow these PS to accumulate in large amounts on their surface, making them more susceptible during the aPDT process (Ghorbani et al., 2018).

In some instances, porphyrins are in the metabolisms of the bacteria and are called endogenous porphyrins. These endogenous porphyrins get excited in the presence of blue light and perform the aPDT process without adding external porphyrins (Cieplik et al., 2014). Since porphyrins are positively charged compounds, they feasibly interact with Gram-positive cells and provide promising aPDT effects. For example, the study has shown that the growth of the *Staphylococcus aureus* was inhibited on aPDT treatment with deuteroporphyrin as PS. In contrast, the development of *E. coli* and *P. aeruginosa* could not be inhibited with deuteroporphyrin alone (Orenstein et al., 1997). Malik and his co-workers could overcome this issue by pre-treating the bacterial cells with EDTA or PMBN (Malik et al., 1992).

Some porphyrins like TMPyP (1-methyl-4-piridium-tetra (p-toluenesulfonate)) have more positive charges that facilitate ambulation onto the Gram-negative cells in large quantities and lead to bacterial inactivation. For example, Hanagova and co-workers investigated the aPDT effect of TMPyP against Gram-positive and Gram-negative bacteria, *S. aureus* and *E. coli*, respectively. The work results demonstrated that TMPyP effectively inhibits the growth of both bacteria (Hanakova et al., 2014). Similarly, other studies have been conducted to analyse the photosensitising capability of aryl cationic porphycenes in aPDT against bacterial and fungal cells. The study results show that aryl cationic porphycenes are promising photosensitisers that potentially inhibit bacterial and fungal cells (Ragàs et al., 2013). Additionally, XF73, a novel porphyrin-based photosensitiser, has been reported. This XF73 photosensitiser effectively destroyed MRSA without harming healthy human cells. This study also claimed that this PS can suppress MRSA infection in hospitals and on open wounds such as burns (T. Maisch et al., 2007). Tricationic porphycene 2,7,12-tris (trimethyl-*p*-tolyl)–17-(*p*-(methoxymethyl)phenyl) porphycene (NMe_3MeO-TBPo,) along with its predecessor, 2,7,12-tris (α-pyridinio-*p*-tolyl)–17-(*p*-(methoxymethyl)phenyl) porphycene (Py_3MeO-TBPo), have been used as PS to photoinactivation of bacterial species including *S. aureus*, *E. coli*, and *P. aeruginosa* (Ruiz-González et al., 2015).

Due to the poor binding efficiency of anionic and neutral porphyrins, they are generally not suggested for targeting Gram-negative bacteria. Conjugating bacterial cell binding materials with anionic and neutral porphyrins can significantly enhance the binding efficacy with Gram-negative bacteria. Different bacterial cell binding materials like cationic antimicrobial peptides and cell-penetrating peptides were conjugated with anionic and neutral porphyrins to enhance the aPDT effectiveness (Dosselli et al., 2013; Bourré et al., 2010). For example, the cationic antimicrobial peptide apidaecin IB was coupled with anionic photosensitiser 5(4′-carboxyphenyl)-10,15,20-triphenylporphyrin (cTPP) and evaluated for its aPDT efficacy against both Gram-positive and Gram-negative bacteria. This study provides a novel PS conjugate with enhanced aPDT activity against Gram bacteria with anionic porphyrins than on conjugated PS (Dosselli et al., 2010). It should be emphasised that human cells also accept porphyrins. Therefore, nonspecific targeting cannot be ruled out, and may reduce the selectivity of APDT (Kessel et al., 2003).

8.4.3 Phthalocyanine, Xanthene, and Fullerene Derivatives

Phthalocyanines have the chemical structure of a heterocyclic macrocycle aromatic similar to porphyrins; however, phthalocyanines have an extra isoindole subunit with secondary amine interconnects. This change in the chemical structure of phthalocyanines brings the difference in the excitation wavelength > 660 nm. For APDT, zinc (II) phthalocyanine-based phthalocyanines have been the subject of most research (Liu et al., 2018; Tunçel et al., 2019; Spesia and Durantini, 2013). In general, native phthalocyanines are not an efficient PS against Gram-negative bacteria. However, in the case of Gram-positive bacteria, phthalocyanines are found to have promising PS with a higher excitation wavelength (>600 nM), which facilitates tissue penetration (Spesia and Durantini, 2013). Phthalocyanine was found more challenging to accumulate in Gram-negative bacteria in large amounts than in Gram-positive bacteria due to the structural differences between both bacteria. This challenge was addressed by combining the phthalocyanine with cell membrane perturbing agents and enhancing the deposition of PS into Gram-negative bacteria (Soncin et al., 2002; Scalise and Durantini, 2005). Furthermore, the addition of a positive charge to the chemical structure of phthalocyanine significantly enhanced the binding affinity and aPDT activity against Gram-positive and Gram-negative bacteria (Ke et al., 2014). Interestingly, the study illustrated that cationic zinc (II) phthalocyanine showed phototoxicity against Gram-negative bacteria (*E. coli*) in the presence of human blood derivatives. This study also claims that the PS mentioned above can be used for blood product sterilisation (Spesia et al., 2010). Among all other phthalocyanines, cationic phthalocyanines are mostly preferred PS due to their significant photoinactivation against bacterial pathogens with lower dark toxicity. The presence of at least one amino group in the native structure of the phthalocyanine has considerably enhanced the binding affinity of negatively charged cells like Gram-negative bacteria (Mikula et al., 2014). Some of the phthalocyanine and fullerene derivatives and their antimicrobial photodynamic properties are listed in Table 8.3.

Eosin Y, Erythrosine, and rose bengal are examples of anionic xanthene dyes and are the derivatives of fluorescein that exhibit strong absorption in the green spectrum with excitation wavelength ~480–550 nm (DeRosa, 2002). The study proved that the PS rose bengal is capable of producing a more significant amount of reactive oxygen species during aPDT that subsequently hinders biofilm

TABLE 8.3
Phthalocyanine and Fullerene PS and Their Antimicrobial Photodynamic Properties

S. No.	Name of the PS	Charge	Excitation Wavelength (nm)	Test Pathogen	aPDT Efficacy	Reference
Phthalocyanine						
1	Zinc phthalocyanine	Neutral	670	*A. hydrophila*	<0.5 log	Bertolini et al. (1990)
2	Zinc (II) phthalocyanine derivative	Tetra Cationic	690	*S. aureus* *P. aeruginosa*	6 log 6 log	Giuliani et al. (2010)
3	Zinc (II) phthalocyanine substituted with a triamino moitety	Cationic	610	MRSA *E. coli* *P. aeruginosa*	~4 log ~4 log ~4 log	Ke et al. (2014)
4	Uncharged zinc(II) pyridyloxyphthalocyanine	Neutral	608	*E. coli*	0.5 log	Scalise and Durantini (2005)
5	Zinc(II) N-methylpyridyloxy-phthalocyanine	Cationic	608		4.5 log	
Fullerene						
6	C_{60} Decaiodide Monoadducts		532	*S. aureus*	3 log	Wang et al. (2012)
7	DTC_{60}^{2+} iodide N,N-dimethyl-2-(4`-N,N,N-trimethylaminophenyl) fulleropyrrolidinium	Cationic	450	*E. coli*	3.5 log	Spesia et al. (2008)
8	5,10,15,20-terekis[3-(N-ethylcarbazoyl)]porphyrin	Tetra Cationic	434	*S. aureus*	2 log	Ballatore et al. (2014)
9	Porphyrin-linked fullerene C_{60} dyad derivatives 1	Cationic	532	*S. aureus*	3.2 log	
10	Fullerenyl malonate quaternary ammonium iodide salt	Cationic	532	*S. aureus* *E. coli* *A. baumannii*	6 log 2 log 2 log	Huang et al. (2014a)

growth forming *Aggregatibacter actinomycetemcomitans* (Goulart et al., 2010). Even though xanthene derivatives mainly function according to type II mechanisms, their negative charge prevents them from being used effectively as PS for aPDT (Wilkinson et al., 1993). For instance, the study assessed the effectiveness of rose bengal and erythrosine-mediated photoinactivation against free-living and biofilm of *S. aureus* and *Salmonella Typhimurium*. The study reported that rose bengal could completely control the growth of the planktonic cells of both bacteria at only five minutes of photosensitisation. At five minutes of photosensitisation, erythrosine also rendered growth of *S. aureus* planktonic cells. Both the PS have significantly controlled the development of the biofilm of *S. aureus* at 30 minutes of photosensitisation. This study concluded that these two xanthene dyes are attractive PS to target foodborne pathogens in both planktonic and biofilm states (Silva et al., 2019).

Fullerene derivatives are other PS groups widely used in antimicrobial photodynamic therapy. The chemical structure of the fullerenes looks like a football. The pentagonal and hexagonal rings of fullerenes can absorb the visible and get excited to mediate the photochemical process (Huang et al., 2014a; 2014b; Ballatore et al., 2015). Fullerenes have exceptional photophysical and photochemical capabilities and are suggested as effective photosensitisers in antimicrobial photodynamic treatment. However, the high insolubility and hydrophobicity of fullerene required chemical modification in their structure to confer water solubility and biological availability (Spesia et al., 2008). The cationic quaternary ammonium groups were functionalised onto the native fullerene compounds, increasing their water solubility and used as PS in aPDT against pathogenic microorganisms (Huang et al., 2014b). Cationic C60-fullerene (LC16) mediated photoinactivation of *Acinetobacter baumannii* and MRSA when iodide or azide was added as an electron donor (Zhang et al., 2015; Yin et al., 2015). It is stated elsewhere that fullerene derivatives are more stable than porphyrins, which may make them more comfortable for pharmacological formulations and clinical applications (Spesia et al., 2008).

8.4.4 Natural Photosensitisers in aPDT

Generally, plant extracts are rich in chlorophyll content, with a higher quantum yield of singlet oxygen, and can absorb visible light (da Silva Souza Campanholi et al., 2020). These intrinsic properties of plant extracts provided a new avenue for natural PS to exert an encouraging photodynamic effect. Natural photosensitisers can be obtained from plants, fungi, and bacteria. Curcuminoids, alkaloids, hypericin, flavins, and bacteriochlorins are the natural PS widely studied for their aPDT applications. Chemical structure of some of the Natural PS are illustrated in Figure 8.3.

8.4.4.1 Curcuminoids

Curcumin has a broad spectrum of biological activities, including antimicrobial and antitumor properties (Xiao et al., 2018). It has been extracted from *Curcuma longa*, generally known as turmeric, which is widely used in food and treatments of wounds in Asian countries. Due to its excitation properties at the wavelength ~405–435 nm and having a broader application in aPDT (Tyagi et al., 2015). Due to its hydrophobic nature, curcumin cannot be used as a PS in PDT applications without chemical functionalisation (Abrahamse and Hamblin, 2016). Curcumin derivatives have been extensively investigated for their aPDT activity against Gram-positive and Gram-negative bacteria, including multidrug-resistant bacteria and fungal species (Abdulrahman et al., 2020; Chang et al., 2020; Gao and Matthews, 2020; Gong et al., 2020). Curcumin was investigated for its photoinactivation effect on the storage quality of fresh-cut fruits. Since the aPDT is a non-thermal method that preserves that freshness and eliminates the surface contaminated microorganism (Tao et al., 2019).

8.4.4.2 Alkaloids

Alkaloids are the other group of natural PS proposed in 1888 (Siewert and Stuppner, 2019). The five natural photoactive alkaloids are quinoline-based alkaloids, pterins, benzylisoquinolines,

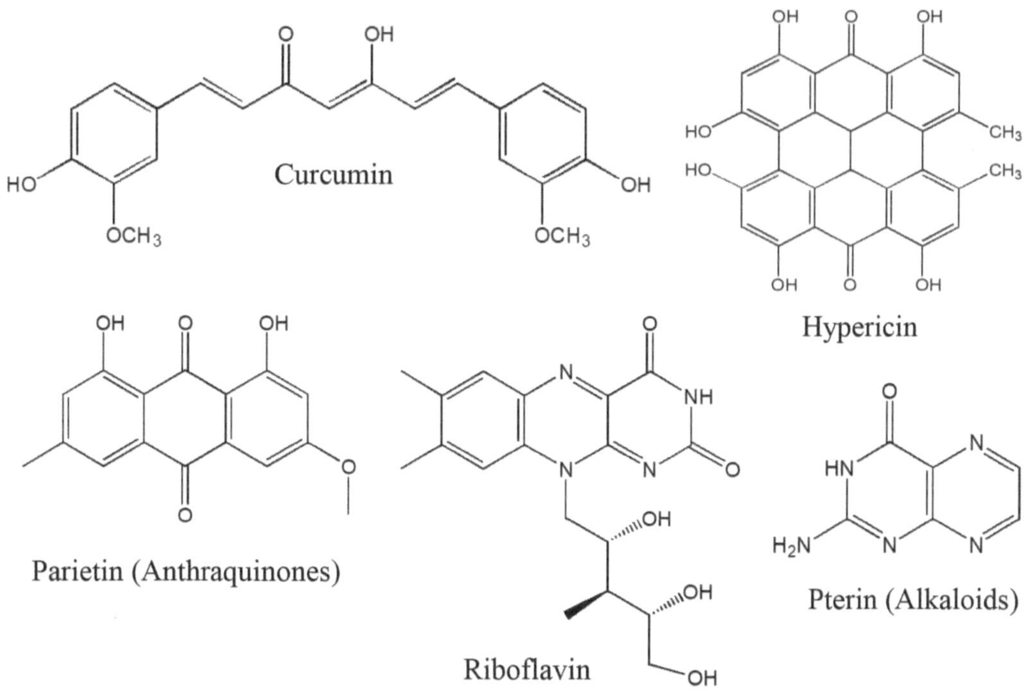

FIGURE 8.3 Chemical structure of some of the commonly used natural PS in aPDT.

beta-carbolines, and indigo alkaloids. The alkaloid from *Indigofera tinctoria* has been evaluated for its phototoxicity properties against bacteria. The indigo alkaloids showed promising photoinactivation against test pathogens (Siewert and Stuppner, 2019). The other four alkaloids also showed promising aPDT activity towards different bacterial pathogens (Siewert and Stuppner, 2019).

8.4.4.3 Hypericin

Hypericin is a commonly occurring natural PS with an excitation maximum at 593 nm. In addition to the photosensitising properties, hypericin is also a potential candidate as a natural drug with antimicrobial and antitumor properties (Engelhardt et al., 2010). Hypericin causes phototoxicity to multidrug-resistant bacteria and other pathogenic microorganisms via a type II mechanism (García et al., 2015; Kashef et al., 2015). Planktonic *S. aureus* was found to be more sensitive to hypericin-mediated photoinactivation than biofilm (Kashef et al., 2015). Polysaccharide in the biofilm prevents penetration of hydrophobic PS such as hypericin. Photoinhibition of bacteria is proportional to the duration of pre-incubation with hypericin; longer pre-incubation resulted in higher aPDT activity against *S. aureus* than shorter pre-incubation time (García et al., 2015). Utilisation of a chemical adjuvant that breaks biofilm integrity along with hypericin could potentially enhance aPDT efficacy (Dinicola et al., 2014). The application of hypericin and N-acetylcysteine combination could improve the intracellular delivery of hypericin in *S. aureus* and increase the aPDT activity against both planktonic and biofilm states *S. aureus* (Kashef et al., 2015).

8.4.4.4 Flavins

Riboflavin, sometimes known as vitamin B2, is a well-known flavin found in many organisms (Xiao et al., 2018; Regensburger et al., 2012). Riboflavin has two excitation wavelengths, 360 nm and 440 nm, and this excitation property assist in application of them as PS in aPDT. In addition, it suppresses the growth of antibiotic-resistant bacteria *E. coli* and MRSA significantly (Xiao et al., 2018). Fewer hydrophilicity properties and lower occurrence of positive charges results in moderate antibacterial activity (Makdoumi and Bäckman, 2016). Recently, riboflavin compounds functionalised with higher amounts of positive charges demonstrated significant enhancements in photodynamic efficiency (Maisch et al., 2014).

8.5 THE BACTERIAL RESPONSE AGAINST aPDT

8.5.1 Oxidative Stress Defence Enzyme

Oxidative stress is pervasive in bacterial life, either as an internal by-product of bacterial metabolism, or as a consequence of interactions with host cells. While molecular oxygen (O_2) is relatively unreactive, it may traverse most cellular membranes rapidly (Imlay 2013). owing to a four-electron reduction sequence with successive electron additions leads to the formation of four different kinds of ROS like superoxide anions ($O_2^{-\bullet}$), hydrogen peroxide (H_2O_2) and free hydroxyl radicals (•OH). It is known that these ROS can oxidise numerous biomolecules and cause significant harm to bacterial cells. The build-up of reactive oxygen species within bacterial cells stimulates them to develop resistance mechanisms against ROS. Superoxide dismutase, catalase, and peroxidase have been used as defence enzymatic pathways to detoxify the ROS inside the bacterial cells (Imlay, 2013).

Theoretically, bacteria can develop resistance mechanisms against ROS produced by the type I mechanism in aPDT by increasing the production of these enzymes (Maisch, 2015; Kashef and Hamblin, 2017). The expression of *SodA* and *SodM* genes were observed in *S. aureus* after photoinactivation with protoporphyrin IX. Although the Sod genes are expressed in *S. aureus*, the photoinactivation completely eliminates the growth of the bacteria (Nakonieczna et al., 2010). This condition shows that activation of Sod genes may not significantly affect the efficiency of the aPDT. Because photoactivation of the PS and ROS generation was occurring at the site of PS accumulation, most often, it is at the cell membrane. The distance between the photoactivation site and Sod enzymes (intracellular) is long and might not permit the defence enzyme to act against generated ROS during aPDT. Moreover, the time utilised to conduct the photoinactivation of the growth bacteria is concise, which is insufficient for the bacteria to express and over produce the defence enzyme to combat generated ROS in aPDT (Maisch, 2015). In contrast, no significant enzyme-mediated resistance mechanisms were found in bacteria against singlet oxygen ($^1O_2$) created by the type II mechanism.

8.5.2 Expression of Efflux Pumps

Transmembrane efflux systems of bacteria potentially block the deposition of PS into the bacterial cell and hinder the efficiency of aPDT (Alekshun and Levy, 2007). For example, a study has been conducted to demonstrate the extensive deposition of PS into bacterial cells by limiting efflux with a covalently linked efflux pump inhibitor and subsequent enhancement of the phototoxicity against bacterial cells. The compound NorA was used as an efflux inhibitor along with PS methylene blue to determine their combined effect on photo-antimicrobial activity against MRSA. The study results argued that hybridising PS and efflux inhibitor is a novel approach for controlling localised MRSA infections (Rineh et al., 2017). Similarly, the hybrid methylene blue and efflux inhibitor have been used as PS and evaluated for their photoinactivation efficacy against *Candida albicans*. The study claims that *C. albicans* infections could be treated effectively by hybridising efflux inhibitor to MB APDI (Prates et al., 2011).

8.6 FUTURE PERSPECTIVES AND CONCLUSIONS

aPDT is a promising alternative therapeutic procedure for antibiotic treatment with significant advantages, including broad-spectrum and non-selective effects on various molecular targets (proteins, lipids, and nucleic acids). The main problem for both conventional medicine and aPDT is the complicated bacterial cell membrane structure, notably that of Gram-negative organisms. Although the uptake of PS is not strictly necessary for aPDT to kill bacteria, selective PS attaching and transiting through the bacterial cell membrane is preferred. The photo-destructive effect of ROS on several intracellular targets, including the membrane, is undoubtedly the most significant advantage of aPDT over antibiotic treatment, which frequently involves a single target. Their therapeutic potential in aPDT for topical and localised infections remains extremely limited, and much needs to be explored.

REFERENCES

Abdulrahman, Hayder, Lama Misba, Shabbir Ahmad, and Asad U. Khan. 2020. "Curcumin Induced Photodynamic Therapy Mediated Suppression of Quorum Sensing Pathway of Pseudomonas Aeruginosa: An Approach to Inhibit Biofilm *in Vitro*". *Photodiagnosis and Photodynamic Therapy* 30 (June): 101645. https://doi.org/10.1016/j.pdpdt.2019.101645

Abrahamse, Heidi, and Michael R. Hamblin. 2016. "New Photosensitizers for Photodynamic Therapy". *Biochemical Journal* 473 (4): 347–64. https://doi.org/10.1042/BJ20150942

Ackroyd, R., C. Kelty, N. Brown, and M. Reed. 2001. "The History of Photodetection and Photodynamic Therapy". *Photochemistry and Photobiology* 74 (5): 656–69. https://doi.org/10.1562/0031-8655(2001)074<0656:thopap>2.0.co;2

Afkhami, Farzaneh, Mahsa Karimi, Abbas Bahador, Paniz Ahmadi, Maryam Pourhajibagher, and Nasim Chiniforush. 2020. "Evaluation of Antimicrobial Photodynamic Therapy with Toluidine Blue against Enterococcus Faecalis: Laser vs LED". *Photodiagnosis and Photodynamic Therapy* 32 (December). https://doi.org/10.1016/j.pdpdt.2020.102036

Alekshun, Michael N., and Stuart B. Levy. 2007. "Molecular Mechanisms of Antibacterial Multidrug Resistance". *Cell* 128 (6): 1037–50. https://doi.org/10.1016/j.cell.2007.03.004

Alenezi, K., A. Tovmasyan, I. Batinic-Haberle, and L. T. Benov. 2017. "Optimizing Zn Porphyrin-Based Photosensitizers for Efficient Antibacterial Photodynamic Therapy". *Photodiagnosis and Photodynamic Therapy* 17: 154–59. https://doi.org/10.1016/j.pdpdt.2016.11.009

Aminov, Rustam I. 2010. "A Brief History of the Antibiotic Era: Lessons Learned and Challenges for the Future". *Frontiers in Microbiology* 1 (December). https://doi.org/10.3389/fmicb.2010.00134

Amos-Tautua, Bamidele, Sandile Songca, and Oluwatobi Oluwafemi. 2019. "Application of Porphyrins in Antibacterial Photodynamic Therapy". *Molecules* 24 (13): 2456. https://doi.org/10.3390/molecules24132456

Ballatore, M. B., M. B. Spesia, M. E. Milanesio, and E. N. Durantini. 2014. Synthesis, Spectroscopic Properties and Photodynamic Activity of Porphyrin–Fullerene C60 Dyads with Application in the Photodynamic Inactivation of Staphylococcus Aureus. *European Journal of Medicinal Chemistry* 83: 685–94. https://doi.org/10.1016/j.ejmech.2014.06.077

Ballatore, M. Belén, Javier Durantini, Natalia S. Gsponer, María B. Suarez, Miguel Gervaldo, Luis Otero, Mariana B. Spesia, M. Elisa Milanesio, and Edgardo N. Durantini. 2015. "Photodynamic Inactivation of Bacteria Using Novel Electrogenerated Porphyrin-Fullerene C 60 Polymeric Films". *Environmental Science & Technology* 49 (12): 7456–63. https://doi.org/10.1021/acs.est.5b01407

Bertolini, G., F. Rossi, G. Valduga, G. Jori, and J. Lier. 1990. "Photosensitizing Activity of Water- And Lipid-Soluble Phthalocyanines on *Escherichia coli*". *FEMS Microbiology Letters* 71(1–2): 149–55. https://doi.org/10.1111/j.1574-6968.1990.tb03814.x

Bourré, Ludovic, Francesca Giuntini, Ian M. Eggleston, Charles A. Mosse, Alexander J. MacRobert, and Michael Wilson. 2010. "Effective Photoinactivation of Gram-Positive and Gram-Negative Bacterial Strains Using an HIV-1 Tat Peptide–Porphyrin Conjugate". *Photochemical & Photobiological Sciences* 9 (12): 1613. https://doi.org/10.1039/c0pp00146e

Carrera, E. T., H. B. Dias, S. C. T. Corbi, R. A. C. Marcantonio, A. C. A. Bernardi, V. S. Bagnato, M. R. Hamblin, and A. N. S. Rastelli. 2016. "The Application of Antimicrobial Photodynamic Therapy (APDT) in Dentistry: A Critical Review". *Laser Physics* 26 (12): 123001. https://doi.org/10.1088/1054-660X/26/12/123001

Chang, Kai-Chih, Ya-Yun Cheng, Meng-Jiun Lai, and Anren Hu. 2020. "Identification of Carbonylated Proteins in a Bactericidal Process Induced by Curcumin with Blue Light Irradiation on Imipenem-resistant Acinetobacter Baumannii". *Rapid Communications in Mass Spectrometry* 34 (S1). https://doi.org/10.1002/rcm.8548

Cieplik, Fabian, Andreas Späth, Christoph Leibl, Anita Gollmer, Johannes Regensburger, Laura Tabenski, Karl-Anton Hiller, Tim Maisch, and Gottfried Schmalz. 2014. "Blue Light Kills Aggregatibacter Actinomycetemcomitans Due to Its Endogenous Photosensitizers". *Clinical Oral Investigations* 18 (7): 1763–69. https://doi.org/10.1007/s00784-013-1151-8

DeRosa, M. 2002. "Photosensitized Singlet Oxygen and Its Applications". *Coordination Chemistry Reviews* 233–234 (November): 351–71. https://doi.org/10.1016/S0010-8545(02)00034-6

Digby, Elyse M., Tianyi Ma, Warren R. Zipfel, Joshua N. Milstein, and Andrew A. Beharry. 2021. "Highly Potent Photoinactivation of Bacteria Using a Water-Soluble, Cell-Permeable, DNA-Binding Photosensitizer". *ACS Infectious Diseases* 7 (11): 3052–61. https://doi.org/10.1021/acsinfecdis.1c00313

Dinicola, S., S. de Grazia, G. Carlomagno, and J. P. Pintucci. 2014. "N-Acetylcysteine as Powerful Molecule to Destroy Bacterial Biofilms. A Systematic Review". *European Review for Medical and Pharmacological Sciences* 18 (19): 2942–48.

Donnelly, R. F., C. M. Cassidy, R. G. Loughlin, A. Brown, M. M. Tunney, M. G. Jenkins, and P. A. McCarron. 2009. "Delivery of Methylene Blue and Meso-tetra (N-methyl-4-pyridyl) Porphine Tetra Tosylate from Cross-Linked Poly(vinyl alcohol) Hydrogels: A Potential Means of Photodynamic Therapy of Infected Wounds". *Journal of Photochemistry and Photobiology B: Biology* 96 (3): 223–31. https://doi.org/10.1016/j.jphotobiol.2009.06.010

Dosselli, Ryan, Cristiano Tampieri, Rubén Ruiz-González, Sonia de Munari, Xavier Ragàs, David Sánchez-García, Montserrat Agut, Santi Nonell, Elena Reddi, and Marina Gobbo. 2013. "Synthesis, Characterization, and Photoinduced Antibacterial Activity of Porphyrin-Type Photosensitizers Conjugated to the Antimicrobial Peptide Apidaecin 1b". *Journal of Medicinal Chemistry* 56 (3): 1052–63. https://doi.org/10.1021/jm301509n

Dosselli, Ryan, Marina Gobbo, Erika Bolognini, Sandro Campestrini, and Elena Reddi. 2010. "Porphyrin-Apidaecin Conjugate as a New Broad-Spectrum Antibacterial Agent". *ACS Medicinal Chemistry Letters* 1 (1): 35–38. https://doi.org/10.1021/ml900021y

Engelhardt, Victoria, Barbara Krammer, and Kristjan Plaetzer. 2010. "Antibacterial Photodynamic Therapy Using Water-Soluble Formulations of Hypericin or MTHPC Is Effective in Inactivation of Staphylococcus Aureus". *Photochemical & Photobiological Sciences* 9 (3): 365. https://doi.org/10.1039/b9pp00144a

Fontana, C. R., A. D. Abernethy, S. Som, K. Ruggiero, S. Doucette, R. C. Marcantonio, C. I. Boussios, R. Kent, J. M. Goodson, A. C. R. Tanner, and N. S. Soukos. 2009. "The Antibacterial Effect of Photodynamic Therapy in Dental Plaque-Derived Biofilms". *Journal of Periodontal Research* 44 (6): 751–59. https://doi.org/10.1111/j.1600-0765.2008.01187.x

Freitas, Laura Marise de, Esteban Nicolás Lorenzón, Norival Alves Santos-Filho, Lucas Henrique de Paula Zago, Marciana Pierina Uliana, Kleber Thiago de Oliveira, Eduardo Maffud Cilli, and Carla Raquel Fontana. 2018. "Antimicrobial Photodynamic Therapy Enhanced by the Peptide Aurein 1.2". *Scientific Reports* 8 (1): 4212. https://doi.org/10.1038/s41598-018-22687-x

Gao, Jingwen, and Karl R. Matthews. 2020. "Effects of the Photosensitizer Curcumin in Inactivating Foodborne Pathogens on Chicken Skin". *Food Control* 109 (March): 106959. https://doi.org/10.1016/j.foodcont.2019.106959

García, Isabel, Sofía Ballesta, Yolanda Gilaberte, António Rezusta, and Álvaro Pascual. 2015. "Antimicrobial Photodynamic Activity of Hypericin against Methicillin-Susceptible and Resistant Staphylococcus Aureus Biofilms". *Future Microbiology* 10 (3): 347–56. https://doi.org/10.2217/fmb.14.114

Ghorbani, Jaber, Dariush Rahban, Shahin Aghamiri, Alireza Teymouri, and Abbas Bahador. 2018. "Photosensitizers in Antibacterial Photodynamic Therapy: An Overview". *Laser Therapy* 27 (4): 293–302. https://doi.org/10.5978/islsm.27_18-RA-01

Giuliani, F., M. Martinelli, A. Cocchi, D. Arbia, L. Fantetti, and G. Roncucci. 2010. "*In Vitro* Resistance Selection Studies of RLP068/Cl, a New Zn(II) Phthalocyanine Suitable for Antimicrobial Photodynamic Therapy". *Antimicrobial Agents and Chemotherapy* 54 (2): 637–42. https://doi.org/10.1128/AAC.00603-09

Gollmer, Anita, Ariane Felgenträger, Wolfgang Bäumler, Tim Maisch, and Andreas Späth. 2015. "A Novel Set of Symmetric Methylene Blue Derivatives Exhibits Effective Bacteria Photokilling—A Structure-Response Study". *Photochemical and Photobiological Sciences* 14 (2): 335–51. https://doi.org/10.1039/c4pp00309h

Gong, Chen, Yujin Li, Ruichang Gao, Feng Xiao, Xiaodong Zhou, Haiyan Wang, He Xu, Ruihong Wang, Pan Huang, and Yuanhui Zhao. 2020. "Inactivation of Specific Spoilage Organism (Pseudomonas) of Sturgeon by Curcumin-Mediated Photodynamic Inactivation". *Photodiagnosis and Photodynamic Therapy* 31 (September): 101827. https://doi.org/10.1016/j.pdpdt.2020.101827

Goulart, Rosangela de Carvalho, Mayte Bolean, Tony de Paiva Paulino, Geraldo Thedei, Sérgio L. S. Souza, António Cláudio Tedesco, and Pietro Ciancaglini. 2010. "Photodynamic Therapy in Planktonic and Biofilm Cultures of Aggregatibacter Actinomycetemcomitans". *Photomedicine and Laser Surgery* 28 (S1): S-53-S-60. https://doi.org/10.1089/pho.2009.2591

Halili, Francisco, Alejandro Arboleda, Heather Durkee, Mukesh Taneja, Darlene Miller, Karam A. Alawa, Mariela C. Aguilar, Guillermo Amescua, Harry W. Flynn, and Jean Marie Parel. 2016. "Rose Bengal- and

Riboflavin-Mediated Photodynamic Therapy to Inhibit Methicillin-Resistant Staphylococcus Aureus Keratitis Isolates". *American Journal of Ophthalmology* 166 (June): 194–202. https://doi.org/10.1016/j.ajo.2016.03.014

Hamblin, Michael R., and Tianhong Dai. 2010. "Can Surgical Site Infections Be Treated by Photodynamic Therapy?" *Photodiagnosis and Photodynamic Therapy* 7 (2): 134–36. https://doi.org/10.1016/j.pdpdt.2010.04.004

Hanakova, Adela, Katerina Bogdanova, Katerina Tomankova, Klara Pizova, Jakub Malohlava, Svatopluk Binder, Robert Bajgar, et al. 2014. "The Application of Antimicrobial Photodynamic Therapy on S. Aureus and *E. Coli* Using Porphyrin Photosensitizers Bound to Cyclodextrin". *Microbiological Research* 169 (2–3): 163–70. https://doi.org/10.1016/j.micres.2013.07.005

Huang, Liyi, Min Wang, Tianhong Dai, Felipe F. Sperandio, Ying-Ying Huang, Yi Xuan, Long Y. Chiang, and Michael R. Hamblin. 2014a. "Antimicrobial Photodynamic Therapy with Decacationic Monoadducts and Bisadducts of [70]Fullerene: *In Vitro* and *in Vivo* Studies". *Nanomedicine* 9 (2): 253–66. https://doi.org/10.2217/nnm.13.22

Huang, Tsun Chin, Chun Ju Chen, Shinn Jyh Ding, and Chun Cheng Chen. 2019. "Antimicrobial Efficacy of Methylene Blue-Mediated Photodynamic Therapy on Titanium Alloy Surfaces *in Vitro*". *Photodiagnosis and Photodynamic Therapy* 25 (March): 7–16. https://doi.org/10.1016/j.pdpdt.2018.11.008

Huang, Ying-Ying, Sulbha K. Sharma, Rui Yin, Tanupriya Agrawal, Long Y. Chiang, and Michael R. Hamblin. 2014b. "Functionalized Fullerenes in Photodynamic Therapy". *Journal of Biomedical Nanotechnology* 10 (9): 1918–36. https://doi.org/10.1166/jbn.2014.1963

Ichinose-Tsuno, Akiko, Akira Aoki, Yasuo Takeuchi, Teruo Kirikae, Takuro Shimbo, Masaichi-Chang-il Lee, Fumihiko Yoshino, et al. 2014. "Antimicrobial Photodynamic Therapy Suppresses Dental Plaque Formation in Healthy Adults: A Randomized Controlled Clinical Trial". *BMC Oral Health* 14 (1). https://doi.org/10.1186/1472-6831-14-152

Imlay, James A. 2013. "The Molecular Mechanisms and Physiological Consequences of Oxidative Stress: Lessons from a Model Bacterium". *Nature Reviews Microbiology* 11 (7): 443–54. https://doi.org/10.1038/nrmicro3032

Kashef, Nasim, and Michael R. Hamblin. 2017. "Can Microbial Cells Develop Resistance to Oxidative Stress in Antimicrobial Photodynamic Inactivation?" *Drug Resistance Updates* 31 (March): 31–42. https://doi.org/10.1016/j.drup.2017.07.003

Kashef, Nasim, Shima Karami, and Gholamreza Esmaeeli Djavid. 2015. "Phototoxic Effect of Hypericin Alone and in Combination with Acetylcysteine on Staphylococcus Aureus Biofilms". *Photodiagnosis and Photodynamic Therapy* 12 (2): 186–92. https://doi.org/10.1016/j.pdpdt.2015.04.001

Kasimova, K. R., M. Sadasivam, G. Landi, T. Sarna, and M. R. Hamblin. 2014. "Potentiation of Photoinactivation of Gram-Positive and Gram-Negative Bacteria Mediated by Six Phenothiazinium Dyes by Addition of Azide Ion". *Photochem. Photobiol. Sci.* 13 (11): 1541–48. https://doi.org/10.1039/C4PP00021H

Ke, M.-R., J. M. Eastel, K. L. K. Ngai, Y.-Y. Cheung, P. K. S. Chan, M. Hui, D. K. P. Ng, and P.-C. Lo. 2014. "Photodynamic Inactivation of Bacteria and Viruses Using Two Monosubstituted Zinc(II) Phthalocyanines". *European Journal of Medicinal Chemistry* 84: 278–83. https://doi.org/10.1016/j.ejmech.2014.07.022

Kessel, David, Raymond Luguya, M. Graça, and H. Vicente. 2003. "Localization and Photodynamic Efficacy of Two Cationic Porphyrins Varying in Charge Distribution". *Photochemistry and Photobiology* 78 (5): 431. https://doi.org/10.1562/0031-8655(2003)078<0431:LAPEOT>2.0.CO;2

Kou, Jiayuan, Dou Dou, and Liming Yang. 2017. "Porphyrin Photosensitizers in Photodynamic Therapy and Its Applications". *Oncotarget* 8 (46): 81591–603. https://doi.org/10.18632/oncotarget.20189

de Kraker, Marlieke E. A., Andrew J. Stewardson, and Stephan Harbarth. 2016. "Will 10 Million People Die a Year Due to Antimicrobial Resistance by 2050?" *PLOS Medicine* 13 (11): e1002184. https://doi.org/10.1371/journal.pmed.1002184

Lambrechts, S. A. G., M. C. G. Aalders, D. H. Langeveld-Klerks, Y. Khayali, and J. W. M. Lagerberg. 2004. "Effect of Monovalent and Divalent Cations on the Photoinactivation of Bacteria with meso-Substituted Cationic Porphyrins". *Photochemistry and Photobiology* 79 (3): 297. https://doi.org/10.1562/SA-03-15.1

Liu, Dafeng, Linsen Li, Jincan Chen, Zhuo Chen, Longguang Jiang, Cai Yuan, and Mingdong Huang. 2018. "Dissociation of Zinc Phthalocyanine Aggregation on Bacterial Surface Is Key for Photodynamic Antimicrobial Effect". *Journal of Porphyrins and Phthalocyanines* 22 (09n10): 925–34. https://doi.org/10.1142/S1088424618500888

Liu, Yao, Rong Qin, Sebastian A. J. Zaat, Eefjan Breukink, and Michal Heger. 2015. "Antibacterial Photodynamic Therapy: Overview of a Promising Approach to Fight Antibiotic-Resistant Bacterial Infections". *Journal of Clinical and Translational Research* 1 (3): 140–67. https://doi.org/10.18053/jctres.201503.002

Mahmoudi, Hassan, Abbas Bahador, Maryam Pourhajibagher, and Mohammad Yousef Alikhani. 2018. "Antimicrobial Photodynamic Therapy: An Effective Alternative Approach to Control Bacterial Infections". *Journal of Lasers in Medical Sciences* 9 (3): 154–60. https://doi.org/10.15171/jlms.2018.29

Maisch, T., C. Bosl, R.-M. Szeimies, B. Love, and C. Abels. 2007. "Determination of the Antibacterial Efficacy of a New Porphyrin-Based Photosensitizer against MRSA Ex Vivo". *Photochemical & Photobiological Sciences* 6 (5): 545. https://doi.org/10.1039/b614770d

Maisch, Tim, Anja Eichner, Andreas Späth, Anita Gollmer, Burkhard König, Johannes Regensburger, and Wolfgang Bäumler. 2014. "Fast and Effective Photodynamic Inactivation of Multiresistant Bacteria by Cationic Riboflavin Derivatives". *PLoS One* 9 (12): e111792. https://doi.org/10.1371/journal.pone.0111792

Maisch, Tim. 2015. "Resistance in Antimicrobial Photodynamic Inactivation of Bacteria". *Photochemical & Photobiological Sciences* 14 (8): 1518–26. https://doi.org/10.1039/C5PP00037H

Makdoumi, Karim, and Anders Bäckman. 2016. "Photodynamic UVA-Riboflavin Bacterial Elimination in Antibiotic-Resistant Bacteria". *Clinical & Experimental Ophthalmology* 44 (7): 582–86. https://doi.org/10.1111/ceo.12723

Malik, Zvi, Hava Ladan, and Yeshayahu Nitzan. 1992. "Photodynamic Inactivation of Gram-Negative Bacteria: Problems and Possible Solutions". *Journal of Photochemistry and Photobiology B: Biology* 14 (3): 262–66. https://doi.org/10.1016/1011-1344(92)85104-3

Masiera, Natalia, Agnieszka Bojarska, Iwona Gawryszewska, Ewa Sadowy, Waleria Hryniewicz, and Jacek Waluk. 2017. "Antimicrobial Photodynamic Therapy by Means of Porphycene Photosensitizers". *Journal of Photochemistry and Photobiology B: Biology* 174 (September): 84–89. https://doi.org/10.1016/j.jphotobiol.2017.07.016

Meire, M. A., T. Coenye, H. J. Nelis, and R. J. G. de Moor. 2012. "Evaluation of Nd:YAG and Er: YAG Irradiation, Antibacterial Photodynamic Therapy and Sodium Hypochlorite Treatment on Enterococcus Faecalis Biofilms". *International Endodontic Journal* 45 (5): 482–91. https://doi.org/10.1111/j.1365-2591.2011.02000.x

Mikula, Premysl, Libor Kalhotka, Daniel Jancula, Stepan Zezulka, Radka Korinkova, Jiri Cerny, Blahoslav Marsalek, and Petr Toman. 2014. "Evaluation of Antibacterial Properties of Novel Phthalocyanines against Escherichia Coli—Comparison of Analytical Methods". *Journal of Photochemistry and Photobiology B: Biology* 138 (September): 230–39. https://doi.org/10.1016/j.jphotobiol.2014.04.014

Mitton, D., and R. Ackroyd. 2005. "History of Photodynamic Therapy in Great Britain". *Photodiagnosis and Photodynamic Therapy* 2 (4): 239–46. https://doi.org/10.1016/S1572-1000(05)00111-0

Mondal, Dhananjoy, and Smritilekha Bera. 2014. "Porphyrins and Phthalocyanines: Promising Molecules for Light-Triggered Antibacterial Nanoparticles". *Advances in Natural Sciences: Nanoscience and Nanotechnology* 5 (3): 033002. https://doi.org/10.1088/2043-6262/5/3/033002

Nakano, M., J. Suehiro, K. Konishi, T. Kikutani, and R. Hamada. 2011. "Development of Rapid Oral Bacteria Detection Apparatus Based on Dielectrophoretic Impedance Measurement Method". *IET Nanobiotechnology* 5 (2): 25–31. https://doi.org/10.1049/iet-nbt.2010.0011

Nakonieczna, Joanna, Ewelina Michta, Magda Rybicka, Mariusz Grinholc, Anna Gwizdek-Wiśniewska, and Krzysztof P Bielawski. 2010. "Superoxide Dismutase Is Upregulated in Staphylococcus Aureus Following Protoporphyrin-Mediated Photodynamic Inactivation and Does Not Directly Influence the Response to Photodynamic Treatment". *BMC Microbiology* 10 (1): 323. https://doi.org/10.1186/1471-2180-10-323

Nyman, Emma S., and Paavo H. Hynninen. 2004. "Research Advances in the Use of Tetrapyrrolic Photosensitizers for Photodynamic Therapy". *Journal of Photochemistry and Photobiology B: Biology* 73 (1–2): 1–28. https://doi.org/10.1016/j.jphotobiol.2003.10.002

O'Connor, Aisling E., William M. Gallagher, and Annette T. Byrne. 2009. "Porphyrin and Nonporphyrin Photosensitizers in Oncology: Preclinical and Clinical Advances in Photodynamic Therapy". *Photochemistry and Photobiology* 85 (5): 1053–74. https://doi.org/10.1111/j.1751-1097.2009.00585.x

Oliveira, Fabiana Sodré de, Thiago Cruvinel, Daniela Alejandra Cusicanqui Méndez, Evandro José Dionísio, Daniela Rios, and María Aparecida Andrade Moreira Machado. 2018. "The *in Vitro* Effect of Antimicrobial Photodynamic Therapy on Dental Microcosm Biofilms from Partially Erupted Permanent Molars: A Pilot Study". *Photodiagnosis and Photodynamic Therapy* 21 (March): 163–67. https://doi.org/10.1016/j.pdpdt.2017.12.005

Orenstein, Arie, Dov Klein, Juri Kopolovic, Eyal Winkler, Zvi Malik, Nathan Keller, and Yeshayahu Nitzan. 1997. "The Use of Porphyrins for Eradication of Staphylococcus Aureus in Burn Wound Infections". *FEMS Immunology & Medical Microbiology* 19 (4): 307–14. https://doi.org/10.1111/j.1574-695X.1997.tb01101.x

Pang, Xin, Dengfeng Li, Jing Zhu, Jingliang Cheng, and Gang Liu. 2020. "Beyond Antibiotics: Photo/Sonodynamic Approaches for Bacterial Theranostics". *Nano-Micro Letters* 12: 1–23. https://doi.org/10.1007/s40820-020-00485-3

Paz-Cristobal, M. P., D. Royo, A. Rezusta, E. Andrés-Ciriano, M. C. Alejandre, J. F. Meis, M. J. Revillo, C. Aspiroz, S. Nonell, and Y. Gilaberte. 2014. "Photodynamic Fungicidal Efficacy of Hypericin and Dimethyl Methylene Blue against Azole-Resistant Candida Albicans Strains". *Mycoses* 57 (1): 35–42. https://doi.org/10.1111/myc.12099

Perni, Stefano, Emily C. Preedy, and Polina Prokopovich. 2021. "Amplify Antimicrobial Photo Dynamic Therapy Efficacy with Poly-Beta-Amino Esters (PBAEs)." *Scientific Reports* 11 (1): 7275. https://doi.org/10.1038/s41598-021-86773-3

Phoenix, D. A., Z. Sayed, S. Hussain, F. Harris, and M. Wainwright. 2003. "The Phototoxicity of Phenothiazinium Derivatives against Escherichia Coli and Staphylococcus Aureus". *FEMS Immunology and Medical Microbiology* 39 (1): 17–22. https://doi.org/10.1016/S0928-8244(03)00173-1

Prates, R. A., I. T. Kato, M. S. Ribeiro, G. P. Tegos, and M. R. Hamblin. 2011. "Influence of Multidrug Efflux Systems on Methylene Blue-Mediated Photodynamic Inactivation of Candida Albicans". *Journal of Antimicrobial Chemotherapy* 66 (7): 1525–32. https://doi.org/10.1093/jac/dkr160

Qiu, Haofeng, Zhangyong Si, Yang Luo, Peipei Feng, Xujin Wu, Wenjia Hou, Yabin Zhu, Mary B. Chan-Park, Long Xu, and Dongmei Huang. 2020. "The Mechanisms and the Applications of Antibacterial Polymers in Surface Modification on Medical Devices". *Frontiers in Bioengineering and Biotechnology* 8 (November). https://doi.org/10.3389/fbioe.2020.00910

Ragàs, X., T. Dai, G. P. Tegos, M. Agut, S. Nonell, and M. R. Hamblin. 2010. "Photodynamic Inactivation of Acinetobacter Baumannii Using Phenothiazinium Dyes: *In vitro* and *in vivo* Studies". *Lasers in Surgery and Medicine* 42(5), 384–90. https://doi.org/10.1002/lsm.20922

Ragàs, Xavier, Xin He, Montserrat Agut, Mónica Roxo-Rosa, António Gonsalves, Arménio Serra, and Santi Nonell. 2013. "Singlet Oxygen in Antimicrobial Photodynamic Therapy: Photosensitizer-Dependent Production and Decay in *E. Coli*". *Molecules* 18 (3): 2712–25. https://doi.org/10.3390/molecules18032712

Regensburger, Johannes, Alena Knak, Tim Maisch, Michael Landthaler, and Wolfgang Bäumler. 2012. "Fatty Acids and Vitamins Generate Singlet Oxygen under UVB Irradiation". *Experimental Dermatology* 21 (2): 135–39. https://doi.org/10.1111/j.1600-0625.2011.01414.x

Rineh, Ardeshir, Naveen K. Dolla, Anthony R. Ball, María Magana, John B. Bremner, Michael R. Hamblin, George P. Tegos, and Michael J. Kelso. 2017. "Attaching the NorA Efflux Pump Inhibitor INF55 to Methylene Blue Enhances Antimicrobial Photodynamic Inactivation of Methicillin-Resistant Staphylococcus Aureus *in Vitro* and *in Vivo*". *ACS Infectious Diseases* 3 (10): 756–66. https://doi.org/10.1021/acsinfecdis.7b00095

Rossoni, R. D., J. O. Barbosa, F. E. de Oliveira, L. D. de Oliveira, A. O. C. Jorge, and J. C. Junqueira. 2014. "Biofilms of Candida Albicans Serotypes A and B Differ in their Sensitivity to Photodynamic Therapy". *Lasers in Medical Science* 29 (5): 1679–84. https://doi.org/10.1007/s10103-014-1570-z

Ruiz-González, Rubén, Montserrat Agut, Elena Reddi, and Santi Nonell. 2015. "A Comparative Study on Two Cationic Porphycenes: Photophysical and Antimicrobial Photoinactivation Evaluation". *International Journal of Molecular Sciences* 16 (11): 27072–86. https://doi.org/10.3390/ijms161125999

Scalise, Inés, and Edgardo N. Durantini. 2005. "Synthesis, Properties, and Photodynamic Inactivation of Escherichia Coli Using a Cationic and a Noncharged Zn(II) Pyridyloxyphthalocyanine Derivatives". *Bioorganic & Medicinal Chemistry* 13 (8): 3037–45. https://doi.org/10.1016/j.bmc.2005.01.063

Scanone, A. C., N. S. Gsponer, M. G. Alvarez, and E. N. Durantini. 2018. "Porphyrins Containing Basic Aliphatic Amino Groups as Potential Broad-Spectrum Antimicrobial Agents". *Photodiagnosis and Photodynamic Therapy* 24: 220–27. https://doi.org/10.1016/j.pdpdt.2018.09.017

Schastak, Stanislaw, Svitlana Ziganshyna, Burkhard Gitter, Peter Wiedemann, and Thomas Claudepierre. 2010. "Efficient Photodynamic Therapy against Gram-Positive and Gram-Negative Bacteria Using THPTS, a Cationic Photosensitizer Excited by Infrared Wavelength". *PLoS One* 5 (7): e11674. https://doi.org/10.1371/journal.pone.0011674

Shang, Yunfeng, Haiwei Li, and Ren Zhang. 2021. "Effects of Pandemic Outbreak on Economies: Evidence from Business History Context". *Frontiers in Public Health* 9 (March). https://doi.org/10.3389/fpubh.2021.632043

Siewert, Bianka, and Hermann Stuppner. 2019. "The Photoactivity of Natural Products—An Overlooked Potential of Phytomedicines?" *Phytomedicine* 60 (July): 152985. https://doi.org/10.1016/j.phymed.2019.152985

Silva Souza Campanholi, Katieli da, Jonas Marcelo Jaski, Ranulfo Combuca da Silva Junior, Ana Beatriz Zanqui, Danielle Lazarin-Bidóia, Claudia Marques da Silva, Edson António da Silva, et al. 2020. "Photodamage on Staphylococcus Aureus by Natural Extract from Tetragonia Tetragonoides (Pall.) Kuntze: Clean Method of Extraction, Characterization and Photophysical Studies". *Journal of Photochemistry and Photobiology B: Biology* 203 (January): 111763. https://doi.org/10.1016/j.jphotobiol.2019.111763

Silva, Alex Fiori, Adriele Rodrigues Santos, Daliah Alves Coelho Trevisan, Edineia Bonin, Camila Fabiano Freitas, Andreia Farias Pereira Batista, Noboru Hioka, Manuel Simões, and Jane Martha Graton Mikcha. 2019. "Xanthene Dyes and Green LED for the Inactivation of Foodborne Pathogens in Planktonic and Biofilm States". *Photochemistry and Photobiology* 95 (5): 1230–38. https://doi.org/10.1111/php.13104

Soncin, Marina, Clara Fabris, Alessandra Busetti, Donata Dei, Daniele Nistri, Gabrio Roncucci, and Giulio Jori. 2002. "Approaches to Selectivity in the Zn(<scp>ii</Scp>)–Phthalocyanine-Photosensitized Inactivation of Wild-Type and Antibiotic-Resistant Staphylococcus Aureus". *Photochemical and Photobiological Sciences* 1 (10): 815–19. https://doi.org/10.1039/B206554A

Sperandio, Felipe, Ying-Ying Huang, and Michael Hamblin. 2013. "Antimicrobial Photodynamic Therapy to Kill Gram-Negative Bacteria". *Recent Patents on Anti-Infective Drug Discovery* 8 (2): 108–20. https://doi.org/10.2174/1574891X113089990012

Spesia, M. B., M. E. Milanesio, and E. N. Durantini. 2008. "Synthesis, Properties and Photodynamic Inactivation of Escherichia Coli by Novel Cationic Fullerene C60 Derivatives". *European Journal of Medicinal Chemistry* 43 (4): 853–61. https://doi.org/10.1016/j.ejmech.2007.06.014

Spesia, Mariana B., and Edgardo N. Durantini. 2013. "Photodynamic Inactivation Mechanism of Streptococcus Mitis Sensitized by Zinc(II) 2,9,16,23-Tetrakis[2-(N,N,N-Trimethylamino)Ethoxy]Phthalocyanine". *Journal of Photochemistry and Photobiology B: Biology* 125 (August): 179–87. https://doi.org/10.1016/j.jphotobiol.2013.06.007

Spesia, Mariana B., Marisa Rovera, and Edgardo N. Durantini. 2010. "Photodynamic Inactivation of Escherichia Coli and Streptococcus Mitis by Cationic Zinc(II) Phthalocyanines in Media with Blood Derivatives". *European Journal of Medicinal Chemistry* 45 (6): 2198–205. https://doi.org/10.1016/j.ejmech.2010.01.058

Stojiljkovic, Igor, Brian D. Evavold, and Veena Kumar. 2001. "Antimicrobial Properties of Porphyrins". *Expert Opinion on Investigational Drugs* 10 (2): 309–20. https://doi.org/10.1517/13543784.10.2.309

Sun, Y., D. Xing, L. Shen, M. Sun, M. Fang, L. Bi, Y. Sui, Z. Zhang, and W. Cao. 2013. "Bactericidal Effects of Hematoporphyrin Monomethyl Ether-Mediated Photosensitization against Pathogenic Communities from Supragingival Plaque". *Applied Microbiology and Biotechnology* 97 (11): 5079–87. https://doi.org/10.1007/s00253-013-4903-0

Tandukar, Madan, Seungdae Oh, Ulas Tezel, Konstantinos T. Konstantinidis, and Spyros G. Pavlostathis. 2013. "Long-Term Exposure to Benzalkonium Chloride Disinfectants Results in Change of Microbial Community Structure and Increased Antimicrobial Resistance". *Environmental Science & Technology* 47 (17): 9730–38. https://doi.org/10.1021/es401507k

Tao, Ran, Fang Zhang, Qing-juan Tang, Chuan-shan Xu, Zhi-Jing Ni, and Xiang-hong Meng. 2019. "Effects of Curcumin-Based Photodynamic Treatment on the Storage Quality of Fresh-Cut Apples". *Food Chemistry* 274 (February): 415–21. https://doi.org/10.1016/j.foodchem.2018.08.042

Tegos, George P., Kayo Masago, Fatima Aziz, Andrew Higginbotham, Frank R. Stermitz, and Michael R. Hamblin. 2008. "Inhibitors of Bacterial Multidrug Efflux Pumps Potentiate Antimicrobial Photoinactivation". *Antimicrobial Agents and Chemotherapy* 52 (9): 3202–9. https://doi.org/10.1128/AAC.00006-08

Teixeira, A. H., E. S. Pereira, L. K. A. Rodrigues, D. Saxena, S. Duarte, and I. C. J. Zanin. 2012. "Effect of Photodynamic Antimicrobial Chemotherapy on *in vitro* and in situ Biofilms". *Caries Research* 46 (6): 549–54. https://doi.org/10.1159/000341190

Tseng, Sung Pin, Wei Chun Hung, Hsiao Jan Chen, Yu Tzu Lin, Hung Sih Jiang, Hao Chieh Chiu, Po Ren Hsueh, Lee Jene Teng, and Jui Chang Tsai. 2017. "Effects of Toluidine Blue O (TBO)-Photodynamic Inactivation on Community-Associated Methicillin-Resistant Staphylococcus Aureus Isolates". *Journal of Microbiology, Immunology and Infection* 50 (1): 46–54. https://doi.org/10.1016/j.jmii.2014.12.007

Tunçel, Ayça, İsmail Öztürk, Mine Ince, Kasim Ocakoglu, Mine Hoşgör-Limoncu, and Fatma Yurt. 2019. "Antimicrobial Photodynamic Therapy against Staphylococcus Aureus Using Zinc Phthalocyanine and Zinc Phthalocyanine-Integrated TiO 2 Nanoparticles". *Journal of Porphyrins and Phthalocyanines* 23 (01n02): 206–12. https://doi.org/10.1142/S1088424619500238

Turbay, M. B. E., V. Rey, N. M. Argañaraz, F. E. Morán Vieyra, A. Aspée, E. A. Lissi, and C. D. Borsarelli. 2014. "Effect of Dye Localization and Self-Interactions on the Photosensitized Generation of Singlet Oxygen by Rose Bengal Bound to Bovine Serum Albumin". *Journal of Photochemistry and Photobiology B: Biology* 141: 275–82. https://doi.org/10.1016/j.jphotobiol.2014.09.014

Tyagi, Poonam, Madhuri Singh, Himani Kumari, Anita Kumari, and Kasturi Mukhopadhyay. 2015. "Bactericidal Activity of Curcumin I Is Associated with Damaging of Bacterial Membrane". *PLoS One* 10 (3): e0121313. https://doi.org/10.1371/journal.pone.0121313

Viens, A. M., and Jasper Littmann. 2015. "Is Antimicrobial Resistance a Slowly Emerging Disaster?" *Public Health Ethics* 8 (3): phv015. https://doi.org/10.1093/phe/phv015

Wainwright, M., D. A. Phoenix, M. Gaskell, and B. Marshall. 1999. "Photobactericidal Activity of Methylene Blue Derivatives against Vancomycin-Resistant Enterococcus Spp". *Journal of Antimicrobial Chemotherapy* 44 (6): 823–25. https://doi.org/10.1093/jac/44.6.823

Wainwright, M., D. A. Phoenix, J. Marland, D. R. A. Wareing, and F. J. Bolton. 1997. "A Study of Photobactericidal Activity in the Phenothiazinium Series". *FEMS Immunology & Medical Microbiology* 19 (1): 75–80. https://doi.org/10.1111/j.1574-695X.1997.tb01074.x

Wang, M., L. Huang, S. K. Sharma, S. Jeon, S. Thota, F. F. Sperandio, S. Nayka, J. Chang, M. R. Hamblin, and L. Y. Chiang. 2012. "Synthesis and Photodynamic Effect of New Highly Photostable Decacationically Armed [60]- and [70]Fullerene Decaiodide Monoadducts To Target Pathogenic Bacteria and Cancer Cells". *Journal of Medicinal Chemistry* 55 (9): 4274–85. https://doi.org/10.1021/jm3000664

Warnes, Sarah L., Callum J. Highmore, and C. William Keevil. 2012. "Horizontal Transfer of Antibiotic Resistance Genes on Abiotic Touch Surfaces: Implications for Public Health". *MBio* 3 (6). https://doi.org/10.1128/mBio.00489-12

Wilkinson, Francis, W. Phillip Helman, and Alberta B. Ross. 1993. "Quantum Yields for the Photosensitized Formation of the Lowest Electronically Excited Singlet State of Molecular Oxygen in Solution". *Journal of Physical and Chemical Reference Data* 22 (1): 113–262. https://doi.org/10.1063/1.555934

Xiao, Qicai, Juan Wu, Xin Pang, Yue Jiang, Pan Wang, Albert W. Leung, Liqian Gao, Sheng Jiang, and Chuanshan Xu. 2018. "Discovery and Development of Natural Products and Their Derivatives as Photosensitizers for Photodynamic Therapy". *Current Medicinal Chemistry* 25 (7): 839–60. https://doi.org/10.2174/0929867324666170823143137

Yin, Rui, Min Wang, Ying-Ying Huang, Giacomo Landi, Daniela Vecchio, Long Y. Chiang, and Michael R. Hamblin. 2015. "Antimicrobial Photodynamic Inactivation with Decacationic Functionalized Fullerenes: Oxygen-Independent Photokilling in Presence of Azide and New Mechanistic Insights". *Free Radical Biology and Medicine* 79 (February): 14–27. https://doi.org/10.1016/j.freeradbiomed.2014.10.514

Yu, Yanlan, Yan Zhao, Yaxiong He, Jiayin Pang, Zengjun Yang, Mengxue Zheng, and Rui Yin. 2022. "Inhibition of Efflux Pump Encoding Genes and Biofilm Formation by Sub-Lethal Photodynamic Therapy in Methicillin Susceptible and Resistant Staphylococcus Aureus". *Photodiagnosis and Photodynamic Therapy* 39 (September). https://doi.org/10.1016/j.pdpdt.2022.102900

Zand, V., A. S. Milani, M. Amini, M. H. S. Barhaghi, M. Lotfi, S. Rikhtegaran, and A. Sohrabi. 2014. "Antimicrobial Efficacy of Photodynamic Therapy and Sodium Hypochlorite on Monoculture Biofilms of Enterococcus faecalis at Different Stages of Development". *Photomedicine and Laser Surgery* 32 (5): 245–51. https://doi.org/10.1089/pho.2013.3557

Zanin, I. C. J., R. B. Gonçalves, A. B. Junior, C. K. Hope, and J. Pratten. 2005. "Susceptibility of Streptococcus mutans biofilms to photodynamic therapy: an *in vitro* study". *Journal of Antimicrobial Chemotherapy* 56 (2): 324–30. https://doi.org/10.1093/jac/dki232

Zanin, I. C. J., M. M. Lobo, L. K. A. Rodrigues, L. A. F. Pimenta, J. F. Hofling, and R. B. Gonçalves. 2006. "Photosensitization of *in vitro* Biofilms by Toluidine Blue O Combined with a Light-Emitting Diode". *European Journal of Oral Sciences* 114 (1): 64–69. https://doi.org/10.1111/j.1600-0722.2006.00263.x

Zhang, Yunsong, Tianhong Dai, Min Wang, Daniela Vecchio, Long Y. Chiang, and Michael R. Hamblin. 2015. "Potentiation of Antimicrobial Photodynamic Inactivation Mediated by a Cationic Fullerene by Added Iodide: *In Vitro* and *in Vivo* Studies". *Nanomedicine* 10 (4): 603–14. https://doi.org/10.2217/nnm.14.131

9 Natural Photosensitisers for Antimicrobial Therapy

G. Ganga
Sree Ayyappa College, Alappuzha, India

CONTENTS

9.1 INTRODUCTION

Oscar Raab first reported the photodynamic effect in 1904. In 1905 it was applied effectively in the treatment of cancerous-cells, and later used as a therapy option for cancer and microbial diseases (Sowa and Voskuhl, 2020). The challenges of antibiotic resistance and antibiotic contamination are widespread. Infections that are resistant to antibiotics run the risk of developing sepsis, irreversible systemic inflammatory disease, and organ failure. Following the Covid-19 outbreak, pandemic overdose usage of antibiotics resulted in an increase in antimicrobial resistance and the formation of multidrug resistance microbes. Hence an alternative effective therapy is needed. Photosensitisers are light-activated chemicals that produce reactive oxygen species (ROS) that can damage cell structures in microbes or sick mammalian cells, resulting in cell death. The external source of light is the major component of this therapy where light activates the medicine only locally, it has a high level of selectivity and no side effects. Antimicrobial photodynamic treatment (aPDT) is an effective technique to stop antibiotic overuse and prevent drug resistance spread. aPDT is considered to be a successful

DOI: 10.1201/9781003345299-9

treatment regimen against drug resistance with strong antibiotic potency. Photosensitisers are high light energy absorbing molecules that can in turn transfer the energy to a nearby molecule with a visible or infrared range which produces electromagnetic radiation with higher energy and can result in the photoelectric effect. They can absorb incident light and can reach an excited state by absorbing photons. Before the use of antimicrobial therapy with antibiotic drugs, various photoemitters were used. Photodynamic therapy (aPDT) can be employed for antimicrobial therapy against various pathogens. The photosensitiser molecule is excited under light irradiation in a typical photosensitisation process. The excited electrons then go through an intersystem crossover (ISC) from the lowest singlet excited state (S1) to the lowest triplet excited state (T1). Photosensitised singlet oxygen ($^1O_2$) formation from the Excited T1 state to its ground state, which is generally unstable and has a high oxidation propensity which results in the killing of a bacterial cell that interacts with it (Wu et al., 2018).

Studies on photosensitisers began in the early 1900s when scientists observed the efficiency of photosensitisation in biological substrates and the treatment of cancer. Toluidine blue was successfully used as a photosensitiser against bacterial pathogens and aPDT with 21–30 mW of light at 632 nm for 30 minutes of irradiation that had 99% efficacy to kill bacteria. Many dyes can be used similarly to PS and are classified into anionic and cationic dyes. Methylene blue, Rose Bengal dye neutral red, crystal violet and rhodamine G, and Acridine orange are found to be good photosensitisers with varied applications (Sabbahi et al., 2013; Gütlich and Goodwin, 2004).

Antibiotic resistance organisms were increased enormously following the Covid pandemic due to extensive antibiotic uptake and the emergence of multidrug-resistant microbes. Hence there is a need for a paradigm shift towards therapeutic alternatives antibiotic regimen now becomes the need of the hour. To that end, the demand for new alternative therapy against bacteria antibiotics is attention towards aPDT (Gütlich and Goodwin, 2004).

9.2 CRITERIA FOR PHOTOSENSITISERS

9.2.1 Photosensitiser Molecule Criteria

Photosensitisers are substances that are susceptible to causing physicochemical change after it exposed to the incident of light by photons as energy so that it shifts their state from ground to its excited state. There is chemical modification occurs between these shifts. Once the chemical change is complete it turns back to its original state. Hence, they are not chemically affected by these processes. Some photosensitisers can also operate as which speeds up light-dependent chemical reactions. The absorption of light causes photoacids (or photo-bases) to become more acidic (or basic). When photoacids absorb light, their acidity rises (Allcock et al., 2003)

Photosensitisers have existed in natural systems as light-sensitive molecules like chlorophyll and other natural pigments. The mechanism of action of photosensitiser is when it reacts with molecular oxygen resulting in photo-oxidisation ion molecular oxygen to form peroxide species. Quantum efficiencies and fluorescence yields at various wavelengths were calculated by spectroscopic methods such as reaction-intermediate studies and luminescence investigations (Kavarnos and Turro, 1986; Daniell and Hill, 1991).

The term "photosensitiser" was used initially as photooxidation or photooxygenation sensitisers. During the early years instead of chemical synthesisers, organic photosensitisers made up of aromatic hydrocarbon molecules were used in studies. As a result of their function in biological and enzymatic processes, it become a popular therapeutic regime in the 1970s and 1980s (Gollnick, 1968; Julliard and Chanon, 1983). Due to its extensive possibilities and potential application in the biomedical field, intense research is being initiated to understand its potential use in domains such as energy harvesting, photo-redox catalysis in synthetic chemistry, and cancer treatment (O'Regan and Grätzel, 1991).

Light-sensitive photosensitisers and use of particular intensity light are used in Antibacterial Photodynamic Therapy (aPDT) in an oxygen-rich milieu. Photosensitiser absorbs the photons' energy, which is then passed to the surrounding molecules and produces reactive oxygen species and free radicals. can destroy bacterial macromolecules like proteins, lipids, and nucleic acids, resulting in bacterial death. aPDT is a unique approach, which does not result in the selection of gene mutant-resistant organisms. This therapy is a promising regime to address multidrug resistance and is beneficial to act against localised infections. It has become a standard tool for modern therapy.

This chapter discusses how photosensitiser-based therapy, the advantages of aPDT-based therapy, its limitation current modifications, and different natural photosensitisers are used in the therapy. Progressive approaches and new avenues in photosensitisers-based therapy. New PDT-based strategies to target and manage microbial resistance are also emphasised.

9.3 TYPES OF PHOTOSENSITISING REACTIONS

There are two major types of photosensitising reactions. There are different forms of photosensitisers that have been described, and it was observed that any light-absorbing molecule with a decent intersystem crossing yield could be a potential photosensitiser. Organic compounds have noteworthy electron systems that allow them to absorb light in the visible portion of the spectrum (Rai et al., 2010).

9.3.1 Type I Photosensitised Oxidation Reactions

In the case of type 1 after excitation with photons from a light source, the photosensitisers get into triplets form, and combine with non-molecular oxygen substrate. The product's chemical nature and photosensitiser is quenched by a chemical substrate other than molecular oxygen in type I photosensitised processes (Sang et al., 2014).

9.3.2 Type II Photosensitised Oxidation Reactions

A light source excites the photosensitiser into a triplet form in type II photosensitised processes. The excited photosensitiser then reacts with a triplet oxygen molecule in its ground state. The oxygen molecule is excited into the singlet state, resulting in a reactive oxygen species. The singlet oxygen molecule reacts with a substrate to generate a product after being excited. A ground state oxygen molecule quenches the photosensitiser in a type II photosensitised process, which subsequently reacts with a substrate to generate a product (Karimi et al., 2017; Jiang and Weiss, 2020).

Photosensitisers can be used efficiently for various applications with specific features. Photosensitiser must be chemically pure and well-defined, soluble in physiological media, non-toxic but highly phototoxic, absorb in the red region of the spectrum (600–800 nm) for deeper penetration, photostable (to perform for long periods), selective to the target cells, and have a quick clearance time to minimise side effects. In addition, if the photosensitiser has a high fluorescence quantum yield (F), the drug can be used as a diagnostic reporter to pinpoint the location of the disease, making it a diagnostic drug (Hally et al., 2018; 2020; Kelkar and Reineke, 2011).

Antibiotic usage during Covid-19 pandemic is a drastic raise and self-medication is one of the greatest threats on this occasion. The usage results in excess contamination of antibiotics in river ponds wells, resulting in the emergence of multidrug-resistant bacteria. Antibiotic contamination in waterbodies is going to be a widespread issue around the world. Infections with multi-drug-resistant bacteria can lead to sepsis, as well as unavoidable systemic inflammation and organ failure (Yagnik et al., 2018). Extensive usage of such antibiotics puts selection pressure on mutation and the creation of antibiotic-resistant bacteria (Pang et al., 2019). Due to the extensive use of antibiotics against microbial infections, there has recently been an increase in the rise of Multidrug resistance such as Methicillin-resistant Staphylococcus (MRSA), Multidrug resistance in other organisms such as mycobacterium, *E.coli* and Enterococcus etc. It was estimated that by the year 2030 majority of nosocomial infection-related death will be due to antimicrobial resistance and a lack of sufficient drugs to treat the infection (St. Denis et al., 2011).

World Health Organization programmes such as the antibiotic stewardship programme and several other initiatives exist to address Global problems due to drug resistance (WHO, 2014). They suggest various methods to rip out drug resistance by various treatment methods. It was suggested that Photodynamic therapy can be used against multidrug-resistant Pathogens and horizontal R gene transfer between pathogens (Polat and Kang, 2021). Photosensitisers used in antimicrobial Photodynamic therapy (aPDT) can produce singlet of oxygen which in turn results in the generation of functional molecules called reactive oxygen species (ROS) which is responsible for microbial cell destruction and death (Polat and Kang, 2021; Sowa, and Voskuhl, 2020). Photosensitisers from different resources are one of the thrust areas of recent research in which natural extract and synthetic compounds are being investigated. The upcoming area of research is the exploration of new photosensitiser candidates, which are cost-effective, have high solubility, and have high specificity. Recently, cell surface imaging, cotreatment with currently available drugs, computational methods, nanomedicine along with photodynamic therapy are the advent technologies used, along with photosensitisers. Novel aPDT with improved properties is hence found to be a promising area of research (Sowa and Voskuhl, 2020).

aPDT can enhance the efficacy of antibiotic treatment against pathogens by reducing the drug concentration and increasing its bioavailability and production of reactive oxygen molecules, photosensitiser combinatorial therapy with minimum drug concentration can eliminate a variety of pathogens (Viera et al., 2019). Another advantage of aPDT is its tissue selectivity. Another advantage of APDT is its tissue selectivity and selectively takes up only the target cells without causing any side effects or destruction to normal cells or the normal flora of the human body (Park et al., 2020).

9.4 MAJOR IMPLICATIONS OF aPDT

9.4.1 Photosensitiser Activation

For the proper activation of photosensitisers energy sources from light at various wavelengths can be used. Hence for this purpose photons are obtained from various light sources for various photosensitisers (PSs) in a dose-dependent manner, sunlight is considered a natural and effective mode of energy resources that can be exploited in this manner. This sort of Photo-biomodulation results in a change in tissue penetration, treats pain, reduce inflammation, and promotes live tissue recovery (Zhang et al., 2014). Light penetration is one of the major problems encountered with photosensitisers. It was observed that alternative energy sources such as laser beams, metal halide-based lamps, laser beam light emitting diodes (LEDs), and fibre optics can be used for deep tissue penetration (Algorri et al., 2021; Kashef et al., 2017). LED become the most promising light source due to its simplicity, safety, and economy. Babu et al. 2020 reported that light intensities of 5–1000 W/m^2 can be used effectively as an antibacterial photosensitiser. However, it was observed that depending on the light intensity, the exposure duration might range from seconds to minutes (Lukšiene and Brovko, 2013). Zhang et al. (2014) reported that in mammalian cells, this singlet oxygen can cause organ damage and promote programmed cell death. Reduction in oxygen production results in absence of oxidative phosphorylation and adenosine triphosphate synthesis which are basic prerequisites of energy production and metabolism (Memar et al., 2018). In photochemical reduction reactions reactive and short-lived singlet oxygen ($^1O_2$) will be produced. In type I reactions •OH is converted into hydroxide ions H_2O_2 and electrons are getting removed and a break in oxygen–oxygen bond of H_2O_2 results in the production of hydroxide ions. In addition. superoxide dismutase enzymes (SOD) cause dismutation to form oxygen and water (Wu et al., 2018).

According to Zhang et al. (2014), the steps involved are:

- Photosensitisers shift energy from the ground state to the excited state in presence of suitable energy sources.
- There is an exchange triplet oxygen called inter-system crossing if the ground state was not turned back.

- Photosensitiser type I: in these PS there is free radical formation occurs when a Hydrogen atom shifts towards between PS to the tissue where the photochemical reaction occurs. PSs get reduced to form reactive oxygen species (ROS) and superoxide anion radical (O2•) upon reaction through an oxidation-reduction reaction. Photosensitiser type II In this PS energy transfer occurs from a triplet excited state directly from an activated photosensitiser and after an excitation, it turns back to the ground state which results in the formation generating singlet oxygen ($^1O_2$) (Benov, 2015).

Types of photosensitisers:

Photosensitisers are categorised into three generations. First-generation PS are water-soluble porphyrins known as "hematoporphyrins". Second-generation PS have pure, highly selective, and high singlet oxygen yield (Sowa and Voskuhl, 2020). Examples of second-generation PS are chemical dyes or their derivatives such as toluidine blue and methylene blue.

Third-generation PS have been developed with less toxicity and have better bioavailability; hence they can be used for easy Drug target transport, protein engineering-based applications, antibody binding receptors, etc.

Characteristics of photosensitisers:

- Absorption spectra between 650 nm and 800 nm in spectral region (Babu et al., 2020).
- Possesses high tissue selectivity and no cytotoxic effect (Kwiatkowski et al., 2018; Abrahamse and Hamblin, 2017).
- Photosensitisers must have quantum yield, (triplet) resulting in the generation of reactive oxygen species on irradiation (Wang and Zhao, 2017).
- Should have the lipophilic capacity to traverse the phospholipid membrane and prevent self-aggregation, have optimal solubility (Xiao et al., 2017).
- Should have better absorption, distribution, metabolism, and excretion capacity (Xiao et al., 2017).
- Display a high level of storage stability (Vázquez-Hernández, 2017).
- Kill bacteria effectively while avoiding harm to eukaryotic host cells with high membrane penetration capacity (Ciepli et al., 2018).
- Cost-effective and easy to synthesise (Abrahamse and Hamblin, 2017).

9.5 NATURAL PHOTOSENSITISERS FOR ANTIMICROBIAL THERAPY

Herbal medicine and compounds derived from herbs are possible therapeutics for many ailments. Nature is a treasure for many medicinal plants and Ps derived from these medicinal plant resources can be expected to be a potent resource for natural photosensitisers for use in Phototherapeutic applications. However, there have been few attempts made by researchers across the globe to develop PS from naturally available plant extracts or their derivatives. Modern techniques and computational analysis will provide a new platform for analysing the suitability PSs in various applications as well In addition to that photosensitisers are also derived naturally from a variety of resources such as fungus, algae, and microorganisms, natural compounds, and biopigments (Bonifácio et al., 2018). Natural compounds such as alkaloids, curcuminoids anthraquinones perylenequinones flavins, and porphyrin precursors are some of the compounds which can be used in PS and are enlisted in Table 9.1 (Bonifácio et al., 2018; Vieira et al., 2019; Souza et al., 2021; Wang et al. 2021) reported the mode of action of photodithazine a photosensitiser in application against *Staphylococcus aureus* (MSSA) a methicillin-sensitive strain and methicillin-resistant *Staphylococcus aureus* (MRSA) strain (Figure 9.1).

Photoactive principles are found in various natural compounds and are thought to be a promising source of novel photosensitisers in the future. PDT may eventually attain its therapeutic

TABLE 9.1
Natural Photosensitisers with Mode of Action

Compound Origin	Mode of Action	Absorption Maximum	References
Curcumin, Curcuminoids	Seafood pathogen	470	Chen et al. (2020)
Curcumin, Curcuminoids	Inactivation of spoilage organisms	470	Gong et al. (2020)
Curcumin	Multidrug resistance	450	Freitas et al. (2019)
Curcumin	Biofilm	405	Abdulrahman et al. (2020)
Curcuminoids	Peridodentitis	450–475	Pan et al. (2020)
Curcumin, Curcuminoids	Poultry diseases Skin infections	420–480 nm	Kubrak et al. (2022)
Pheophorbide A	Anti-cancer antimicrobial	670 nm	Kubrak et al., 2022
Alkaloids	Multidrug resistance	350–750	Tosato et al. (2020)
Anthraquinones	Multidrug-resistant infection	435	Wang et al. (2021)
Porphyrin precursor	Trichosporonosis	635	Lan et al. (2020)
Perylenequinones	Antifungal effect	590	Yang et al. (2019)
Hypericin	Acne vulgari	660	Barroso et al. (2020)
Chlorophyllin sodium salt Chlorin-type compounds	Plant pathogens	405	Glueck et al. (2019); Žudyte and Lukšiene (2019)
Riboflavin	Listeriosis	460	Kim et al. (2021)
Aloe emodin	MDR candida	400–780	Ma et al. (2020)
Parietin	Local infection	470	Comini et al. (2017)

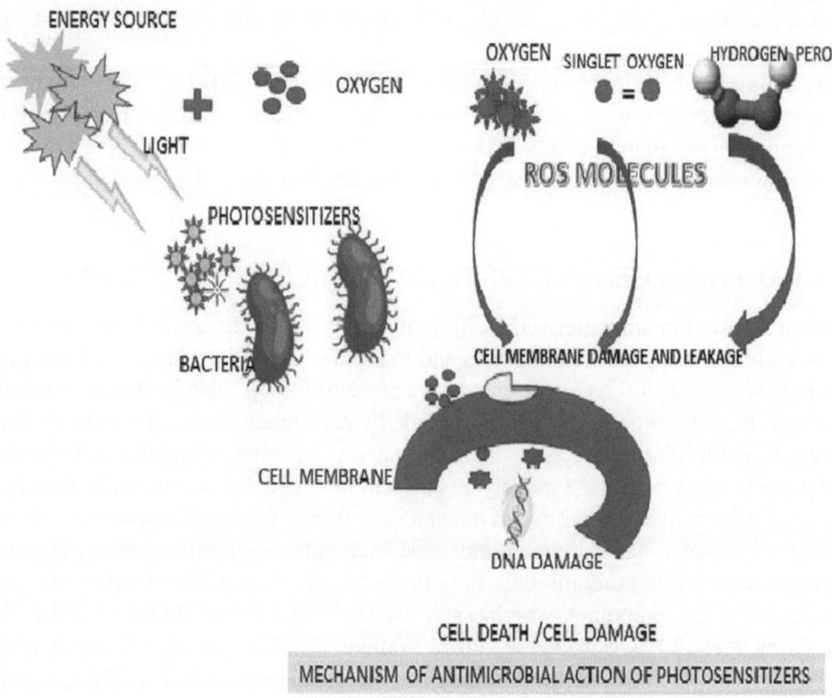

FIGURE 9.1 Meshanism of Actionof Photosensitisers as Antimicrobials

TABLE 9.2
Various Natural Extracts as Photosensitisers and Their Possible aPDT Applications

Extracts	Wavelength (nm)	Infections	Reference
B. orellana leaf	395–480	Bad breath treatment	Gonçalves et al. (2020)
C. longa leaf	420–480	Periodontitis	Saitawee et al. (2018)
H. sabdariffa plant	420	Photodynamic sanitation	Majiya, and Galstyan (2020)
C. xanthorrhiza plant	405	Dental caries	Lee et al. (2017)
T. minuta	315–400	Antivirulence Candida sp.	Giacone et al. (2020)
Essential oils	660.	microbial films	Dascalu et al. (2020)
G. blepharophylla	660	Antifungal effect	Andreazza et al. (2015)
P. obscurum plant	315–400	Oral candidiasis	Postigo et al. (2017)
E. microcorys leaves	tungsten filament lamps	Antibacterial effect	Arfao et al. (2020)
I. truxillensis	660	MDR microbial infections	Andreazza et al. (2015)
T. wilfordii fraction of extracts	660	Skin infection	Alam et al. (2021)
H. perforatum (St. John's wort)	sunlight	Antiviral compounds	Ryberg et al. (2020)
Wort plant	570–1,400	Oral biofilms	Vollmer et al. (2019)

breakthrough as a viable alternative treatment against multi-resistant bacteria and cancer therapy and potential antifungal therapeutic application. Table 9.2 shows representative samples of natural extracts utilised in aPDT (Saitawee et al., 2018).

9.5.1 Synthetic Derivatives of Natural Products as Photosensitisers

Photosensitisers made of synthetic compounds derived either from natural sources or semi-synthetic and synthetic processes can be utilised from PSs (Siewert and Stuppner, 2019; Dias and Mfouo-Tynga, 2020; Dias et al., 2020)

Methods which can be used as synthetic PS include:

- Hydrogenation, annulation, cycloaddition, and breaking are used to make chlorin-based PS, as well as the repair of porphyrin precursors (Dias and Mfouo-Tynga, 2020; Dias et al., 2020)
- Natural chlorophylls are used in the synthesis of chlorin-based PSs (Dias and Mfouo-Tynga, 2020; Dias et al., 2020)
- In antimicrobial photodynamics, both organic plant-based natural and chemical-based synthetic PSs can be utilised. Organic natural PSs can be directly employed in aPDT without much synthesis process in a cost-effective manner. Many natural PS are derived from plant-based compounds and do not require the use of hazardous organic chemicals that are used in the production of synthetic PS; as a result, natural PS are both user-friendly and non-hazardous. On the other hand, the natural PS, have some limitations such as low triplet quantum yields, limited aqueous solubility, and poor ADME properties when irradiated. Synthetic compounds are the derivatives of natural compounds with improved bioactive characteristics. Hence, the discovery of new cost-effective non-hazardous PS is necessary for the development of new photosensitisers for aPDT (Dias and Mfouo-Tynga, 2020; Dias et al., 2020)

9.6 LIMITATIONS IN ANTIMICROBIAL THERAPY

aPDT using various photosensitisers is reported to have a significant effect on bacterial pathogens. However, the application of photosensitisers depends on the nature of Photosensitizes and their inducing light sources. Those PS have poor water solubility, poor UV light penetration does not consider a good photosensitiser. Some PS molecules are ineffective against other microbes, including Gram-negative bacteria due to their membrane-based chemical characteristics (Zhuang et al., 2020).

9.6.1 Limitation of aPDT against Different Bacteria

Zhuang et al., 2020 reported various photosensitisers with ionic nature such as anionic or neutral range are suitable to remove bacteria stained as Gram-positive nature, however, according to the reports they are not appropriate to act upon against Gram-negative bacteria. The action of PS on the selective organism is greatly influenced by parameters such as charge, lipophilicity, molecular size, concentration, and cell wall components of the targeted organism (Jia et al., 2017).

Gram-negative bacteria's outer membrane is made up of a bilayer of phospholipids and lipopolysaccharides with phosphatidylglycerol, phosphatidylethanolamine, and cardiolipin-lactams, and other small hydrophilic medications use porins pores to get into the cell. (Zhuang et al., 2020) Amphipathic compounds with both lipophilic backbones and polar/charged sides make them good photosensitisers that will be permeable to the membrane bilayer. The natural photosensitiser enolic and ketonic forms of curcumin, alkaloids, and plant-based compounds are the most promising PSs for aPDT antimicrobial activities. Amphiphilic nature makes them effective against different microbes (Park et al., 2020). It was observed that hypericin in conjugation with antibiotics can be detrimental to the growth of *P. aeruginosa* due to its high membrane capacity (Zhuang et al., 2020).

9.6.2 Failure in Target Selection

Normal cells along with infectious cells may die if aPDT is not target-specific. Hence, it is crucial to eliminate infectious cells without harming normal cells (Simões et al., 2020). One of the major constraints of aPDT is that it is nonselective against some cellular proteins and molecular targets, including proteins, lipids, and nucleic acids. This could be due to the electrostatic interactions between cells and poor target selection (Ciepli et al., 2018).

9.6.3 Solubility and Cell Target Penetration

Solubility, degradation with some membrane-bound enzymes and aggregation, and quenching are some of the issues related to the reliability and sensitivity of PSs. It was observed that some of the natural PSs have low solubility and very less penetration capacity. which limits their bioavailability to the target cells. Membrane-bound enzymes are able to degrade such PSs if they are not properly protected with certain biopolymers and are susceptible to hydrolytic breakdown. Agglomeration and aggregation are other limitations that increase their molecular size which results in interaction with the target cell. Fluorescent quenching is another drawback in aPDT efficiency. Aggregation can result in a change in fluorescent shift from one wavelength to another or fluorescence quenching. This agglomeration also affects ROS, formation which in turn can reduce aPDT efficiency (Polat and Kang, 2021).

9.7 NEW APPROACHES TO OVERCOME LIMITATIONS

In several applications, including photodynamic treatment, photocatalytic organic waste breakdown, and synthetic chemistry, efficient photosensitisers are widely desired. The development of aPDT is currently hampered by several problems, including those involving light, tissue penetration, enzymatic degradation, cellular perforation and oxygenation, and other integral features of the photosensitisers. New techniques are being introduced to overcome this limitation.

Researchers have introduced several Nano-engineering and cell surface engineering technologies to promote hydrophobic interaction. Chitosan and cholesterol-assisted modification with PSs was found to be an improved method to enhance aPDT efficiency (Jia et al., 2017; Ciepli et al., 2018). PS can be attached with amphiphilic copolymer and were found to be effective to improve the solubilisation efficiency without any photoactive degradation (da Silva Souza Campanholi et al., 2020).

aPDT efficiency is by using polymer-based conjugate from various resources. Conjugated-polymer-based photosensitisers have significantly greater in the lowest excited state of dioxygen molecule called singlet oxygen molecule (o=o) generation in a process known as polymerisation-enhanced photosensitisation. Polymer-based conjugates generated from cyclodextrins can be encapsulated with PSs to enhance their efficiency and ROS activities (Contreras et al., 2019).

Cyclohexdrins based conjugates can be made on cellulose film and are efficient PS against MRSA (methicillin-resistant bacteria) (Sharma et al., 2020). These polymer-based biofilms are effective antimicrobial therapies and applications in skin lesions and wound infections. Chitosan can be used as conjugate PS and can be used as an efficient aPDT due to its antimicrobial activities. Hydrogel-based PS were developed and are proven to be good PS against Candida sp. (Chandna et al., 2020). Water-soluble chlorine derivatives with tannic acid coatings are reported to be good photosensitisers for aPDT (Liu et al., 2017).

Freitas et al. 2019 made antimicrobial peptides with 12 to 50 amino acids in length to combat antimicrobial resistance. The action of antimicrobial peptides in bacterial killing is done by membranous leakages, which can be further enhanced by antimicrobial peptide complexes with PS and making the membrane more sensitive to photons (Torra et al., 2015).

Protein engineering was made to improve the mode of action of antimicrobial peptides. This protein-conjugated peptide was proven to be used as conjugated photosensitisers against pathogens, including pathogens from the *Enterobacteriaceae* family *Staphylococcus aureus*, *Acinetobacter baumannii* (Polat and Kang, 2021; Freitas et al., 2019). This engineered PS can be made with target peptides or antibodies, as well as with other target proteins in cellular compartments. Tagging with specific proteins against target cells and the development of singlet oxygen species will have bactericidal activity, and reduce the dark toxicity of normal cells. fluorescent proteins and flavin-binding proteins are the widely used protein which is used for tagging (Hally et al., 2020). In addition, these proteins are proven to have good photosensitising abilities and antimicrobial properties (Torra et al., 2015).

9.7.1 Nanotechnology-Assisted Photosensitisers

PS can be used with nanomaterials to lessen the adverse effects of conventional PSs and aPDT activity. PS conjugated with nanomaterials or nanoparticles have various applications (Pang et al., 2019). PS conjugated photosensitising NP can act for a longer period and protect PS from deactivation agents. This NP conjugated Ps have improved penetration and retention effect (EPR effect) when exposed to 50–100 nm particles, NPs also improve cells (Ghorbani et al., 2018). In addition to group modification, aPDT uses targeted hydrophobic drug delivery via nanoparticles. Modifications to the functional group can enhance the biochemical properties of nanoparticles and has been used in drug delivery (Pezzuoli et al., 2018). Enhanced rate of cellular absorption, photogeneration of singlet oxygen and bioavailability, and regulated drug delivery are the advantages of NPs with PS conjugated aPDT (Youf et al., 2021; Ghorbani et al., 2018; Vázquez-Hernández et al., 2018).

9.7.2 Carbon Quantum Dots as Photosensitisers

Traditional photosensitisers for photodynamic therapy (PDT) don't produce singlet oxygen ($^1O_2$), and hence have low quantum yield. In addition, the problem of photobleaching will limit activity. Researchers still seek improved photosensitisers, and carbon quantum dots are proven to act against drug-resistant microbes effectively. Green synthesised carbon quantum dots (CQDs) are obtained from various biomass as a sole source of carbon via a hydrothermal synthesis and can be used as an excellent photodynamic agent. Due to their unique photophysical properties, they can effectively be used in Photosentisers with various applications. Synthesis of Carbon quantum dots (CQDs) is made cost-effectively with a one-pot solvothermal/hydrothermal procedure (Pavel et al., 2022).

Preparation CDs is crucial, and the properties of CDs depend on these preparation methods. Bottom-up and top-down approaches are the two methods used for the synthesis of CDs. Physiochemical characteristics, bioavailability, biocompatibility, solubility, antibacterial activity, cell penetration ability, etc are greatly influenced by the methods of synthesis. Synthesis of CDS can be done with minimal waste, and with simple purification methods that can be performed with a simple filtration technique (Nie et al., 2020).

9.7.3 Photosensitiser-Loaded Hydrogels and Composite Biopolymer Film

Hydrogel-based photosensitisers are now available for aPDT against multidrug-resistant bacteria (Chandna et al., 2020) In this photodynamic therapy uses mechanically stable and transparent electron beam-polymerised hydrogels based on poly (ethylene glycol) diacrylate were used. The photosensitisers were released in significant amounts (up to 300 mol/l) by the hydrogels, which was important for efficiency. This was tested for antibacterial efficacy against multidrug-resistant *Escherichia coli* in both a tissue-like and a liquid environment. Light-induced elimination was possible in both systems. The loaded hydrogels were also evaluated against seven multidrug-resistant bacterial strains, including *Enterococcus faecium*, *Staphylococcus aureus*, *Klebsiella pneumoniae*, *Acinetobacter baumannii*, *Pseudomonas aeruginosa*, *Escherichia coli*, and were successfully eradicated (Corrêa et al., 2019; Gao and Matthews, 2020). Glass et al. (2021) noted that photosensitiser-loaded hydrogels are a promising therapy for treating wounds infected with multidrug-resistant microbes. PS-loaded hydrogels such as poly(-vinyl alcohol)-borate and poly(ethylene glycol) diacrylate are found to have good antibacterial activity (Sharma et al., 2020).

Photosensitisers made from polymers or conjugated polymers and conjugated polymer nanoparticles (CPNs) are found to have an extensive advantage for PS production in antimicrobial therapy. They are small molecules combined with nanoparticles and used in photodynamic and photothermal applications (Wei et al., 2020)

According to Sharma et al. (2020), administration of an acceptable concentration of photosensitiser (PS) to the bacterially infected site is an essential step; as a result, PS must be administered there in an appropriate formulation. The photo bactericidal and biocompatibility properties of PS-embedded composite biopolymer films make them effective PS Deliverables for treating methicillin-resistant *Staphylococcus aureus* (MRSA) (Sharma et al., 2020; Kwiatkowski et al., 2018).

9.7.4 Sonodynamic Therapy (SDT) for Enhanced Efficiency

Sonodynamic therapy (SDT) with ADPT was found to be produced enhanced efficiency in antimicrobial therapy. In this technology, Ultrasonic energy-based activation of specific MHz and with specific density can be used to activate certain photosensitisers (Alves et al., 2018; Alves et al., 2019; Corrêa et al., 2019). It was also observed that when compared to light, ultrasound has a greater ability to permeate tissues and pass through cell membranes. According to Costley's research, SDT using ultrasound also kills microorganisms. In that investigation, Rose Bengal PS was found to be a powerful son dynamic technique for eliminating Gram-negative *Pseudomonas aeruginosa* and Gram-positive *Staphylococcus. aureus*. SDT can be utilised alone as well as in conjunction with aPDT (Alves et al., 2018; Alves et al., 2019; Corrêa et al., 2019).

9.7.5 *In Silico* Computational Analysis

Pre-clinical *in vivo* experiments are time-consuming and need more trials Computer-based *in silico* technology can be used to evaluate the efficacy of certain drugs. Many software programs and *in silico* models are in use, which will validate and help in further *in vitro* studies. Computer-based simulations are effective for predetermining the effect of photosensitisers in aPDT (Wu et al., 2020). The effect of porphyrin-derivative PSs as aPDT was predicted using Monte Carlo simulation with the aid of simulated human skin. This Simulations method produces the most efficient aPDT settings, including the wavelength of light (Žudyte and Lukšiene, 2019).

9.8 CONCLUSION

There are several advantages of aPDT against other therapeutic approaches that this sort of approach will never bring antimicrobial resistance and resistance genes. In such therapy, diatomic molecular oxygen reacts with photosensitisers, and certain photons from energy sources to produce harmful reactive oxygen species (ROS). ROS produce oxidative damage which helps in the elimination of various pathogens. Various oxidative mechanisms such as autophagy, apoptosis, and phagocytic activation are some of the mechanisms and activation of several enzyme mechanisms and signalling of various pathways involved with it. Modified PS with high tissue specificity and target specificity can be used more effectively than standard antibiotic therapy. *In vitro* computational simulation will provide insight to further *In vivo* study in Animal models and will be preoptimised and cost-effective procedure. computational simulations also will aid in the development of new aPDT regimes in future. Natural extracts and chemicals, as well as their synthesised derivatives and nano-processing, are currently the focus of many PSs investigations. Research on novel PS systems needs to be intensified so that more natural systems can be evaluated, and new PS structures elucidated with advent technologies and procedures such as nanotechnology, quantum dots, biopolymers synthesis, and protein engineering should be simulated for photosensitisation therapy to cure human infectious diseases more effectively and with fewer side effects and drug resistance.

REFERENCES

Abrahamse, H. and M.R. Hamblin. 2017. New photosensitizers for photodynamic therapy. *Biochem. J.* 473: 347–364. doi: 10.1042/BJ20150942

Abdulrahman, H., L. Misba, S. Ahmad, and A.U. Khan. 2020. Curcumin induced photodynamic therapy mediated suppression of quorum sensing pathway of *Pseudomonas aeruginosa*: An approach to inhibit biofilm *in vitro*. *Photodiagnosis Photodyn. Ther.* 30: 101645. doi: 10.1016/j.pdpdt.2019.101645

Andreazza, N.L., C.C. de Lourenço. M.É.A. Stefanello, T.D.Z. Atvars, and M.J. Salvador. 2015 Photodynamic antimicrobial effects of bis-indole alkaloid indigo from *Indigofera truxillensis* Kunth (Leguminosae). *Lasers Med. Sci.* 30: 1315–1324. doi: 10.1007/s10103-015-1735-4

Alam, S.T., H. Hwang, J.D. Son, U.T.T. Nguyen, J.S. Park, H.C. Kwon, J. Kwon, and K. Kang. 2021. Natural photosensitizers from *Tripterygium wilfordii* and their antimicrobial photodynamic therapeutic effects in a *Caenorhabditis elegans* model. *J. Photochem. Photobiol. B Biol.* 218: 112184. doi: 10.1016/j.jphotobiol.2021.112184

Algorri, J.F., M. Ochoa, P. Roldán-Varona, L. Rodríguez-Cobo, and J.M. López-Higuera. 2021. Light technology for efficient and effective photodynamic therapy: A critical review. *Cancers (Basel)* 13(14):3484. doi: 10.3390/cancers13143484.

Alves, F., N. Mayumi Inada, V. Salvador Bagnato, C. Kurachi, and Sonophotodynamic. 2019. Therapy for the Inactivation of *Staphylococcus aureus* Biofilm. In *Proceedings of the 17th International Photodynamic Association World Congress, Proceedings of the SPIE 11070: 110708 S 5*, Cambridge, MA, USA. doi: 10.1117/12.2528214

Alves, F., A.C. Pavarina, E.G.D.O. Mima, A.P. McHale, and J.F. Callan. 2018. Antimicrobial sonodynamic and photodynamic therapies against *Candida albicans*. *Biofouling* 34: 357–367. doi:10.1080/08927014.2018.1439935.

Allcock, H.R., F.W. Lampe, and J.E. Mark. 2003. Contemporary polymer chemistry (3rd ed.). Upper Saddle River, NJ: Pearson/Prentice Hall. ISBN 0-13-065056-0. OCLC 51096012

Arfao, A.T., O.V.N. Ewoti, M.F. Onana, C.L. Djimeli, S Tchakonté, N.K. Dama, and M. Nola. 2020. The combined effect of eucalyptus aqueous extract and light on pathogenic *Escherichia coli* survival in aquatic microcosm. *J. Microbiol. Biotechnol. Food Sci.*, 9: 1003–1008. doi: 10.15414/jmbfs.2020.9.5.1003-1008

Babu, B., Mack, J., and Nyokong, T. 2020 Sn(iv) N-confused porphyrins as photosensitizer dyes for photodynamic therapy in the near IR region. *Dalton Trans.* 49: 15180–15183. doi: 10.1039/D0DT03296D

Barroso, R.A., R. Navarro, C.R. Tim, L. de Paula Ramos, L.D. de Oliveira, Â.T. Araki, K.G.C. Fernandes, D. Macedo, and L. Assis. 2020. Antimicrobial photodynamic therapy against Propionibacterium acnes biofilms using hypericin (Hypericum perforatum) photosensitizer: *In vitro* study. *Lasers Med. Sci.* doi: 10.1007/s10103-020-03163-3

Benov, L. 2015. Photodynamic therapy: Current status and future directions. *Med. Princ. Pract.* 24(Suppl 1): 14–28. doi: 10.1159/000362416.

Bonifácio, D., C. Martins, B. David, C. Lemos, M.G.P.M.S. Neves, A. Almeida, D.C.G.A. Pinto, M.A.F. Faustino, and Â. Cunha. 2018. Photodynamic inactivation of *Listeria innocua* biofilms with food-grade photosensitizers: A curcumin-rich extract of Curcuma longa vs. commercial curcumin. *J. Appl. Microbiol.* 125: 282–294. doi: 10.1111/jam.13767

Ciepli, K.F., D. Deng, W. Crielaard, W. Buchalla, E. Hellwig, A. Al-Ahmad, and T. Maisch. 2018. Antimicrobial photodynamic therapy– what we know and what we don't. *Crit. Rev. Microbiol.* 44: 571–589. doi: 10.1080/1040841X.2018.1467876

Comini, L.R., F.E. Morán Vieyra, R.A. Mignone, P.L. Páez, M. Laura Mugas B.S. Konigheim, J.L. Cabrera, S.C. Núñez Montoya, and C.D. Borsarelli Parietin. 2017. An efficient photo-screening pigment *in vivo* with good photosensitizing and photodynamic antibacterial effects *in vitro*. *Photochem. Photobiol. Sci.* 16: 201–210. https://doi.org/10.1039/c6pp00334f

Contreras, A., M.J. Raxworthy, S. Wood, J.D. Schiffman, and G. Tronci. 2019. Photodynamically active electrospun fibers for antibiotic-free infection control. *Acs Appl. Bio Mater.* 2: 4258–4270. doi: 10.1021/acsabm.9b00543.

Corrêa, T.Q., K.C. Blanco, J.M. Soares, N.M. Inada, C. Kurachi, M.D.A. Golim, E. Deffune, and V.S. Bagnato. 2019. Photodynamic inactivation for *in vitro* decontamination of *Staphylococcus aureus* in whole blood. *Photodiagnosis Photodyn. Ther.* 28: 58–64. http://doi.org/10.1016/j.pdpdt.2019.08.013

Chandna, S., N.S. Thakur, R. Kaur, and J. Bhaumik. 2020. Lignin-bimetallic nanoconjugate doped pH-responsive hydrogels for laser-assisted antimicrobial photodynamic therapy. *Biomacromolecules* 21: 3216–3230. https://doi.org/10.1021/acs.biomac.0c00695

Chen, B.J., H. Huang Q.H. Li, J.J. Zeng, H. Wang, Y.P. Liu, and Y. Zhao. 2020. Eradication of planktonic *Vibrio parahaemolyticus* and its sessile biofilm by curcumin-mediated photodynamic inactivation. *Food Control* 113: 107181. https://doi.org/10.1016/j.foodcont.2020.107181

Dascalu, L.M., M. Moldovan, D.I. Prodan Ciotlaus, V. Popescu, I. Baldea, R. Carpa, S. Sava, R. Chifor, and M.E. Badea. 2020. Assessment and characterization of some new photosensitizers for antimicrobial photodynamic therapy (aPDT). *Materials* 13: 3012. doi: 10.3390/ma13133012

Daniell, M.D., and J.S. Hill. 1991. A history of photodynamic therapy. *Aust. N. Z. J. Surg.* 61(5): 340–348. doi: 10.1111/j.1445-2197.1991.tb00230.x.

da Silva Souza Campanholi, K., J.M. Jaski, R.C. da Silva Junior, A.B. Zanqui, D. Lazarin-Bidóia, C.M. da Silva, E.A. da Silva, N. Hioka, C.V. Nakamura, and L. Cardozo-Filho, and W. Caetano. 2020 Photodamage on *Staphylococcus aureus* by the natural extract from *Tetragonia tetragonoides* (Pall.) Kuntze: Clean method of extraction, characterization, and photophysical studies. *J. Photochem. Photobiol. BBiol.* 203: 111763. doi: 10.1016/j.jphotobiol.2019.111763

Dias, D.L., and I.S. Mfouo-Tynga. 2020. Learning from nature: Bioinspired chlorin-based photosensitizers immobilized on carbon materials for combined photodynamic and photothermal therapy. *Biomimetics* 5: 53. doi: 10.3390/biomimetics5040053

Dias, L.D., K.C. Blanco, I.S. Mfouo-Tynga, N.M. Inada, and V.S. Bagnato. 2020. Curcumin as a photosensitizer: From molecular structure to recent advances in antimicrobial photodynamic therapy. *Jphotochemrev* 45: 1389–5567. doi: https://doi.org/10.1016/j.jphotochemrev.2020.100384

Freitas, M.A., A.H. Pereira, J.G. Pinto, A. Casas, and J. Ferreira-Strixino. 2019. Bacterial viability after antimicrobial photodynamic therapy with curcumin on multi-resistant *Staphylococcus aureus*. *Future Microbiol.* 14:739–748. doi: 10.2217/fmb-2019-0042

Gao, J., and K.R. Matthews. 2020. Effects of the photosensitizer curcumin in inactivating foodborne pathogens on chicken skin. *Food Control.* 109: 106959. doi: 10.1016/j.foodcont.2019.106959.

Ghorbani, J., D. Rahban, S. Aghamiri A. Teymouri, and A. Bahador. 2018. Photosensitizers in antibacterial photodynamic therapy: An overview. *Laser Ther.* 27(4): 293–302. doi: 10.5978/islsm.27_18-RA-01.

Giacone, L., E. Cordisco M.C. Garrido, E. Petenatti, and M. Sortino. 2020. Photodynamic activity of Tagetes minutes extracts against superficial fungal infections. *Med. Mycol.* 58: 797–809. doi: 10.1093/mmy/myz114

Glass, S., M. Kühnert, N. Lippmann, J. Zimmer, R. Werdehausen, B. Abel, V. Eulenburg, and A. Schulze. 2021. Photosensitizer-loaded hydrogels for photodynamic inactivation of multi-resistant bacteria in wounds. *RSC Adv*. 13: 7600–7609. doi: 10.1039/D0RA09786A

Glueck, M., C. Hamminger, M. Fefer, J. Liu, and K. Plaetzer. 2019. Save the crop: Photodynamic Inactivation of plant pathogens I: Bacteria. *Photochem. Photobiol. Sci.* 18: 1700–1708. doi: 10.1039/C9PP00128J

Gong, C., Y. Li, R. Gao, F. Xiao, X. Zhou, H. Wang, H. Xu, R. Wang, P. Huang, and Y. Zhao. 2020. Inactivation of specific spoilage organism (Pseudomonas) of sturgeon by curcumin-mediated photodynamic inactivation. *Photodiagnosis Photodyn. Ther*. 31: 101827. doi: 10.1016/j.pdpdt.2020.101827

Gollnick, K. 1968. Type II Photooxygenation Reactions in Solution. In *Advances in photochemistry*. John Wiley & Sons, Ltd, pp. 1–122. doi: 10.1002/9780470133361.ch1. ISBN 978-0-470-13336-1

Gonçalves, M.L.L., A.C.C. da Mota, A.M. Deana, L.A.D.S. Cavalcante, A.C.R.T. Horliana, C. Pavani, L.J. Motta, K.P.S. Fernandes, R.A. Mesquita-Ferrari, D.F.T. da Silva, P. de Barros Motta, R.A. Prates, and S.K. Bussadori. 2020. Antimicrobial photodynamic therapy with Bixa orellana extract and blue LED in the reduction of halitosis—A randomized, controlled clinical trial. *Photodiagnosis Photodyn. Ther.* 30: 101751. doi: 10.1016/j.pdpdt.2020.101751.

Gütlich, P., and H.A. Goodwin. 2004. *Spin crossover in transition metal compounds*. Berlin: Springer. ISBN 978-3-540-40394-4. OCLC 56798940.

Hally, C., P. Delcanale, S. Nonell, C. Viappiani, and S. Abbruzzetti. 2020. Photosensitizing proteins for antibacterial photodynamic inactivation. *Transl. Biophotonics* 2: e201900031. doi: 10.1002/tbio.201900031

Hally, C., B. Rodríguez-Amigo, R. Bresolí-Obach, O. Planas, J. Nos, E. Boix-Garriga, R. RuizGonzález, and S. Nonell. 2018. *Theranostics and image guided drug delivery*, ed. M. Thanou, vol. 4. Cambridge: Royal Society of Chemistry, 86–122, doi: 10.1039/9781788010597

Jia, H.R., Y.X. Zhu, Z. Chen, and F.G. Wu, 2017. Cholesterol-assisted bacterial cell surface engineering for photodynamic inactivation of gram-positive and gram-negative bacteria. *ACS Appl. Mater. Interfaces* 9: 15943–15951. doi: 10.1021/acsami.7b02562

Jiang, Y., and E.A. Weiss. 2020. Colloidal quantum dots as photocatalysts for triplet excited state reactions of organic molecules. *J. Am. Chem. Soc.* 36: 15219–15229. doi: 10.1021/jacs.0c07421.

Julliard, M., and M. Chanon. 1983. Photoelectron-transfer catalysis: Its connections with thermal and electrochemical analogs. *Chem. Rev*. 83(4): 425–506. doi: 10.1021/cr00056a003

Kavarnos, G.J., and N.J. Turro. 1986. Photosensitization by reversible electron transfer: Theories, experimental evidence, and examples. *Chem. Rev*. 86: 401–449. doi: 10.1021/cr00072a005. ISSN 0009-2665.

Kelkar, S.S., and T.M. Reineke. 2011. Theranostics: Combining imaging and therapy. *Bioconjug. Chem.* 22: 1879–1903. https://doi.org/10.1021/bc200151q

Karimi, M., P. Sahandi Zangabad, S. Baghaee-Ravari, M. Ghazadeh, H. Mirshekari, and M.R. Hamblin. 2017. Smart nanostructures for cargo delivery: Uncaging and activating by light. *J. Am. Chem. Soc*. 13: 4584–4610. doi: 10.1021/jacs.6b08313.

Kashef, N., Y.Y. Huang, and M.R. Hamblin. 2017. Advances in antimicrobial photodynamic inactivation at the nanoscale. *Nanophotonics* 6: 853–879. doi: 10.1515/nanoph-2016-0189

Kim, M.-J., D.K. Lianto, G.H. Koo, and H.-G. Yuk. 2021. Antibacterial mechanism of riboflavin-mediated 460 nm light emitting diode illumination against Listeria monocytogenes in phosphate-buffered saline and on smoked salmon. *Food Control* 124: 107930. https://doi.org/10.1016/j.foodcont.2021.107930

Kwiatkowski, S., B. Knap, D. Przystupski, J.K. Saczko, E Kędzierska, K. Knap-Czop, J. Kotlińska, O. Michel, K. Kotowski, and J. Kulbacka. 2018 Photodynamic therapy—Mechanisms, photosensitizers and combinations. *Biomed. Pharmacother*. 106: 1098–1107. doi: 10.3390/molecules25214964

Kubrak, T.P., P. Kołodziej, J. Sawicki, A. Mazur, K. Koziorowska, and D. Aebisher. 2022 Some natural photosensitizers and their medicinal properties for use in photodynamic therapy. *Molecules* 27(4): 1192. doi: 10.3390/molecules27041192. 555

Lan, Y., S Lu, B. Zheng, Z. Tang, J. Li, and J. Zhang. 2020. Combinatory effect of ALA-PDT and itraconazole treatment for *Trichosporon asahii*. *Lasers Surg. Med*. 25(21): 4964. doi: 10.1002/lsm.23343

Lee, H.J., S.M. Kang, S.H. Jeong, K.H. Chung, and B.I. Kim. 2017. Antibacterial photodynamic therapy with curcumin and Curcuma xanthorrhiza extract against *Streptococcus mutans*. *Photodiagnosis Photodyn. Ther*. 20: 116–119. doi: 10.1016/j.pdpdt.2017.09.003

Liu, Y., K. Ma, T. Jia, R. Xing, G. Shen, and X. Yan. 2017. Water-insoluble photosensitizer nanocolloids stabilized by supramolecular interfacial assembly towards photodynamic therapy. *Sci. Rep*. 7: 1–8. doi: 10.1038/srep42978

Lukšiene, Z.. and L. Brovko. 2013. Antibacterial photosensitization-based treatment for food safety. *Food Eng. Rev*. 5: 185–199. doi: 10.1007/s12393-013-9070-7

Ma, W.P.C.C., J. Liu, M. Li, Y.H. Hao, Y. Ji, and X.Y. Zeng. 2020. The effects of aloe emodin-mediated antimicrobial photodynamic therapy on drug-sensitive and resistant *Candida albicans*. *Photochem. Photobiol. Sci.* 19: 485–494. doi: /10.1039/C9PP00352E

Majiya, H., and A. Galstyan. 2020. Dye extract of calyces of Hibiscus sabdariffa has photodynamic antibacterial activity: A prospect for sunlight-driven fresh produce sanitation. *Food Sci. Nutr.* 8: 3200–3211. doi: 10.1002/fsn3.1580

Memar, M.Y., R. Ghotaslou, M. Samiei, and K. Adibkia. 2018. Antimicrobial use of reactive oxygen therapy: Current insights. *Infect. Drug Resist.* 11: 567–576. doi: 10.2147/IDR.S142397

Nie, X., C. Jiang, S. Wu, W. Chen, P. Lv, Q. Wang, J. Liu, C. Narh, X. Cao, R.A. Ghiladi, and Q. Wei. 2020. Carbon quantum dots: A bright future as photosensitizers for *in vitro* antibacterial photodynamic inactivation. *J. Photochem. Photobiol. B*. doi: 10.1016/j.jphotobiol.2020.111864

O'Regan, B., and M. Grätzel. 1991. A low-cost, high-efficiency solar cell based on dye-sensitizedcolloidal-TiO2films. *Nature* 353(6346): 73774. doi: 10.1038/353737a0.

Pan, H., D. Wang, H. Li, and F. Zhang. 2020. *In vitro* antimicrobial effect of curcumin-based photodynamic therapy on *Porphyromonas gingivalis* and *Aggregatibacter actinomycetemcomitans*. *Photodiagnosis Photodyn. Ther.* 32: 102055. doi: 10.1016/j.pdpdt.2020.102055

Pang, Z., R. Raudonis, B.R. Glick, T.J. Lin, and Z. Cheng. 2019. Antibiotic resistance in *Pseudomonas aeruginosa*: Mechanisms and alternative therapeutic strategies. *Biotechnol. Adv.* 37: 177–192. doi: 10.1016/j.biotechadv.2018.11.013

Park, D., M. Kim, J.W. Choi, J.H. Baek, S.H. Lee, and K. Baek. 2020. Antimicrobial photodynamic therapy efficacy against specific pathogenic periodontitis bacterial species. *Photodiagnosis Photodyn. Ther.* 30: 101688. doi: 10.1016/j.pdpdt.2020.101688

Pavel, L., K. Pavel, B. Zuzana, and M. Jiří. 2022. Antibacterial nanoparticles with natural photosensitizers extracted from spinach leaves. *ACS Omega* 7: 1505–1513. doi: 10.1021/acsomega.1c06229

Pezzuoli, D., M. Cozzolino, C. Montali, L. Brancaleon, P. Bianchini, M. Zantedeschi, S. Bonardi, C. Viappiani, and S. Abbruzzetti. 2018 Serum albumins are efficient delivery systems for the photosensitizer hypericin in photosensitization-based treatments against *Staphylococcus aureus*. *Food Control* 94: 254–262. doi: 10.1016/j.foodcont.2018.07.027

Polat, E., and K. Kang. 2021. Natural photosensitizers in antimicrobial photodynamic therapy. *Biomedicine* 9(6): 1–30. https://doi.org/10.3390/biomedicines9060584

Postigo, A., M. Funes, E. Petenatti, H. Bottai, A. Pacciaroni, and M. Sortino. 2017. Antifungal photosensitive activity of *Porophyllum obscurum* (Spreng.) DC. Correlation of the chemical composition of the hexane extract with the bioactivity. *Photodiagnosis Photodyn. Ther.* 20: 263–272. 10.1016/j.pdpdt.2017.10.023

Rai, P., S. Mallidi, X. Zheng, R. Rahmanzadeh, Y. Mir, S. Elrington, A. Khurshid, and T. Hasan. 2010. Development and applications of photo-triggered theranostic agents. *Adv. Drug Deliv. Rev.* 62(11): 1094–1124. doi: 10.1016/j.addr.2010.09.002.

Ryberg, E.C., J. Knight, and J.H. Kim. 2020. Farm-to-tap water treatment: Naturally sourced photosensitizers for enhanced solar disinfection of drinking water. *ACS EST Eng.* 206: 111864. https://doi.org/10.1016/j.jphotobiol.2020.111864

Sabbahi, S., L.B. Ayed, and A. Boudabbous. 2013. Cationic, anionic and neutral dyes: Effects of photosensitizing properties and experimental conditions on the photodynamic inactivation of pathogenic bacteria. *J. Water Health* 4: 590–599. doi: 10.2166/wh.2013.219

Sang, X., J. Li, L. Zhang, Z. Wang, W. Chen, Z. Zhu, S. Zhogmin, and W. Enbo. 2014. A novel carboxyethyltin functionalized sandwich-type germanotungstate: Synthesis, crystal structure, photosensitivity, and application in dye-sensitized solar cells. *ACS Appl. Mater. Interfaces* 6(10): 7876–7884. doi: 10.1021/am501192f

Saitawee, D., A. Teerakapong, N.P. Morales, P. Jitprasertwong, and D. Hormdee. 2018. Photodynamic therapy of *Curcuma longa* extract stimulated with blue light against *Aggregatibacter actinomycetemcomitans*. *Photodiagnosis Photodyn. Ther.* 22: 101–105. doi: 10.1016/j.pdpdt.2018.03.001

Sharma, M., A. Dube, and S.K. Majumder. 2020. Antibacterial photodynamic activity of photosensitizer-embedded alginate-pectin-carboxymethyl cellulose composite biopolymer films. *Lasers Med. Sci.* 36: 763–772. https://doi.org/10.1007/s10103-020-03083-2

Siewert, B., and H. Stuppner. 2019. The photoactivity of natural products—An overlooked potential of phytomedicines? *J. Phymed.* 60: 152985. doi: 10.1016/j.phymed.2019.15298

Simões, J.C.S., S. Sarpaki, P. Papadimitroulas, B. Therrien, and G. Loudos. 2020. Conjugated photosensitizers for imaging and PDT in cancer research. *J. Med. Chem.* 63: 14119–14150. doi: 10.1021/acs.jmedchem.0c00047

Sowa, A., and J. Voskuhl. 2020. Host-guest complexes—Boosting the performance of photosensitizers. *Int. J. Pharm.* 586: 119595. doi: 10.1016/j.ijpharm.2020.119595

Souza, B.M.N., J.G. Pinto, A.H.C. Pereira, A.G. Miñán, and J. Ferreira-Strixino. 2021. Efficiency of antimicrobial photodynamic therapy with Photodithazine® on MSSA and MRSA strains. *Antibiotics (Basel)* 10: 869. doi: 10.3390/antibiotics10070869.

St. Denis, T.G., T. Dai, L. Izikson, C. Astrakas, R.R. Anderson, M.R. Hamblin, and G.P. Tegos. 2011. All you need is light, antimicrobial photoinactivation as an evolving and emerging discovery strategy against infectious disease. *Virulence* 2: 509–520. doi: 10.4161/viru.2.6.17889

Tosato, M.G., P. Schilardi, M.F.L. de Mele, A.H. Thomas, C. Lorente, and A. Miñán. 2020. Synergistic effect of carboxypterin and methylene blue applied to antimicrobial photodynamic therapy against mature biofilm of *Klebsiella pneumoniae*. *Heliyon* 6(3): e03522. doi: 10.1016/j.heliyon.2020.e03522.6

Torra, J., A. Burgos-Caminal, S. Endres, M. Wingen, T. Drepper, T. Gensch, R. Ruiz-González, and S. Nonell. 2015. Singlet oxygen photosensitisation by the fluorescent protein Pp2FbFP L30M, a novel derivative of *Pseudomonas putida* flavin-binding *Photochem. Photobiol. Sci.* 14: 280–287. 10.1039/C4PP00338A

Vázquez-Hernández, F., D.A. Granada-Ramírez, J.S. Arias-Cerón, P. Rodriguez-Fragoso, J.G. Mendoza-Álvarez, E. RamónGallegos, A. Cruz-Orea, and J.P. Luna-Arias. 2018. Use of nanostructured materials in drug delivery. In *Nanobiomaterials*; R. Narayan, Ed.; Sawston, UK: Woodhead Publishing, pp. 503–549. ISBN: 9780081007167

Vieira, C., A. Santos, M.Q. Mesquita, A.T.P.C. Gomes, M.G.P.M.S. Neves, and M.A.F. Faustino, and A. Almeida. 2019. Advances in aPDT based on the combination of a porphyrinic formulation with potassium iodide: Effectiveness on bacteria and fungi planktonic/biofilm forms and viruses. *J. Porphyr. Phthalocyanines* 23: 534–545. doi: 10.1142/s1088424619500408

Vollmer, A., A. Al-Ahmad, A. Argyropoulou, T. Thurnheer, E. Hellwig, T. Attin, K. Vach, A. Wittmer, K. Ferguson, A.L. Skaltsounis, and L. Karygianni. 2019. Antimicrobial photoinactivation using visible light plus water-filtered infrared-A (VIS + wIRA) and *Hypericum Perforatum* modifies *in situ* oral biofilms. *Sci. Rep.* 9: 1–15. doi: 10.1038/s41598-019-56925-7

Wang, Y., J. Li, S. Geng, X. Wang, Z. Cui, W. Ma, M. Yuan, C. Liu, and Y. Ji. 2021. Aloe-emodin-mediated antimicrobial photodynamic therapy against multidrug-resistant *Acinetobacter baumannii*: An *in vivo* study. *Photodiagnosis Photodyn. Ther.* 34: 102311. doi: 10.1016/j.pdpdt.2021.102311

Wang, Z., and J. Zhao. 2017. Bodipy–anthracene dyads as triplet photosensitizers: effect of chromophore orientation on triplet-state formation efficiency and application in triplet–triplet annihilation upconversion. *Org. Lett.* 19: 4492–4495. doi: 10.1021/acs.orglett.7b0204

Wei, G., G. Yang, Y. Wang, H. Jiang, Y. Fu, G. Yue, and R. Ju. 2020. Phototherapy-based combination strategies for bacterial infection treatment. *Theranostics* 30: 12241–12262. doi: 10.7150/thno.52729.

Wu, W., D. Mao, S. Xu, H. Kenry, F. Hu, X. Li, D. Kong, and B. Liu. 2018. Polymerization-enhanced photosensitization. *Chem* 4(8): 1937–1951. doi: 10.1016/j.chempr.2018.06.003

World Health Organization. 2014. *Antimicrobial resistance: Global report on surveillance*. World Health Organization. https://apps.who.int/iris/handle/10665/112642

Wu, F., Y. Zhou, L. Li, X. Shen G. Chen, X. Wang X. Liang, M. Tan, and Z. Huang. 2020. Computational approaches in preclinical studies on drug discovery and development. *Front. Chem.* 11(8): 726. doi: 10.3389/fchem.2020.00726.

Xiao, Q., J. Wu, X. Pang, Y. Jiang, P. Wang, A. Leung, W.L. Gao, S. Jiang, and C. Xu. 2017. Discovery and development of natural products and their derivatives as photosensitizers for photodynamic therapy. *Curr. Med. Chem.*, 839–860. doi: 10.2174/0929867324666170823143137

Yagnik, D., V. Serafin, and A.J. Shah. 2018. Antimicrobial activity of apple cider vinegar against *Escherichia coli*, *Staphylococcus aureus* and *Candida albicans*; downregulating cytokine and microbial protein expression. *Sci. Rep.* 8: 1–13. doi: 10.1038/s41598-017-18618-x

Yang, Y., C. Wang, Y. Zhuge, J. Zhang, K. Xu Q. Zhang, H. Zhang, H. Chen, M. Chu, and C. Jia. 2019. Photodynamic antifungal activity of hypocrellin A against *Candida albicans*. *Front. Microbiol.* 10: 1–13. doi: 10.3389/fmicb.2019.01810

Youf, R., M. Müller, A. Balasini, F. Thétiot, M. Müller, A. Hascoët, U. Jonas, H. Schönherr, G. Lemercier, T. Montier, and T. Le Gall. 2021. Antimicrobial photodynamic therapy: Latest developments with a focus on combinatory strategies. *Pharmaceutics* 13: 1995. doi: 10.3390/pharmaceutics13121995.

Zhang, X., T. Liu, Z. Li, and X. Zhang. 2014. Progress of photodynamic therapy applications in the treatment of musculoskeletal sarcoma. *Oncol. Lett.* 2014(8): 1403–1408. 10.3892/ol.2014.2332

Zhuang, J., H. Yang, Y.B. Li, N. Wang, N. Li, and N. Zhao. 2020. Efficient photosensitizers with aggregation-induced emission characteristics for lysosome- And Gram-positive bacteria-targeted photodynamic therapy. *Chem. Commun.* 56: 2630–2633. doi: 10.1039/D0CC00394H

Žudyte, B., and Ž. Lukšiene. 2019. Towards better microbial safety of wheat sprouts: Chlorophyllin-based photosensitization of seeds. *Photochem. Photobiol. Sci.* 18: 2521–2530. doi: 10.1039/c9pp00157c [illegible] 2022. [illegible] imaging with [illegible] and [illegible]. [illegible]

[illegible] Imaging [illegible]. [illegible] 2017.

[illegible]

[illegible] Y. [illegible] 2019. [illegible] for [illegible] images [illegible]

Wang, [illegible] Y. [illegible] and Y. [illegible] 2023. [illegible] multi-[illegible] [illegible] for [illegible] [illegible] 2023: 121.

[illegible]

Wang, G., [illegible] Y. [illegible] [illegible]

[illegible]

Zhang, Y. [illegible] for [illegible] 2022.

[illegible]

10 Nanostructures in Antimicrobial Photodynamic Therapy

Santhi Latha Pandrangi and Nikhilesh Ghantala
GITAM (Deemed to be) University, Visakhapatnam, India

Suseela Lanka
Krishna University, Machilipatnam, India

CONTENTS

10.1 INTRODUCTION

In the last three decades, the WHO has noted increased resistance of microorganisms to drugs and reduced development of new antibiotics. High doses of antibiotics and vaccines cannot stop the spread of microbial infections. Microbial infections are the leading contributors to morbidity and mortality worldwide. Sensitivity to infections depends on the person's immune system. People who are very young, old, and immunocompromised are most susceptible. The epithelial surface and circulatory system, immune cells (macrophages, phagocytes) are conquered by microorganisms, then they evoke oxygen-dependent and oxygen-independent pathogenic killing reactions (Jori, 2006). There is an intense demand to develop new antibacterial regimens for treating microbial infections. Therapies based on light, such as laser modifiers and PDT, have significant ability in tackling microbial infections. In contrast, the use of PDT has previously been shown to not result in microbial resistance due to its numerous cell targets including nucleic acids, proteins, and lipids. There is

DOI: 10.1201/9781003345299-10

TABLE 10.1
Examples of Nanoparticles in PDT

Ser	Nanoparticle	Photosensitisers	Size
1	Poly Lactic-Co-Glycolic Acid	Zinc(II) Phthalocyanine	Not Assessed
2	Gold	Phthalocyanine	2–4
3	Poly Lactic-Co-Glycolic Acid	Bacteriochlorophyll-a	660
4	Organically Modified Silica	2-(1-Hexyloxyethyl)-2-Devinyl Pyropheophorbide	30
5	Polyacrylamide	Methylene Blue	30
6	Silica-Based Magnetic Nanoparticles	2,7,12,18- Tetramethyl-3,8-di-(1-propoxyethyl)-13,17-bis-(3-hydroxypropyl)porphyrin	20–30
7	Silica Nanoparticles	Hypocrellin A	110

dominance for popular and fundamental antibiotics that likely induce microbial resistance because they are restricted to fewer targets. PDT includes the use of photosensitisers and light sources. After photo-activation of the photosensitiser, a chemical reaction takes place to produce ROS, which produce damage to cells. In PDT, photosensitisers are particularly localised to microbial infection affected areas, either locally or by injecting. This procedure restricts damage to adjacent tissue and makes photodynamic therapy very appealing. However, photodynamic therapy is hydrophobic and aggregates in a physiological environment, so most photosensitisers have the problem of low water solubility. This makes it even more difficult to determine the optimal dose of photosensitiser and light irradiation, which can reduce ROS production, quenching the elevated energy state of the cluster photosensitiser (Nancy, 2020). Use of nanoparticlescan label the modern challenges of aPDT and improve the production. Their wide contact area allows for good attribute through the cell membrane and enhance more efficient reactive oxygen species production by restricting photosensitiser accumulation. The nanoparticle system will increase the water solubility of the photosensitiser, more effectively transfer the desired dose to the specific site, and minimise loss to adjacent tissue. In addition, nanoparticles will minimise photo-degradation and rapid release at the target site during delivery of the photosensitiser. Nanoparticles are accumulated near specific cells because of its improved permeability and retention (Balhaddad et al., 2020; Fekrazad et al., 2017; Ghorbani et al., 2018; Bloise, 2017) (Table 10.1).

10.2 ANTIMICROBIAL CONCEPTS OF PHOTODYNAMIC THERAPY

There must be three key components to perform PDT: visible light, PS, and oxygen. Note that PS is harmless unless exposed to visible light. The visible light at a particular wavelength is exposed, PS is excited to a higher energy triplet state. The combination of triplet PS and biomolecules will give an outcome of cytotoxicity and antibacterial activity of photodynamic therapy. The visible light changes the photosensitiser material from the lower energy level to the excited level. The absorbed energy is emitted by the excited PS, the single excited PS is transferred to the triple excited state before coming to the ground state. Triple excited photosensitisers start one out of the two probable paths that describe photodynamic effects (type I or type II). Type I mechanisms include transfer of electrons within excited photosensitisers and the natural constituents within cell (Chandna et al., 2020). The outcome with the combination is a free radical that is highly reactive and interacts with O_2 molecules to produce harmful ROS such as H_2O_2, O_2^-, HO •. Reactive oxygen species molecules strike the cell membrane and cause rupture through peroxidation of lipid in the cell membrane components. In the two pathways, type II pathway has the photosensitiser that elevates the reaction straight with oxygen molecules to produce highly reactive singlet oxygen. Singlet oxygen molecules react with different types of biomolecules, causing damage due to oxidation of cell membranes or cell walls. Singlet oxygen can kill a variety of microbes like viruses, bacteria, fungi, and protozoa. Gràm-negative bacteria are difficult to destroy, as triplet PS should induce its action when interacted with the outer wall (Bekmukhametova et al., 2020; Perni et al., 2011). Although PDT can act through all types of pathways, singlet oxygen is noted to be the major toxic to the cell involved in the process of photodynamic therapy. Therefore, aPDT is mainly affected with the type II reaction. The lifetime of ROS and singlet oxygen produced was short. A single molecule of oxygen is observed in the biological environment for < 0.04 ms and has a radius of action of 0.02 mm. Therefore, for efficacy and security, the PS must be transported locally to the destination. Since visible light must have a specific wavelength of comparatively high intensity to excite PS chemicals, lasers can significantly improve photodynamic therapy and provide powerful monochromatic light with a small amount of energy (Pagonis et al., 2010; Huang et al., 2012a;2012b).

10.2.1 Advantages of Photodynamic Therapy

The safety and effectiveness of photodynamic therapy in local applications, procedural selectivity is one of the benefits of PDT. In addition, the use of photodynamic therapy does not try to induce tolerance as seen in the administration of antibacterial agents. PDT has several advantages over traditional chemical antibiotics. PDT can be used to:

i. Eradicate different bacteria that cause different oral disorders.
ii. Remove bacteria in a short time (Silvestre et al., 2021).

10.2.2 Limitations in Photodynamic Therapy

Precautions and effectiveness are one of the advantages of PDT, but few key obstacles threaten these two key factors. First, the photosensitiser molecule has a wide range of p-conjugations that flatten the 3D structure. In addition, these particles are too hydrophobic. As the effect, photosensitiser molecules try combining with the aqueous media, reducing the potency at the specific site (because photosensitiser had the strongest potency when it is in singlet state). Second, PS has been observed to not move or stick to specific target cells, making it hard to affect only lesioned tissue. Therefore, much effort has been devoted towards developing the delivery system to overcome these deficiencies (Ghorbani et al., 2018; Bloise et al., 2017).

10.3 NANOMATERIALS IN aPDT

Although the above methods have proven to be new alternatives for eradicating bacterial infections, it is possible that bacteria can establish a mechanism of resistance to these methods. As bacterial pathogens evolve at a faster rate, it is a bacterial species through a variety of mechanisms, including alteration of membrane permeability, evolution of multidrug efflux pumps, enzymatic degradation of drugs, self-mutations, and covalent changes (Mirzahosseinipour et al., 2020; Fathi, 2013), facilitates the acquisition of resistance to antibacterial agents. These factors greatly complicate the situation and indirectly enhance the pathogen resistance mechanism. Recent developments in nanobiotechnology mediate current research that focuses on these complications and provides appropriate solutions to these problems, and nanoparticle as a delivery system to facilitate targeted therapeutic practices. When used, it potentially increases the effectiveness of available antimicrobials (Lucky et al., 2015). Over the last decade, nanobiotechnology has grown tremendously, offering many uses in the health industry. Nanomaterials are primarily seen as multitasking products in managing the diseases that may emerge as possible antibacterial agents against resistant bacteria. In addition, nanomaterials are widely used as drug carriers that can deliver traditional antibiotics to the correct location, facilitating a specific target therapeutic approach. Nanoparticles are utilised as a proper tool in the identification of infectious diseases (Perni, 2011; Paramanantham et al., 2019; Sun et al., 2018). Many other diagnostic tools based on nano systems are currently available for accurate detection of infectious microorganisms in the early stages of infection. The nano-based identification tools have been discovered to particularly find pathogens and to differentiate between Gram-positive and Gram-negative bacteria. In future the nanoscale diagnostic tools are being developed and widely used in antibacterial applications. In other aspect, nanomaterials such as liposomes, solid lipid, polymer, silica, and gold are used widely as potential drug delivery systems (Revuelta-Maza et al., 2021; Mroz et al., 2007). From the above-described nanomaterials, carbon-based have their extraordinary structural, physical properties, relatively good biocompatibility, and morphology. The trend of carbon NPs has produced a novel variety of nanocarriers that can effectively deliver the drug molecules to the desired site of infection. The use of nanomaterials as a delivery system is accepted because it has the potential to recruit therapeutic molecules to the bacterial cytoplasm. Interestingly, the nanocarriers can be combined with one or more bioactive molecules such as therapeutic compounds, antibacterial peptides, biosensing proteins, and nucleic acids to deliver antibacterial agents to cells or organs. The undesired absorption referred in the point is the attachment of materials from the system to the CNPs during drug delivery, which minimises the initial effects of the nano formulated drugs and their release. Impact is high on profile and target specificity. Among different nano-based therapeutic ways, CNPs have proven to be more effective and cheaper. For example, using antibiotics to make carbon nanomaterials is cheaper than using antibiotics to make the nanomaterial liposomes. In addition, antibiotics administered with carbon nanocarriers have proven to be more effective with both the steps *in vitro* and *in vivo* than free antibiotics (Hah et al., 2011). Carbon is widely available element on Earth, as it has the capability to combine with different carbons that can transform into multiple carbon allotropes. This property of carbon offers the potential to power a wide variety of carbon nanomaterials including fullerenes, dots (quantum) and nanotubes. It was found that the orbital hybrids of all CN materials were likely to be same due to its dimensional changes, but each CN material showed differences in its antibacterial process and the action mechanism. Interestingly, carbon nanomaterials of similar dimensions show changes in antibacterial activity with other parameters like shape, size, number of layers, particle charge, and surface material. The specific physicochemical activity of a CN material may depend primarily on the process that follows to produce that particular carbon nanomaterial (Lu et al., 2018; Paramanantham et al., 2019; Prasad et al., 2020).

10.3.1 Carbon Nanotubes

The discovery of antimicrobial resistant bacteria in the past few years has created the gap to develop new and effective antimicrobials through the integration of nanotechnology. Increased microbial drug resistance has been discovered for possible and other therapeutic strategies. In addition, antibacterial PDT can be enhanced as an antibacterial agent or by using nanoparticles for effective delivery of photosensitisers. In the different available groups of nanoparticles, carbon-based nanomaterials are very interesting and efficient at antibacterial photodynamic therapy. The possible use of carbon nanomedicines in antibacterial PDT is due to its good visual properties, bioadaptability, low toxicity, and superior tensile strength. The nanomaterials and its derivatives have received considerable attention in several areas due to their great physio-chemical properties. CNT is a one-dimensional carbon-based material. CNTs are of two types: single-walled carbon nanotubes (SWCNTs) in which a single layer of graphene is wrapped, and multi-walled carbon nanotubes (MWCNTs) in which another layer is wrapped concentrically. CNTs are gaining increasing interest in the field of biomedicine due to their unique structure and characteristics (huge surface area, numerous functionalities, size stability, etc.). The functioning of CNTs with different molecules will improve biocompatibility and take into cells. Binding or encapsulating carbon nanotubes with other photosensitisers is the vital tool for photodynamic therapy in infection causing diseases. Carbon nanotubes are not soluble in aqueous solution, so different surface improvements, especially non-covalent, are often used to retain the actual properties and enhance their natural applications. Single-walled carbon nanotubes are phototoxic to bacteria than multi-walled carbon nanotubes, describing the nano diameter and nano size are two vital parameters in photodynamic therapy treatment. The interaction among bacterial cells and CNTs causes destruction in the membrane and ultimate bacterial senescence by increasing ROS production. Single-walled carbon nanotubes have been applied as a possible sensitiser for photodynamic therapy in cancers (Lu et al., 2018; Paramanantham et al., 2019; Sah et al., 2018).

10.3.2 Fullerenes

Work continues on the on the development of novel and effective photosensitisers to improve the efficiency of PDT. Fullerenes have entered the market due to its high drug carrying capacity. It has a unique structure in which 60–100 carbon bioactive molecules are arranged in the shape of a soccer ball. The major applications of fullerenes are the electronics, composites and drug delivery in nanotechnology due to its tensile strength, high electrical conductivity and thermal properties. With the drastic increase in interest of fullerene in medical applications, these materials have decent priority as potential photosensitiser mediators of PDT in a variety of diseases. C60 is widely used in biomedicine, because modifying the surface of C60 with a few functional groups attached to these molecules improves water solubility, improves biocompatibility, and reduces toxicity (Lu et al., 2010, 2018). A variety of hydrophilic or amphipathic branches or a connected ring structures were fixed to the spherical core. The functions depict a high capability to generate singlet oxygen, O_2^-, HO • upon irradiation and the active species are the effective photodynamic therapy bridges for several applications. The advantages in fullerenes for PDT are:

- Highly photostable, and less photobleaching when compared to tetrapyrrole and synthetic dyes.
- Exhibition of two types of photochemistry, including type I (free radicals) and type II (singlet oxygen), where tetrapyrrole mainly shows type 2 photochemistry (Sharma et al., 2011).

10.3.2.1 *In vitro* PDT with Fullerenes

The toxicity of fullerenes with lighting help to kill various pathogens. It was Tokuyama et al., primarily released the possible phototoxicity of carboxylic acid made live fullerenes in cervical cancer cells (HeLa) of human. A 10 μM neat C60 that has a mercury lamp with visible light in it, was used to cause phototoxicity to Erich cancer cells. In other observation, the tris-malonic acid fullerene derivatives are having efficient phototoxicity than dendritic derivatives in destroying the immortalised line of human T lymphocyte cells when exposed to UV-A and B light (Paramanantham et al., 2019; Lu et al., 2010).

10.3.2.2 *In vivo* PDT with Fullerenes

In PS, Fullerenes acts efficient to kill microbial cells and also cancer cells. The important point to be discussed is whether or not this high-level *in vitro* function can be converted into *in vivo* anti-cancer or antibacterial activity for a suitable animal model. The first report for the utilisation of fullerenes in performing photodynamic therapy on real tumours was in 1997. Fullerenes were chemically changed to develop the molecule which are water soluble and increased the molecular size by PEGylation of C60. When PS was injected intravenously into mice with subcutaneous tumours on the back, C60-PEG conjugates showed higher aggregation and longer retaining in malignant tissue than in ordinary tissue. Conjugates were flushed without accumulating in specific organs. Increase in C60 dose and fluence has affected the fullerene conjugates in antitumor (Perni, 2011; Sharma et al., 2011).

10.4 APPLICATIONS OF METAL NANOPARTICLES

The use of nanotechnology has become popular and is affecting more and more scientific disciplines. Nanomaterials can easily produce compounds that are complex with different substances, including organic materials. In addition, the surface developed with nanoparticles has increased chemical activity, allowing the use of oxidised NPs to reduce the progression of pathogens and also those with antibiotic resistance. Irradiating the nanoparticles with low-power radiation of nanoparticles will result in photo stimulatory responses on the surface (Sun et al., 2018). There are several types of nanoparticles, including gold, silver, and copper; we shall also cover other applications.

10.4.1 Gold

The most popular nanoparticle in medicine is gold. One approach for cancer treatment is gold particles heated with infrared laser radiation. The unique property is the electromagnetic energy absorption in metal nanoparticles, where nanoparticle size is less than wavelength. Here, the absorption that is related with surface plasmon resonance, as the oscillations are collectively conducted with electron gas on the plane of nanoparticles. The wavelength of plasmon resonance is visible in metals. When gold nano star placed in water, the wavelength is near 520 nm for the surface plasmon resonance, because wavelength required for heating nanometres for gold is 520 nm, these properties prevent light from penetrating deep into the tissue (Maliszewska et al., 2021). The optimum wavelength is 800–900 nm. The light absorption properties of gold nanoparticles can be adjusted with control in the physical properties of the nanostructure during the synthesis. Hydrophobic thiol group is used with PS containing in it and also the phthalocyanine, as the stabiliser, 2–4 nm sized gold nanoparticles were obtained. The photosensitisers are gold nanoparticles that are used notably in the field of targeted delivery, which has significantly modified the efficiency of PDT. In addition, conjugates are not shown as toxic when injected into the body at therapeutic doses (Khan, 2012).

10.4.2 Silver

Silver is a well-known antimicrobial agent in nature and has long been used by humans to destroy various microorganisms. Colloidal nano silver is composed of fine silver nanoparticles that are suspended in water which is demineralised and deionised. In general silver nanoparticles are 20–25 nm in size. Since the specific surface area is very large, the contact surface between silver and bacteria and viruses becomes large, and the bactericidal effect is greatly improved. Therefore, using silver nanoparticles can reduce the silver concentration by a factor of 100 with preserving the bactericidal properties. The metal nanoparticles are widely used as a technique for targeted transport of PS to tissues (Sun et al., 2018). The function of silver NPs with the binding of photosensitisers in such nano vessels is significant for the productivity of microbial defect identification and separation with the bimodal action of such composites, may lead to a prominent development. The nano-sized metal particles formation and stabilisation requires careful consideration of manufacturing conditions and the stabilisers performance. Nanoparticles such as silver, gold, platinum, and copper are produced in different ways, but only sphere shape is available. Finally, silver NPs are consistent and productive approach for improving the photodynamic (PD) effects on microbes and their safety on mammalian cells for the PDT of tumours (Król et al., 2018).

10.4.3 Copper

Copper is an element that is indispensable to the human. With the recommendations, the daily required amount of copper is approximately 900 μg. Immunodeficiency is occurred with the deficiency in copper that reduces granulocyte phagocytosis and immunoglobulin synthesis. Copper has the property of cell proliferation and differentiation in the biological process. Copper gluconate

(Cu^{2+}) has been experimentally proven to help increase IgG levels in immunodeficiency, controls the emergence of malignant cells and improve anticancer effects. On the contrary, copper deficiency opens up new growth potential. Copper sulphide is the mostly used nanoparticle in photodynamic therapy. Copper sulphide nanoparticles are characterised by low cost, uncomplicated manufacturing, and small size for surface modification, with small size being prominent for use as a cancer drug treatment. Photodynamic therapy is another treatment for cancer. Photosensitiser produces ROS such as hydroxyl radical, singlet oxygen, and peroxide, and destroys cancer cells by photo stimulation, and isused because human tissues are permeable to this energy, so it is ideal that PS should occur at wavelengths of 700–1000 nm for clinical applications (Firezuk et al., 2011; Yin et al., 2022).

10.4.4 Other Applications

aPDT also known as photodynamic inactivation (PDI) is a promising strategy to eradicate pathogenic microorganisms such as Gram-positive and Gram-negative bacteria.

i. The polyphenol that has less molecular weight is curcumin (diferuloymethane) isolated from *Curcuma longa*, also has antimicrobial effects but not soluble in water. It is not stable and less bioavailability in the clinical process (Mirzahosseinipour et al., 2020). When curcumin is encapsulated into drug delivery system that will enhance the properties. Polylactic acid (PLA) is encapsulated in hydrophobic drugs as it is naturally stable, biodegradable. The polymeric nanoparticles (PLGA and dextran sulphate) are embedded with curcumin that are water soluble were used for aPDT of *C. albicans* and for oral candidiasis (Gutierrez et al., 2017; Eduard et al., 2019).
ii. The discovery of novel antibiotics is decreasing the requirement for alternative treatment for bacterial infections. For respiratory infections there is inhalable nano composition for PDT, the curcumin NPs are embedded in matrix made with mannitol. These are easily miscible in aqueous medium that has NPs properties without any change (Lipovsky, 2011).
iii. Chitosan (CS) has good efficiency in aPDT and studied the capability of erythrosine (ER) action against bacteria and yeast. By using novel gelation method chitosan NPs embedded with ER and observed its potential on planktonic cells and biofilms of *S. mutans*, *P. aeruginosa*, and *C. albicans* (Chen et al., 2012; Zhang et al., 2022).
iv. When the biofilms and planktonic cells were treated with NPs in the dark, toxicity was not present, but when they were exposed to light after the treatment, a prominent phototoxicity is seen. So, the antimicrobial activity of NPs was more than the free form. The factors affecting the PDT is the incubation time and the size of particle against *S. mutans* and *C. albicans* (Prasad et al., 2020).
v. In the present medical care system, the existing multidrug-resistant bacteria (MDRB) is the serious threat. With the traditional antibiotics MDRB cannot be treated because of its rapid mutations and resistance. The types of assimilations between the NPs and PS are divided as(Dai et al., 2010) (Figure 10.1).
vi. To enhance the effectiveness of PDI of bacteria and fungi there is a biocompatible and stable carboxy methyl CS polyethyleneimine protoporphyrin IX NPs (Hah et al., 2011; Maisch et al. 2009).
vii. The silica nanospheres are supported with the improvement of the silver NPs on the surface for APDT. The presence of Photolon in silica nano powders was tested with doping the silver NPs. It is observed that the optical absorption of Photolon is maximum which is dependent with the concentration and it can be improved with the presence of silver NPs, also used as the PS (Król et al., 2018; De Freitas et al., 2016).
viii. Liposomes have a nano-structured sac, made of bilayered lipid around the liquid core, with size varying from 50 to 1000 nm, and has a wide range of applications in nano carriers, as they can take hydrophilic drugs in liquid core and hydrophobic drugs in lipid bilayer that

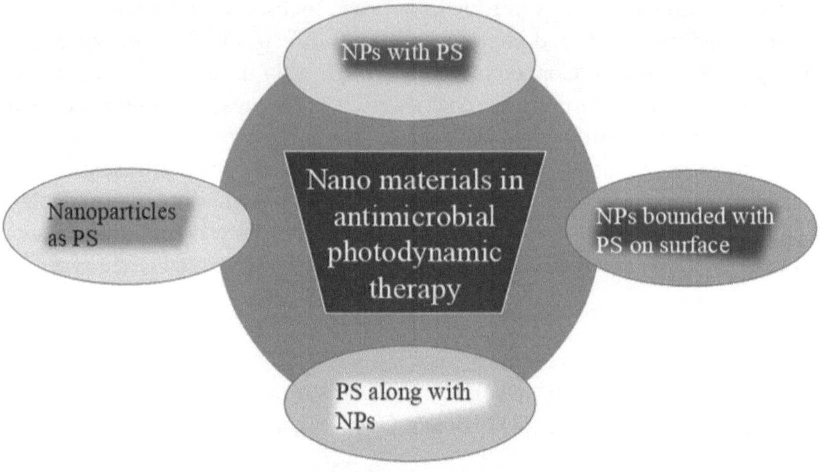

FIGURE 10.1 Different nano materials used in aPDT.

has fatty acid chain. When PS is embedded in liposomes for targeted delivery against different bacteria, it is efficient *in vitro*. The liposomes with positive charge attach firmly to the negatively charged surface of the bacteria (Derycke, 2004).

ix. Nanoemulsions (NEs) consists of incompatible liquids, as one is continuous and other is noncontinuous droplets, also as skin percolate inducers. In aPDT for peripheral skin infections, NEs that have skin percolate capability is used to stratum corneum that reaches the inner side of the epidermis. NEs can reach inner epidermis where other drugs are not able to do it (Hah et al., 2011).

x. For the delivery of photosensitisers, nano micelles are very efficient than liposomes as they have good solubility of PS and also prevent the segregation of photosensitisers in liquid solution (Pagonis et al., 2010).

xi. To have efficient aPDT against biofilms, thymol was embedded in chitosan micelles. Light dependent micelles are used for sustained response of thymol that can produce ROS with strong bactericidal effect on biofilms (Prasad et al., 2020).

xii. Chitosan micelles encapsulated with thymol were synthesised for the effective aPDT against biofilms. Light controllable micelles are formulated for controlled release of thymol which generated ROS with strong bactericidal effect on biofilms of *L. monocytogenes* and *S. aureus* (Chen et al., 2012; Zhang et al., 2022).

10.5 FUTURE PERSPECTIVES AND CONCLUSIONS

Carbon in nanomedicine is widely recommended by the research community, but use in the clinical field is not yet proven. The adjustable form of carbon nanomaterials helps to overcome barriers to go through the biological process due to its insoluble nature. The unique property of light is to produce various types of cytotoxic species provides nanostructures to be integrated into photodynamic therapy as photosensitisers or carriers. The current development of carbon nanomedicines opens up extraordinary possibilities for discovery of novel biomaterials. The action mechanism of these nanomaterials in antimicrobial PDT is described as cell membrane damage, cytoplasmic separation, ROS formation, and death due to oxidation. The antibacterial effect is influenced by the type of nanostructure, and also different conditions such as temperature, light, shape, size, and type of pathogen. In general, carbon nanostructures are potential antibacterial candidates for multiple biological uses because of its capability to destroy pathogenic bacteria and restrict its attachment and biofilm formation. The history of CNPs for aPDT is not yet completely elucidated. The application of nanostructures for effective PDT with various groups of antibiotic-resistant microorganisms and their use in clinical medicine is currently underway. Carbon nanostructures are reliable photosensitisers, carrying PDT to restrict the development of microorganisms that are antibiotic-resistant, and as well biofilms and act as effective alternatives to traditional antibiotics.

REFERENCES

Balhaddad AA, Garcia IM, Ibrahim MS, Rolim JPML, Gomes EAB, Martinho FC, Melo MAS. 2020. Prospects on Nano-Based Platforms for Antimicrobial Photodynamic Therapy against Oral Biofilms. *Photobiomodulation, Photomedicine, and Laser Surgery*: 1–16. Available from: doi:10.1089/photob.2020.4815.

Bekmukhametova A, Ruprai H, Hook JM, Mawad D, Houang J, Lauto A. 2020. Photodynamic Therapy with Nanoparticles to Combat Microbial Infection and Resistance. *Nanoscale*. Available from: doi:10.1039/d0nr04540c.

Bloise N, Minzioni P, Imbriani M, Visai L.2017. Can Nanotechnology Shine a New Light on Antimicrobial Photodynamic Therapies? *Photomedicine-Advances in Clinical Practice*. Available from: doi:10.5772/65974.

Chandna S, Thakur NS, Kaur R, Bhaumik J. 2020. Lignin-Bimetallic Nanoconjugate Doped pH-Responsive Hydrogels for Laser-Assisted Antimicrobial Photodynamic Therapy. *Biomacromolecules*. 21(8): 3216–3230. Available from: doi: 10.1021/acs.biomac.0c00695.

Chen CP, Chen CT, Tsai T. 2012. Chitosan Nanoparticles for Antimicrobial Photodynamic Inactivation: Characterization and *In Vitro* Investigation. *Photochemistry and Photobiology* 88(3): 570–576. Available from: doi:10.1111/j.1751-1097.2012.01101.x.

Dai T, Tegos GP, Zhiyentayev T et al. 2010. Photodynamic Therapy for Methicillin-Resistant Staphylococcus Aureus Infection in a Mouse Skin Abrasion Model. *Lasers in Surgery and Medicine* 42: 38–44. Available from: doi:10.1002/lsm.20887.

Derycke A. 2004. Liposomes for Photodynamic Therapy. *Advanced Drug Delivery Reviews* 56: 17–30. Available from: doi:10.1016/j.addr.2003.07.014.

De Freitas LM, Calixto GMF, Chorilli M, et al. 2016. Polymeric Nanoparticle-Based Photodynamic Therapy for Chronic Periodontitis*in vivo*. *International Journal of Molecular Sciences* 17: 769. Available from: doi:10.3390/ijms17050769.

Eduard P, Elias B, Michael AR, Thomas A, Marcel P, Marc S, Udo B. 2019. Spray Dried Curcumin Loaded Nanoparticles for Antimicrobial Photodynamic Therapy. *European Journal of Pharmaceutics and Biopharmaceutics* 142: 531–539. Available from: doi:10.1016/j.ejpb.2019.07.023.

Fathi A, Rasekaran R et al. 2013. Influence of Hydrogen Peroxide or Gold Nanoparticles in Protoporphyrin IX Mediated Antimicrobial Photodynamic Therapy on Staphylococcus Aureus. *African Journal of Microbiology Research* 7: 4617–4624. Available from: doi:10.5897/2013.5885.

Fekrazad R, Nejat A, Kalhori KAM. 2017. Antimicrobial Photodynamic Therapy with Nanoparticles versus Conventional Photosensitizer in Oral Diseases. *Nanostructures for Antimicrobial Therapy*: 237–259. Available from: doi:10.3390/molecules26030623.

Firczuk M, Winiarska M, Szokalska A, Jodlowska M, Swiech M, Bojarczuk K, Salwa P, Nowis D. 2011. Approaches to Improve Photodynamic Therapy of Cancer. *Frontiers in Bioscience-Landmark* 16: 208–224. Available from: doi:10.3390/molecules16054140.

Ghorbani J, Rahban D, Aghamiri S, Teymouri A, Bahador A. 2018. Photosensitizers in Antibacterial Photodynamic Therapy: An Overview. *Laser Therapy* 27(4): 293–302. Available from: doi:10.5978/islsm.27_18-RA-01.

Gutierrez TJK, Zanatta GC, Ortega ALM, Balastegui MIC, Sanitá PV, Pavarina AC, et al. 2017. Encapsulation of Curcumin in Polymeric Nanoparticles for Antimicrobial Photodynamic Therapy. *PLoS One* 12(11): e0187418. Available from: doi:10.1371/journal.pone.0187418.

Hah HJ, Kim G, Lee Y-EK et al. 2011. Methylene Blue-Conjugated Hydrogel Nanoparticles and Tumor-Cell Targeted Photodynamic Therapy. *Macromolecular Bioscience* 11: 90–99. Available from: doi:10.1002/mabi.201000231.

Huang L, Xuan Y, Koide Y et al. 2012a. Type I and Type II Mechanisms of Antimicrobial Photodynamic Therapy: An*in vitro*Study on Gram-Negative and Gram-Positive Bacteria. *Lasers in Surgery and Medicine* 44: 490–499. Available from: doi:10.1002/lsm.22045.

Huang Y-Y, Sharma SK, Dai T et al. 2012b. Can nanotechnology Potentiate Photodynamic Therapy? *Nanotechnology Reviews* 1: 111–146. Available from: doi:10.1515/ntrev-2011-0005.

Jori G, Fabris C, Soncin M et al. 2006. Photodynamic Therapy in the Treatment of Microbial Infections: Basic Principles and Perspective Applications. *Lasers in Surgery and Medicine* 38: 468–481. Available from: doi:10.1002/lsm.20361.

Khan S, Khan AU, Azam A, Alam F. 2012. Gold Nanoparticles Enhance Methylene Blue-Induced Photodynamic Therapy: A Novel Therapeutic Approach to Inhibit Candida Albicans Biofilm. *International Journal of Nanomedicine*: 3245–3257. Available from: doi:10.2147/IJN.S31219.

Król WK, Janus OS, Plesch G, Plecenik A, Podbielska H, Bauer J. 2018. Nano-Silver Modified Silica Particles in Antibacterial Photodynamic Therapy. *Applied Surface Science* 461: 260–268. Available from: doi:10.1016/j.apsusc.2018.05.014

Lipovsky A, Gedanken A, Nitzan Y, Lubart R. 2011. Enhanced Inactivation of Bacteria by Metaloxide Nanoparticles Combined with Visible Light Irradiation. *Lasers in Surgery and Medicine* 43: 236–240. Available from: doi:10.1002/lsm.21033.

Lucky SS, Soo KC, Zhang Y. 2015. Nanoparticles in Photodynamic Therapy. *Chemical Reviews* 115: 1990–2042. Available from: doi:10.1021/cr5004198.

Lu D, Tao R, Wang Z. 2018. Carbon-Based Materials for Photodynamic Therapy: A Mini-Review. *Frontiers of Chemical Science and Engineering*. Available from: doi:10.1007/s11705-018-1750-7.

Lu Z, Dai T, Huang L, Kurup DB, Tegos GP, Jahnke A et al. 2010. Photodynamic Therapy with a Cationic Functionalized Fullerene Rescues Mice from Fatal Wound Infections. *Nanomedicine (London, England)* 5(10): 1525–1533. Available from: doi:10.2217/nnm.10.98.

Maisch T, Santarelli F, Schreml S et al. 2009. Fluorescence Induction of Protoporphyrin IX by a New 5-Aminolevulinic Acid Nano Emulsion Used for Photodynamic Therapy in a Full-Thickness Ex Vivo Skin Model. *Experimental Dermatology* 19: e302–e305. Available from: doi:10.1111/j.1600-0625.2009.01001.x.

Maliszewska I, Wanarska E, Thompson AC, Samuel IDW, Matczyszyn K. 2021. Biogenic Gold Nanoparticles Decrease Methylene Blue Photobleaching and Enhance Antimicrobial Photodynamic Therapy. *Molecules* 26(3): 623. Available from: doi:10.3390/molecules26030623

Mirzahosseinipour M, Khorsandi K, Hosseinzadeh R, Ghazaeian M, Shahidi FK. 2020. Antimicrobial Photodynamic and Wound Healing Activity of Curcumin Encapsulated in Silica Nanoparticles. *Photodiagnosis and Photodynamic Therapy* 29: 101639. Available from: doi:10.1016/j.pdpdt.2019.101639.

Mroz P, Tegos GP, Gali H et al. 2007. Photodynamic Therapy with Fullerenes. *Photochemical & Photobiological Sciences* 6: 1139–1149. Available from: doi:10.1039/b711141j.

Nancy S, Vincent S, Tan-Sothea O, Christophe TM, Gilles G. 2020. Encapsulation of a Ru(II) Polypyridyl Complex into Polylactide Nanoparticles for Antimicrobial Photodynamic Therapy. *Pharmaceutics* 12(10): 961. Available from: doi:10.3390/pharmaceutics12100961.

Pagonis TC, Chen J, Fontana CR, Devalapally H, Ruggiero K, Song X, Foschi F, Dunham J, Skobe Z, Yamazaki H, Kent R, Tanner AC, Amiji MM, Soukos NS. 2010. Nanoparticle-Based Endodontic Antimicrobial Photodynamic Therapy. *Journal of Endodontia* 36(2): 322–328. Available from: doi:10.1016/j.joen.2009.10.011.

Paramanantham P, Anju VT, Dyavaiah M, Siddhardha B. 2019. Applications of Carbon-Based Nanomaterials for Antimicrobial Photodynamic Therapy. *Microbial Nanobionics*: 237–259. Available from: doi:10.1007/978-3-030-40337-9.

Perni S, Prokopovich P, Pratten J, Parkin IP, Wilson M. 2011. Nanoparticles: Their Potential Use in Antibacterial Photodynamic Therapy. *Photochemical & Photobiological Sciences* 10(5): 712. Available from: doi:10.1039/C0PP00360C.

Prasad R, Siddhardha B, Dyavaiah M. 2020. Nanostructures for Antimicrobial and Antibiofilm Applications. *Nanotechnology in the Life Sciences*. Available from: doi:10.1007/978-3-030-40337-9.

Revuelta-Maza MA, Heras E, Agut M, Nonell S, Torres T, Torre G. 2021. Self-Assembled Binaphthyl-Bridged Amphiphilic AABB Phthalocyanines: Nanostructures for Efficient Antimicrobial Photodynamic Therapy. *Chemistry—A European Journal* 27(15): 4955–4963. Available from: doi:10.1002/chem.202005060.

Sah U, Sharma K, Chaudhri N, Sankar M, Gopinath P. 2018. Antimicrobial Photodynamic Therapy: Single-Walled Carbon Nanotube (SWCNT)-Porphyrin Conjugate for Visible Light Mediated Inactivation of Staphylococcus Aureus. *Colloids and Surfaces. B, Biointerfaces* 162: 108–117. Available from: doi:10.1016/j.colsurfb.2017.11.046.

Sharma SK, Chiang LY, Hamblin MR. 2011. Photodynamic Therapy with Fullerenesin Vivo: Reality or a Dream? *Nanomedicine* 6(10): 1813–1825. Available from: doi:10.2217/nnm.11.144.

Silvestre ALP, Di Filippo LD, Besegato JF, de Annunzio SR, Furquim A, de Camargo B, de Melo PBG, Rastelli ANS, Fontana CR, Chorilli M. 2021. Current Applications of Drug Delivery Nanosystems Associated with Antimicrobial Photodynamic Therapy for Oral Infections. *International Journal of Pharmaceutics* 592: 120078. Available from: doi:10.1016/j.ijpharm.2020.120078.

Sun J, Kormakov S, Liu Y, Huang Y, Wu D, Yang Z. 2018. Recent Progress in Metal-Based Nanoparticles Mediated Photodynamic Therapy. *Molecules* 23(7): 1704. Available from: doi:10.3390/molecules23071704.

Yin R, Agrawal T, Khan U, Gupta GK, Rai V, Huang YY, Hamblin MR. 2022. Antimicrobial Photodynamic Inactivation in Nanomedicine: Small Light Strides against Bad Bugs. *Nanomedicine* 10(15): 2379–2404. Available from: doi:10.2217/nnm.15.67.

Zhang J, Yang Z, Li YH, Durrani S, Pang AP, Gao Y, Wu FG, and Lin F. 2022. Super-Stable Chitosan-Based Nanoparticles for Broad-Spectrum Antimicrobial Photodynamic Therapy. *ACS Applied Polymer Materials* 4(1): 425–434. Available from: doi:10.1021/acsapm.1c01339.

11 Photodynamic Antimicrobial Chemotherapy

Advancements and Applications

Suseela Lanka
Krishna University, Machilipatnam, India

Shaik Hamad Sharif and Santhi Latha Pandrangi
GITAM (Deemed to be) University, Visakhapatnam, India

CONTENTS

DOI: 10.1201/9781003345299-11

11.1 INTRODUCTION

In this expeditiously growing world economy and healthcare system, infectious diseases are known to be one of the greatest challenges across the globe. The influence correlated with diseases like HIV/AIDS, tuberculosis, skin, soft tissue infections, and sepsis remains highly predominant and among these bacterial infections play a vital role associated with a number of deaths in many countries (Laxminarayan et al., 2020). The emanation of antimicrobial resistance (AMR) had intensified the situation. The introduction of penicillin and other antibiotics had a great impact on waning the prevalence and mortality rate associated with microbial infections (Frieri et al., 2017). WHO had recognised the significance of antimicrobial resistance as global precedence which requires a new method to fight against infections (Oyim et al., 2021).

The most common hallmark of AMR had been the unregulated use of antibiotics and antiseptics. Different circumstances like the exposure of normal flora along with infection-causing agents to the drugs, poor adherence due to frequent administration of prescribed drugs, and inadequate amount of bacterial flora at the site of infection are some unenviable events that have led to the development of resistance towards orthodox agents of some bacterial species makes less triumphant of the treatment regimens (Omolo et al., 2018). The ubiquitous misuse of antimicrobial agents results in the development of antibiotic/antiseptic resistance in bacterial communities (Rice, 2012). Most of the antimicrobial classes of antibiotics such as β-lactams, fluoroquinolones, and glycopeptides have reportedly developed resistance, also the newly introduced antimicrobial agents lack a different mechanism of action from the existing once, besides these newly introduced drugs in the market are alterations to the existing conventional antimicrobial agents (Theuretzbacher et al., 2020). Therefore, new agents and methods are required with a novel mechanism of action against AMR.

11.2 PHOTODYNAMIC ANTIMICROBIAL CHEMOTHERAPY (PACT)

PACT is an optimistic approach to fight against pathogenic bacterial infections. To produce reactive oxygen species (ROS), light responded photosensitisers (PS) are employed. PACT is a minimal invasion technique. The mechanism involves the action that occurs through the cytotoxic ROS, produced by the photosensitive moieties of light irradiation through a photochemical process that causes invariable damage and cell death (Vatansever et al., 2013; Wainwright, 1998). PACT is effective against a broad range of bacteria, i.e., Gram-positive, Gram-negative, multidrug-resistant strains, and antibiotic sensitive. PACT gains a lot of attention in the research area as an alternative strategy to challenge antimicrobial resistance (Sobotta et al., 2019; Zeina et al., 2001). PS gets excited from the ground state to the singlet state by the absorption of the light, followed by the intersystem crossing of the excited photosensitiser reach to the excited triplet state. To generate the toxic ROS for eliminating the microorganisms, molecular oxygen will quench the electrons (Coitiño et al., 2014; Mai et al., 2017; Oyim et al., 2021). Photosensitisers like porphyrins, acridines, phenothiazines, phthalocyanines, and chlorins, are studied for use as suitable PACT photosensitisers (Skwor et al., 2016).

11.3 PREVALENCE OF INFECTIOUS DISEASES

Infectious disease is one of the substantial challenges faced by mankind. The influence of infectious diseases is highly predominant, especially microbial infections, that play a prominent role in the increasing number of deaths occurring in the world. The advancement in novel diagnostic methods had expanded our knowledge of epidemiology and pathophysiology of disease-causing agents which occur more frequently in children and adults. The emergence of pathogens that reappeared in recent years had raised alerts among healthcare officials because of ineffective conventional methods. The discovery of new anti-infective treatments had given some relief but it is crucial to know how and when to use them to prevent misuse of drugs to avoid the highest possible negative antimicrobial resistance outcomes (Esposito, 2016). Emerging infectious disease is defined as an infectious disease that is either newly emerged in a population or had already existed and quickly risen with incidence. At least 15% of all human pathogens are the outcome of emerging infections. Antimicrobial resistance crashed the prevailing health measures because of the evolution of an existing organism and the spreading of new infections or existing infections into new topographical areas or the areas that are under continuous geographical transformation.

Emerging infections are mostly either zoonotic or synoptic, requiring an animal host as a repository to incubate and transmit among the human population. These emerging infectious diseases are airborne, foodborne, or vector-borne. The infection-causing agent must be introduced into the population to initiate, and the agent should have the ability to spread from one person to another to cause infection (McArthur, 2019). There is a high impact of burden of infections around the globe due to the outbreak of historical and contemporary infections as emerging greatly.

11.4 ANTIMICROBIAL DRUG RESISTANCE (AMR)

Antimicrobial resistance can be defined as after a course of time organisms like bacteria, viruses, and fungi will no longer respond to the medicines or get resistant and making infections harder to treat and increasing the risk of spreading of infection and death rate. The use of antimicrobial agents has been in existence for decades to treat infectious diseases and reduce the infection and death rate caused by infectious microbial species. But regrettably, it was observed that microorganisms like bacteria, viruses, and fungi are capable of evolving resistance which makes the antimicrobial agents less productive. The root cause contributing to the emerging resistance mechanism among the microbial population is the misuse of antibiotics. Excessive administration of broad-spectrum antibiotics in response to minimal or minute infections and incognizance of the use of broad-spectrum antibiotics leads to the generation and spread of antimicrobial resistance genes (ARGs). ARGs can be attained by genetic recombination like horizontal gene transfer, transformation, transduction, and conjugation in addition to this inter- and intra-species competition, exotic antibiotic pressure, and immune recognition responses induce genetic mutations (Brinkac et al., 2017).

11.4.1 Causes of AMR

Bacteria, viruses, and fungi transform over time. The essence of microbes is to thrive, reproduce, spread quickly and efficiently. In order to survive, microbes adapt to their surrounding environment to ensure their survival. Genetic changes may occur to enable the survival of microbes against any agents which prevent microbial growth.

In case of selective pressure, the presence of antimicrobial agents, microbes will be either killed or develop resistant genes to survive and in mutation, during replication, mutation may arise and help an individual microbe to survive against the effect of antimicrobial agent. As they divide quickly and adapt to new environmental conditions, remnant microbes will continue to divide, and their lineage will quickly become presiding among the microbial population. Gene transfer method help to indulge antimicrobial-resistant gene copied from resistant bacteria to non-resistant bacteria. As a result, in presence of antimicrobial agents, the drug-resistant bacteria will survive and thrive. The inappropriate use of antimicrobial drugs to treat viral infections or in undiagnosed conditions and self-medication makes the drug less effective when the infection reappeared. Inadequate information to diagnose a particular infection and prescribing a broad-spectrum antimicrobial drug where a specific antibiotic can cure the infection. These conditions lead to selective pressure and accelerate antimicrobial resistance.

Microbial resistance to third-generation antibiotics is increasing significantly, as is the resistance to third-generation cephalosporin, due to the acquisition of extended-spectrumβ-lactamase (ESBL) genes. Most of the human population is colonised by ESBL-producing enteric bacteria which are growing annually at approximately14% around the globe (Samreen et al., 2021).

For example, tuberculosis (TB) is an airborne bacterial infection that affects the lungs of humans. TB was catastrophic before the existence of antibiotics. Due to the emergence of drug-resistant TB worldwide, it takes longer to cure, as standard antibiotic drugs will not affect the organism's ability to spread the infection as a result it takes more time for drug administration and requires close observation of patients.

11.4.2 Newer Treatment Strategies to Combat AMR

Due to the emergence and spread of antimicrobial resistance in both nosocomial and community, environments are the main cause behind the urgent need to develop alternative strategies to combat AMR. The concern regarding AMR is raising significantly in the medical field because these drug-resistant microbes will not respond to the medicine as a result, infection will spread more quickly and effectively which can increase the death rate. Hence it is important to focus on new approaches to effectively control the multidrug-resistant microorganism. One of the best non-conventional methods to combat AMR is Photodynamic antimicrobial therapy (PACT). Either PACT alone or in combination with conventional methods is the best approach to control the AMR.

11.5 PACT AND AMR

In the incidence of an increasing number of pathogen resistance to antibiotics and antiseptics, there needs an immediate requirement for new antimicrobial approaches with efficient pathogen inactivation without any risk of inducing resistance. Photodynamic antimicrobial therapy (PACT) is an evolved method to replace or compete with antibiotics, to reduce the burden of antibiotics and it is an alternative approach to conventional methods. PACT involves the combination of light and photosensitiser that can be applied in various fields, such as in the treatment of cancers, and the fight against microbial infections. When exposed to appropriate light, the photosensitiser drug gets excited and generates reactive oxygen species which have the ability to eliminate microbes within the vicinity of the drug. For efficient ROS production by PACT, different kinds of biocompatible photosensitisers, light sources, and biconjugate methods are developed (Anas et al., 2021).

In a photodynamic reaction, by the absorption of a photon, the photosensitizer (PS) gets excited to the singlet state (S_1). After excitation, the electrons will relapse to the ground state (S_0) either radiative or nonradiative and undergoes in a biological substrate having a target molecule or it gets excited to triplet state (T_1) by intersystem crossing (ISC). The photophysical process and chemical reactions involving electron and energy transfer in S_1 and T_1 states are categorised under type I and type II mechanisms. In photodynamic therapy, the most dominant pathway is type II where triplet state (T_1) energy transfers to oxygen which leads to the generation of $^1O_2$ but not type I which involves the transfer of an electron from S_1 to a biomolecule. In the excited state process, S_1 and T_1 states will generate ROS eventually, such as O_2, H_2O_2, HO, and $^1O_2$.

11.5.1 Type I Mechanism

In type 1 mechanism, transfer of charge (electron) from singlet (S_1) or triplet (T_1) excited PS to a biomolecule is the basic principle involved. A photoexcited PS is more oxidising or reducing in nature than when it is in the ground state (Anas et al., 2021). When the PS is excited from S_1 toT_1

state it cannot relapse to the ground state, and as a result the transfer of electrons into surrounding substrates takes place, giving rise to the formation of superoxide or radicals which undergoes dismutation into H_2O_2. Through Fenton-like reactions, the precursors of highly reactive free hydroxyl radicals are formed. In PACT, the type I mechanism is preferred. Nucleic acids, amino acids, and proteins are susceptible to radical ions or ROS formed by the excited PS. The type I reactions will be triggered by these radical ions or ROS, which can lead to significant damage of biomolecules and subcellular organelles (Maisch et al., 2007).

11.5.2 Type II Mechanism

In the type II mechanism, in contrast to the type I mechanism where electrons are transferred, the transfer of energy takes place in the type II mechanism immediately to the ground state where $^3O_2$ leads to the formation of $^1O_2$ which is highly unstable. The transfer of energy from this condition leads to oxidative stress by increasing the production of ROS and again, the relapsed PS in its ground state is ready to absorb a new photon and produce more ROS. Thus, the cycle of ROS generation and absorption is repeated. An excited PS can yield thousands of $^1O_2$ molecules before being destroyed (Alves et al., 2014; Baptista et al., 2017). Cell death is induced by $^1O_2$ in PACT by interaction and harming biomolecules in the cell membrane, mitochondria, nucleus, etc. as the $^1O_2$ is involved in several biological processes in the photooxidation of xenobiotics, enzymes, and lipids, the $^1O_2$ generation by excitation of PS is a key step in PACT. Mechanisms of action differ for each PS depending upon the structure. The ability to generate singlet oxygen quantum yield describes the amount of type II mechanism. In PDT or PACT, direct transfer of charge in the type I mechanism may lead to PS damage, thus the type II mechanism is preferred (Anas et al., 2021).

To develop a successful PACT application, factors like physicochemical properties of the PS, dosage applied, singlet oxygen, localisation of PS delivery along with cell structure and organisation, physiological state, and type of organism, are considered. The exposure time and wavelength of the light source contribute to the productivity of PACT. Even though the generation of ROS is due to electron transfer or energy transfer processes, the determination of reaction mechanisms in a biological system is challenging. But based on the reaction of products and interaction between biomolecules or oxygen with PS we may speculate on the path of the mechanism. The phototherapeutic pathway is determined by factors like the lifetime or stabilities of the intermediates and the environment surrounding PS. In high oxygen concentrations, a $^3\Sigma_g - O_2$ efficiency complex in a triplet excited PS, forces the energy transfer pathway to produce $^1O_2$. In hypoxic conditions, the type I mechanism is followed by PDT/PACT.

11.5.3 Interactions between Oxidative Stress Induced by PDT and Bacteria

Elevated levels of ROS inside a cell defined as oxidative stress, are omni present in the bacterial lifecycle due to intracellular bacterial metabolism or the interaction with the host of other species of bacteria. The reduction of electrons leads to the metabolic distress with the consecutive addition of electrons generating hydrogen peroxide, superoxide anions, and free hydroxyl radicals leading to the formation of ROS. Molecular oxygen is highly unreactive and able to cross biological membranes whereas various biomolecules can be oxidised by potential ROS known to cause substantial damage (Imlay, 2013). Bacteria have developed some defence mechanisms with the help of specific enzymatic pathways which can be catalysed by peroxidase, superoxide dismutase (Sod), catalase, etc., for the detoxification of ROS. Bacteria may overcome the ROS effect induced by the type I mechanism by upregulating the expression of these enzymes. But the enzymes are ineffective to help the bacteria to withstand the effect of ROS despite having some defence mechanism due to the fact that oxidative damage occurs only at the site where the PS had been restrained. Therefore, it is difficult for bacteria to survive the PACT challenge (Maisch, 2015). There is no particular defence system present in microbes to resist the $^1O_2$ which is energised molecular oxygen generated by the type II mechanism (Baptista et al., 2017).

11.5.4 Development of Resistance towards PACT

The development of resistance to antibiotics and antiseptics is a common approach in pathogens that focus on one common structure according to the key-lock principle. Pathogens may overcome the effect of antimicrobial challenge by means of upregulation of defence-associated enzymes, small mutations, or expression of efflux pumps. Whereas in the case of PACT, which is a non-selective procedure with multiple target sites in focus, developing resistance towards PACT will be difficult for bacteria (Mcdonnell and Russell, 1999; Wainwright et al., 2017).

11.6 LIGHT RESOURCES FOR PACT

The activation or excitation of PS requires a proper light source. To achieve successful photodynamic therapy, choosing an appropriate light source is imperative. For light-activation of PS, gas-discharge lamps, light emitting diodes (LEDs), and lasers are the main light sources available so far. Argon, aluminium, neodymium-doped: yttrium, and diode are examples of lasers. Lasers are used mainly because of their monochromaticity, a result of which is that they can be integrated into a single optical fibre and be easily mounted to any light device rather than for properties like parallel beam propagation, narrow spatial intensity, and coherence. Whereas LEDs costs little compared to laser, they deliver a wide range of the spectrum. Halogen lamps have the advantage which they can be spectrally filtered to match any PS. Halogen lamps like xenon-discharge lamps and quartz-tungsten-halogen cannot be efficiently combined into a single optical fibre bundle, also they cause more heat in contrast to lasers and LEDs (Nagata et al., 2012; Wilson and Patterson, 2008). Evaluation of light source should be done regarding the release of heat before opting for any kind of clinical applications of PACT in humans. Factors like the energy dosage used, period of irradiation, and the light source chosen, can lead to a rise in temperature which can significantly damage tissue (Spranley et al., 2012). For instance, irradiation of teeth for greater than or equal to 60 seconds using a halogen lamp increases the temperature of dentin by around 5°C. whereas an increase in temperature of 5.5°C will show adverse effects on dental pulp (Nammour et al., 2010). Crucial factors to be considered for irradiation of any type of PS are the type of light source, intensity of the corresponding light source, spectral emission, and mode of delivery.

11.7 SOURCES OF PHOTOSENSITISERS

The pioneering of the photodynamic process leads to the evolution of relevant light-activable dyes as a source of PS. In the last two decades, to improve the properties of PS several paths were considered either by introducing new classes or altering the chemical structure of PS. Some of the most used PS derivatives are phenothiazinium, porphyrin, xanthene, fullerene, riboflavin, curcumin, etc. (Wainwright et al., 2017).

11.7.1 Phenothiazinium Derivatives

Methylene blue consisting of a three-ring-π-system with an auxochrome side group, and a single positive charge was the first phenothiazinium derivative developed in 1870. It exhibits a singlet quantum yield below 0.5. They act predominantly according to the type I mechanism and strong absorption is seen at the red spectrum (600–680 nm) with the advantage of penetration of light from a longer wavelength into the tissue. It was the first clinically approved PS for dentistry, and it is a standard PS *in vitro*. Methylene Blue and Toluidine Blue are the examples of phenothiazinium derivatives (Gollmer et al., 2015; Wilkinson et al., 1993).

11.7.2 Porphyrin Derivatives

These are tetrapyrrole molecules with four pyrrole cycles or three pyrrole and one pyrroline subunits which are structurally similar to chlorins and phthalocyanines which are heterocyclic-macrocyclic compounds. Porphyrin derivatives are beneficial PS to use. The maximum absorption is seen strongly in UV-Vis near the IR region. The ability to generate triplet quantum yield is the main factor that makes them unique ROS generators (Nam et al., 2000). As a result, the porphyrin derivatives are remarkably considered as antimicrobial agents (Vzorov et al., 2002). They act mainly according to the type II mechanism with the production of singlet oxygen quantum with yields between 0.5 to 0.8 (Fernandez et al., 1997). The exhibition of concentrated absorption maximum is seen at 405 nm and smaller peaks are seen at a wavelength longer than 500 nm (Gouterman, 1961). Due to their hydrophobicity, the bioavailability of porphyrins is very low and are poorly water-soluble. Modifications in their structures by the addition of polar subunits or by coupling with hydrophilic components such as proteins, peptides, and amino acids with enhanced water solubility is the key factor in the application of PACT. For example TMPyP [5,10,15,20-tetrakis(1-methyl-4-pyridinium)- porphyrin tetra-(p-toluene sulfonate] is a fourfold derivative which is positively charged. TMPyPis widely used *in vitro* for this purpose.

11.7.3 Fullerene Derivatives

These are made up of carbon atoms that resemble the shape of a soccer ball, the most studied compound is fullerene C60. They are functionalised in many ways due to the addition of positively charged moieties (Mizunoet al. 2011; Sharma et al., 2011). Fullerenes exhibit strong absorbance in UV and visible range. They react predominantly to both type I and type II mechanisms depending upon the type of solvent and chemical modifications. It is a special class of PS used to study its effect on biofilms.

11.7.4 Riboflavin Derivatives

Riboflavin also known as vitamin B_2 is a known molecule that generates singlet oxygen up on UVA irradiation. But poor hydrophilicity and paucity of positive charges made its antimicrobial efficacy unexceptional (Makdoumi and Bäckman, 2016). Whereas riboflavin derivatives which are substituted with positive charges had shown improved photodynamic efficacy. These derivatives show maximum absorption in UVA/blue spectrum (350–470 nm) with a quantum yield between 0.7–0.8 (Maisch et al., 2014) as the decomposition of flavin molecules is universal, it is safe to use on humans.

11.7.5 Curcumin Derivatives

Curcumin is a crude, yellow-coloured dye extracted from the roots of *C. longa* and is an approved food supplement. In spite of having poor hydrophilicity, curcumin had gone through several chemical modifications by which it carries positive charges has been developed (Spaeth et al., 2018; Tortik et al., 2014). Curcumin derivatives quantitatively act accordingly to the type I mechanism and show maximum absorption at UV/blue spectrum (300–500 nm) (Araújo et al., 2014). Curcumin is an approved food additive, and its derivatives of PACT are positively charged with improved water solubility.

11.8 BIOCONJUGATE METHODS TO PRODUCE EFFICIENT ROS

Besides being a by-product of aerobic respiration, ROS also plays a key role in metabolism regulation and signal transduction. Radicals and non-radical molecules like superoxide, H_2O_2, and $^1O_2$

have been considered deleterious for cellular metabolism (Tripathy and Oelmüller, 2012; Zhou et al., 2016). In mitochondria, chloroplast, and peroxisomes, ROS is produced constantly during photosynthesis and aerobic respiration. In maintaining the equilibrium of many physiological processes such as apoptosis, autophagy, cell cycle, and immune response, ROS acts as a vital cellular signalling molecule (Covarrubias et al., 2008; Ren and Ai, 2013). The disparity between production and depletion levels of ROS leads to diseased conditions (Casetta et al., 2005). In recent studies, genetically encoded ROS generators have gained a lot of interest. Some fluorescent proteins have been considered effective PS; also, certain enzymes produce ROS effectively through biochemical interactions. To understand the biological functions of ROS, protein-based ROS generators are considered a fine research tool (Gaietta et al., 2002; Toyokuni, 1999).

Certain groups of genetically fused fluorescent proteins target proteins through the generation of ROS which then diffuses into surrounding tissue and inactivate that region. The selective inactivation of proteins within a cell can be done using the technique of Chromophore-assisted light inactivation (CALI). The genetically encrypted PS is significant to fuse into almost any target protein. These genetically encrypted PS are used to target the mitochondria, plasma membrane, and chromatin to eliminate cells selectively in both *in vivo* and in culture models. Also, fluorescent proteins have been investigated as phototoxic cancer therapeutic agents (Ryumina et al., 2013; Serebrovskaya et al., 2009).

NADPH Oxidase (NOX), is a known enzyme to produce ROS without the need for light, also these are first identified in mitochondria of neutrophils, and it is a vital source for external production of ROS in the form of superoxide and hydrogen peroxide. Components of the respiratory chain in mitochondrial can be regulated to produce superoxide in huge amounts (Murphy, 2009). Myeloperoxidase (MPO) and Xanthine oxidase (XOR) are considered as effective ROS producers (Kuppusamy and Zweier, 1989). Flavin adenine dinucleotide-containing enzymes are excellent ROS-generating components that involve the conversion of D-amino acid in presence of O_2 into α-keto acid which leads to the enzyme and D-amino acid-dependent H_2O_2 production.

Synthetic PS along with genetically encoded elements are developed to control the localisation of synthetic PS Hybrid ROS generators. Chemical photosensitisers can be combined with proteins through genetically encrypted tags using FISH, and halo tag technologies (Takemoto et al., 2011). The availability of different synthetic photosensitisers is an advantage. Excitation with NIR light regulates proteins and cells in the tissue of a living organism. The main disadvantage associated with hybrid ROS generators is the ingestion of external PS leads to penetration, localisation, and distribution which could be unfavourable for live cell, tissue, and *in vivo* applications (Xiong et al., 2019).

11.9 ANTIBIOTICS IN PACT

The photodynamic inactivation of microorganisms by approaching PACT to eliminate bacteria is an exceptional method to prevent the growth of AMR pathogens. But there is a possible limitation of regrowth of bacteria that were not inactivated during irradiation (Wozniak and Grinholc, 2018). To overcome such limitations, PACT in combination with antibiotics can be used to eliminate the regrowth of such microbes which were not inactivated during irradiation. For example, 2-aminothiazolo[4,5-c]-2,7,12,17-tetrakis (methoxyethyl) porphycene (ATAZTMPo)- gentamicin conjugate a porphycene derivative which is effective in to suppress the antimicrobial effect of *S. aureus* and *E. coli* (Nieves et al., 2020). 5-aminolevulinic acid (5-ALA) a combination of protoporphyrin IX (PpIX), and a low concentration of gentamycin and vancomycin in combination with meso-tetra (N-methyl-4-pyridyl) porphine tetra-tosylate (TMP), a cationic porphyrin is effective against *S. aureus* biofilm formation (Iluz et al., 2018). The mechanism involved is PACT will disrupt the cell membrane so that gentamycin can penetrate better inside the membrane and eliminate the microbe. Deuteroporphyrin-PS combined with oxacillin has the potential as a cell wall synthesis inhibitor. Methylene blue (MB) a PS in combination with chloramphenicol and MB with ciprofloxacin showed an additive effect against planktonic *S. aureus* and also in biofilm formation (di Poto et al., 2009).

11.10 APPLICATIONS OF PACT

11.10.1 Dentistry

There are several conditions in the field of dentistry, where the PACT is highly applicable both in restricting the use of antibiotics and also to improve treatment modalities. The use of antibiotics is the most frequently used method for treating critical cases followed in periodontology to ensure favourable treatment for the patient. PACT is a more effective method to control dental infections and implants, with the advantage of narrowing the dawn of antibiotic-resistant microorganisms. Patients who are allergic to some types of antibiotics can opt for PACT, which is highly advantageous in this group of patients (Meimandi et al., 2017; Zhang et al., 2019). It was shown as a secondary action procedure with nonsurgical handling. The concomitant effect of PACT remains argumentative as it was unproductive in removing microbial load in perplexing critical situations such as patients with a history of smoking (Queiroz et al., 2014). The situation can be possible to overcome by choosing an effective and appropriate PS by optimising their concentrations and light sources and exposure time, and chemical variables like pH, and exudates of the area undergoing treatment (Vohra et al., 2014). PACT offers advanced microbial disinfection that ensures successful long-term outcomes. *In vitro* studies had shown that with the help of photodynamic therapy, 90–98% bacterial killing efficiency was achieved compared to triple antibiotic paste. PACT is one of the best methods for He/O_2 cold jet plasma treatment in disinfecting root canals (Armand et al., 2019). The potential of PACT in dentistry varies in the efficiency of killing microbes depending upon PS concentration. The intensity of laser depends upon interactions between PS with various bacterial infections (Shrestha and Kishen, 2012).

11.10.2 Dermatology

PACT is a valuable application for treatment in dermatology, to treat chronic wounds and all kinds of infections that arise from either bacteria or fungal origin (Baltazar et al., 2015; Brown, 2012; Maisch et al., 2004). Stimulation of different growth factors and immune responses leads to better wound healing, and it is an important application of PACT treatment of skin ulcers and chronic wounds. The application of phenothiazinium derivative as PS in the case of diabetic foot ulcers and chronic leg ulcers had shown accelerated wound healing along with broad-spectrum bacterial cell elimination (Brown, 2012; Morley et al., 2013). For the treatment against the infection of *P. aeruginosa* in lower limbs, O-aminolevulinic acid-mediated PACT had shown significant healing of infection. PACT can be suggested to prevent amputations due to diabetic foot ulcers. PACT can be considered an efficient and affordable treatment method to prevent surgical debridement (Tardivo et al., 2014).

11.10.3 Microbial Inactivation and Biofilm Formation

The formation of polymeric matrix called biofilms acts as a blockade to reducing the effect of antibiotics entering inside the matrix and thus acts as a first line of defence for the cells (Taraszkiewicz et al., 2013). As a result, bacteria that are capable of biofilm formation are hard to kill using conventional antibiotics. The biofilm consists of different proteins, lipids, and polysaccharides that are responsive to oxidative stress generated by PACT (Hu et al., 2018). Microbial infections of mucosa and skin are easily treated by PACT using an appropriate light source and PS. The disruption of biofilm integrity leads to better diffusion of antibiotics. The formation of biofilm by *S. haemolyticus*is affected by 5-aminolevulinic acid PS and the subsequent gentamycin treatment killed bacterial cells (Barra et al., 2015).

11.10.4 Implants

Titanium is the most widely used implant material due to its biomechanical and biocompatibility. The formation of biofilms on these surfaces leads to implant failure. *S. aureus* and *S. epidermidis* are one of the most commonly occurring infections on implants. The formation of abscesses due to such infections will severe complications. Generally, dental implants get infected. PACT is an efficient method to treat such kinds of bacterial infections with more efficacies.

11.10.5 Fungal Infections

Chemicals that are used to eliminate fungal contamination can be vulnerable to humans so it is difficult to treat fungal infections. This condition can be overcome with the help of PACT because of the fact that the oxidative stress tolerance mechanism of humans is greater compared to microbial pathogens (Lam et al., 2011; Snell et al., 2012). Using hydrogels having a PS, when exposed to LED light had shown better results to reduce *C. albicans* infection. Invitro studies have shown that aloe-emodin an aloe vera extracted PS have the potential to damage cell wall and nucleus of *C. albicans* (Carmello et al., 2016).

11.10.6 Viral Infections

PACT shows a significant effect on viral infections like curing localised lesions of herpes (Wainwright, 2003; Wiehe et al., 2019). Because of their simple structure, viruses are amenable to ROS. ROS, can induce peroxidation of lipids, mediates the withdrawal of hydrogen from proteins, and genetic material is damaged which leads to loss of structural integration of viral proteins and enzyme inactivation. By considering the theoretical applications, bacterial, viral and other microbial respiratory tract infections can be controlled with recent advances in PACT (Dias and Bagnato, 2020).

11.10.7 Skin Burns

In the case of severely burned patients, infection prevention and skin growth PACT can be employed by using the combinations of PSs along with fibroblast growth factors which is a promising method. Poly (lactic-co-glycolic acid) (PLGA)-encapsulated basic fibroblast growth factor (bFGF) is an efficient hydrogel for this dual function. This hydrogel can eradicate 99.99% of *S. aureus* (Mai et al., 2020).

11.10.8 Disinfection and Surface Sterilisation

PACT acts as an excellent preventive strategy in disinfecting patients' surfaces in hospital areas and instruments to regulate the spread of infections. Photodynamic surface coating in laboratories and hospitals has shown a significant reduction below the mark of 2.5 CFU cm^{-2} in bacterial loads with the efficacy of a longer period (Nezhadi et al., 2019).

The escape of microorganisms from sewage plant treatment to natural habitat had raised concern for the spread of diseases like cholera. Sewage plant waste is discharged into canals and oceans. High microbial load is seen in subordinate treatment, which should be a concern. The orthodox methods of treating sewage/water treatment with chemicals or antibiotics can be replaced with photodynamic disinfection which can be a two fold advantage of degrading organic pollutants and killing pathogenic microbes (Bartolomeu et al., 2017; Sathyendranath et al., 2020). Therefore, technological advancements should be made to eliminate harmful bacteria from sewage before releasing them into the natural ecosystem. The photodynamic method is safer to treat food materials like cheese, meat, and fruits to decontaminate compared to UV treatment (Lin et al., 2012; Penha et al., 2017; Silva et al., 2018).

11.10.9 Water Purification

Purification of drinking water will be an advantage to people having limited access to clean water using PACT potential. The sustainable development goal of the UN is to put a stop to waterborne diseases and provide safe drinking water. PACT can be an effective method that can be used in treatment systems to disinfect drinking water (Eichner et al., 2020).

11.10.10 Aquaculture

The arbitrary use of antibiotics widely leads to the emanation of several antimicrobial-resistant bacteria to commonly used antimicrobial agents in the field of aquaculture (Thornber et al., 2020). PACT can be an appropriate alternative to these methods. The emergence of antimicrobial-resistant pathogens, and due to lack of awareness among farmers lead to the deterioration of the environment. As a result, it becomes a health hazard to humans. PACT is the safest approach to ensure efficiency in eliminating bacteria without inducing any antimicrobial-resistant pathogens (Cabello, 2006).

11.10.11 PACT and Cancer

Photodynamic therapy (PDT) is a clinically approved procedure. Factors like exerting selective cytotoxic activity on cancer cells with minimally invasive in nature made PACT more compatible. The administration of a photosensitiser followed by irradiation of a suitable wavelength leads to absorbance and activates the photosensitising agent. In presence of oxygen, a chain reaction is initiated that leads to tumour cell death. Tumour that are detected at an early stage can be cured by the application of PDT. It can improve the quality of life and extend the survival of patients suffering from inoperable tumour. The lack of acquired resistance mechanism with minimal normal tissue toxicity, and greatly declined long-term morbidity makes PDT a valuable therapeutic option. It can be a great asset to use the potential of PDT and integrate it into main stream cancer treatment (Agostinis et al., 2011). PDT can be used to treat various types of cancers such as oropharyngeal, oesophageal, skin, etc. Many attempts are made to combine PDT with conventional cancer therapies to improve its therapeutic effect.

The irradiation of laser on the selective PS leads to alterations in biochemical functions causing ROS generation by utilising the oxygen molecules present in the tumour tissue which induce toxic effects leading to apoptosis and necrosis of tumour cells (Chen et al., 2002; Allison and Sibata, 2010). Lesions, Microvascular injury, and thrombosis in tumour tissue can be persuaded by light irradiation at maximum concentrations of PS. Depending upon the PS concentrations in the vascular endothelial cells of the tumour, these lesions will significantly lead to tumour ischaemia and necrosis (Garg et al., 2010). The proliferation of immunocytes such as macrophages, neutrophils, and lymphocytes is stimulated by PDT. These cells infiltrate into tumour tissue to trigger specific immune responses against tumour cells provoked by strong inflammatory reactions (de Rosa and Bentley, 2000).

PS can be engaged by both normal and tumour cells, but normal tissues are able to eliminate PS over time whereas due to the absence of existing lymphatics, tumour tissue cannot clear the PS. The PS is preferentially accumulated into tumour tissue by an unknown mechanism but a popular theory that can be applied is the morphological difference between normal and tumour tissue (Hamblin and Luke Newman, 1994). Abnormal vasculature with defective inner lining, rapid and uncontrolled growth of tumour cells are observed in cancer conditions and the phenomenon known as Enhanced permeability and retention effect (EPR) is utilised in cancer therapeutics. Also, due to impaired lymphatic drainage in tumour tissue, healthy tissue cannot be retained longer period compared to tumor tissue (Maeda et al., 2013).

11.11 PDT IN COMBINATION WITH CHEMOTHERAPY

In general, chemotherapeutic drugs may act on cancer cells by inhibiting cell division and preventing DNA replication that leads to cell death which is facilitated by drugs that bind to the DNA of the tumour cells (Faber et al., 1995; Sederholm et al., 2005). Though the drugs are effective in eliminating cancer cells to a certain extent because of non-specific drugs and are susceptible to drug resistance with limited clinical applications, still chemotherapy exhibits severe side effects on the body.

To overcome these adverse side effects, and drug resistance to chemotherapy, several attempts have been made to combine PDT with conventional chemotherapy. PDT and chemotherapy combination may result in coordinated antitumour effects and reduce the dosage of drugs. Photodynamic combination therapy has an advantage in improving therapeutic effects for late-stage patients. The combination of doxorubicin (DOX) and nickel oxide nanoparticles (NOPs) as NOP-DOX@BSA-FA is a good example which is a potential PDT agent that increased the rate of cell death compared to when only NOP is used. Using $^1O_2$ NOP-DOX@BSA-FA in PDT and chemotherapy combination enabled the effectiveness of DOX to reach tumour sites effectively and increase the rate of cell death and decreased the side effects of chemotherapy (Wang et al., 2016). The combination of PDT with chemotherapy may overcome multidrug resistance which is associated with the upregulation of P-glycoprotein on the tumour surface which is caused due to chemotherapeutic drugs. Downregulation of P-glycoprotein and the mass accumulation of ROS may result in necrosis or apoptosis of drug-resistant cancer cells (Q. Zhang and Li, 2018). The combination of meso-tetra (di-adjacent sulphonatophenyl) porphine and vincristine or taxol had shown better antitumour effects in mouse models (Ma et al., 1996). The combination of doxorubicin and MB-PS had shown great therapeutic potential in the mouse mammary adenocarcinoma tumour model (Khdair et al., 2010). Also, the combination of heme oxygenase I along with poly (ADP-ribose) polymerase inhibitors may enhance the productivity of PACT (Kästle et al., 2011).

11.12 PDT IN COMBINATION WITH RADIOTHERAPY

The application of X-rays in radiotherapy is an effective method against the local tumour. Radiation therapy is received by most of patients suffering from malignant tumour. In nearly 90% of patients, the therapeutic rate of only radiotherapy is seen with early detection of cancer. Radiotherapy in combination with PDT can significantly alleviate the symptoms, relieve pain, sustain survival in patients with advanced tumour, and improve the quality of life. PDT has low toxicity, high efficiency, and selectivity in the treatment of tumour. PDT combined with radiotherapy can make tumour more sensitive to radiation, improve the efficacy of treatment, and it can also decrease the exposure time to radiation or dosage of radiation (Ghoodarzi et al., 2016; Takahashi et al., 2015).

11.13 FUTURE DIRECTIONS

The concept of PACT as a substituent for antibiotics to kill microbes has been progressed in the last few decades. The immediate action and advantage over antibiotic resistance make PACT an interesting technique. More research should be done to develop new light sources, photosensitisers, and light intensity which is more important for developing PACT. Implementation of PACT in sanitisation is an encouraging concept to develop especially in these pandemic situations. Development of new PS derivatives either alone or in combination with chemotherapy, and radiotherapy in the field of cancer therapeutics to improve the quality of life for the patients. Better antimicrobial activity-without any undesirable reactions on patients will be the perfect combination for the PACT system. It should be easy to handle. PS should possess features like low molecular weight which should facilitate better penetration through the polymeric matrix (biofilm) structures, high singlet oxygen yield, high photostability during irradiation, and it should not have any kind of toxicity or mutagenicity towards eukaryotic cells, and a better therapeutic window to efficiently kill microorganisms without damaging host cell.

11.14 CONCLUSION

The emergence of antimicrobial drug resistance accompanied by its quick and active spread had raised concern among clinicians. The conventional methods are not efficient to fight AMR. The development of PACT changed the scenario to combat the AMR and decreased the load of antibiotic usage. The principle behind PACT is by using an appropriate light source and efficient photosensitiser will generate ROS using the oxygen molecule present in the tissue. Bacteria and fungi will not be able to tolerate these conditions and thus they are eliminated from the system. PACT alone or in combination with antibiotics gives the best result. And it is highly impossible for microbes to develop resistance to PACT. There are many types of PS derivatives. Choosing an appropriate PS and light source will have significant impact on the microbial population. Also, PACT is used in several fields such as cancer therapeutics, where a solid tumour detected early can be treated easily by the application of PACT. PACT along with the combination of conventional chemotherapy and radiotherapy had shown promising therapeutic outcomes and also it helps in treating multi-drug-resistant cancer cells. With the implementation of PACT in cancer therapy, drug usage might be decreased resulting in alleviation of chemotherapeutic side effects. PACT can also be applied in water filtration, and sewage waste treatment to control the microbial population. Hence PACT has a privilege over conventional methods in numerous fields.

REFERENCES

Agostinis, P., Berg, K., Cengel, K. A., Foster, T. H., Girotti, A. W., Gollnick, S. O., Hahn, S. M., Hamblin, M. R., Juzeniene, A., Kessel, D., Korbelik, M., Moan, J., Mroz, P., Nowis, D., Piette, J., Wilson, B. C., & Golab, J. (2011). Photodynamic therapy of cancer: An update. *CA: A Cancer Journal for Clinicians, 61*(4). https://doi.org/10.3322/caac.20114

Allison, R. R., & Sibata, C. H. (2010). Oncologic photodynamic therapy photosensitizers: A clinical review. *Photodiagnosis and Photodynamic Therapy, 7*(2).

Alves, E., Faustino, M. A. F., Neves, M. G. P. M. S., Cunha, A., Tome, J., & Almeida, A. (2014). An insight on bacterial cellular targets of photodynamic inactivation. *Future Medicinal Chemistry, 6*(2). https://doi.org/10.4155/fmc.13.211

Anas, A., Sobhanan, J., Sulfiya, K. M., Jasmin, C., Sreelakshmi, P. K., & Biju, V. (2021). Advances in photodynamic antimicrobial chemotherapy. *Journal of Photochemistry and Photobiology C: Photochemistry Reviews*, 49. https://doi.org/10.1016/j.jphotochemrev.2021.100452

Araújo, N. C., Fontana, C. R., Bagnato, V. S., & Gerbi, M. E. M. (2014). Photodynamic antimicrobial therapy of curcumin in biofilms and carious dentine. *Lasers in Medical Science, 29*(2). https://doi.org/10.1007/s10103-013-1369-3

Armand, A., Khani, M., Asnaashari, M., Ali Ahmadi, A., & Shokri, B. (2019). Comparison study of root canal disinfection by cold plasma jet and photodynamic therapy. *Photodiagnosis and Photodynamic Therapy, 26*. https://doi.org/10.1016/j.pdpdt.2019.04.023

Baltazar, L. M., Ray, A., Santos, D. A., Cisalpino, P. S., Friedman, A. J., & Nosanchuk, J. D. (2015). Antimicrobial photodynamic therapy: An effective alternative approach to control fungal infections. *Frontiers in Microbiology*, 6(Mar). https://doi.org/10.3389/fmicb.2015.00202

Baptista, M. S., Cadet, J., di Mascio, P., Ghogare, A. A., Greer, A., Hamblin, M. R., Lorente, C., Nunez, S. C., Ribeiro, M. S., Thomas, A. H., Vignoni, M., & Yoshimura, T. M. (2017). Type I and type II photosensitized oxidation reactions: Guidelines and mechanistic pathways. *Photochemistry and Photobiology, 93*(4). https://doi.org/10.1111/php.12716

Barra, F., Roscetto, E., Soriano, A. A., Vollaro, A., Postiglione, I., Pierantoni, M. G., Palumbo, G., & Catania, M. R. (2015). Photodynamic and antibiotic therapy in combination to fight biofilms and resistant surface bacterial infections. *International Journal of Molecular Sciences, 16*(9). https://doi.org/10.3390/ijms160920417

Bartolomeu, M., Reis, S., Fontes, M., Neves, M. G. P. M. S., Faustino, M. A. F., & Almeida, A. (2017). Photodynamic action against wastewater microorganisms and chemical pollutants: An effective approach with low environmental impact. *Water (Switzerland), 9*(9). https://doi.org/10.3390/w9090630

Brinkac, L., Voorhies, A., Gomez, A., & Nelson, K. E. (2017). The threat of antimicrobial resistance on the human microbiome. *Microbial Ecology, 74*(4). https://doi.org/10.1007/s00248-017-0985-z

Brown, S. (2012). Clinical antimicrobial photodynamic therapy: Phase II studies in chronic wounds. *JNCCN Journal of the National Comprehensive Cancer Network, 10*(Supp. 2). https://doi.org/10.6004/jnccn.2012.0182

Cabello, F. C. (2006). Heavy use of prophylactic antibiotics in aquaculture: A growing problem for human and animal health and for the environment. *Environmental Microbiology, 8*(7). https://doi.org/10.1111/j.1462-2920.2006.01054.x

Carmello, J. C., Alves, F., Basso, F. G., de Souza Costa, C. A., Bagnato, V. S., Mima, E. G. D. O., & Pavarina, A. C. (2016). Treatment of oral candidiasis using Photodithazine®-mediated photodynamic therapy *in vivo*. *PLoS One*, *11*(6). https://doi.org/10.1371/journal.pone.0156947

Casetta, I., Govoni, V., & Granieri, E. (2005). Oxidative stress, antioxidants and neurodegenerative diseases. *Current Pharmaceutical Design*, *11*(16). https://doi.org/10.2174/1381612054065729

Chen, B., Roskams, T., & de Witte, P. A. M. (2002). Antivasculartumor eradication by hypericin-mediated photodynamic therapy. *Photochemistry and Photobiology*, *76*(5). https://doi.org/10.1562/0031-8655(2002)076<0509:atebhm>2.0.co;2

Coitiño, E. L., Mella, A., & Cárdenas-Jirón, G. I. (2014). Theoretical assessment of the photosensitization mechanisms of porphyrin-ruthenium(II) complexes for the formation of reactive oxygen species. *Journal of Photochemistry and Photobiology A: Chemistry*, *294*. https://doi.org/10.1016/j.jphotochem.2014.08.003

Covarrubias, L., Hernández-García, D., Schnabel, D., Salas-Vidal, E., & Castro-Obregón, S. (2008). Function of reactive oxygen species during animal development: Passive or active? *Developmental Biology*, 320(1). https://doi.org/10.1016/j.ydbio.2008.04.041

de Rosa, F. S., & Bentley, M. V. L. B. (2000). Photodynamic therapy of skin cancers: Sensitizers, clinical studies and future directives. *Pharmaceutical Research*, *17*(12). https://doi.org/10.1023/A:1007612905378

di Poto, A., Sbarra, M. S., Provenza, G., Visai, L., & Speziale, P. (2009). The effect of photodynamic treatment combined with antibiotic action or host defence mechanisms on Staphylococcus aureus biofilms. *Biomaterials*, *30*(18). https://doi.org/10.1016/j.biomaterials.2009.02.038

Dias, L. D., & Bagnato, V. S. (2020). An update on clinical photodynamic therapy for fighting respiratory tract infections: A promising tool against COVID-19 and its co-infections. *Laser Physics Letters*, *17*(8). https://doi.org/10.1088/1612-202X/ab95a9

Eichner, A., Holzmann, T., Eckl, D. B., Zeman, F., Koller, M., Huber, M., Pemmerl, S., Schneider-Brachert, W., & Bäumler, W. (2020). Novel photodynamic coating reduces the bioburden on near-patient surfaces thereby reducing the risk for onward pathogen transmission: a field study in two hospitals. *Journal of Hospital Infection*, *104*(1). https://doi.org/10.1016/j.jhin.2019.07.016

Esposito, S. (2016). Infectious diseases: Pathophysiology, diagnostics and prevention. *International Journal of Molecular Sciences*, *17*(9). https://doi.org/10.3390/ijms17091464

Faber, M., Coudray, C., Hida, H., Mousseau, M., & Favier, A. (1995). Lipid peroxidation products, and vitamin and trace element status in patients with cancer before and after chemotherapy, including adriamycin—A preliminary study. *Biological Trace Element Research*, *47*(1–3). https://doi.org/10.1007/BF02790108

Fernandez, J. M., Bilgin, M. D., & Grossweiner, L. I. (1997). Singlet oxygen generation by photodynamic agents. *Journal of Photochemistry and Photobiology B: Biology*, *37*(1–2). https://doi.org/10.1016/S1011-1344(96)07349-6

Frieri, M., Kumar, K., & Boutin, A. (2017). Antibiotic resistance. *Journal of Infection and Public Health*, 10(4). https://doi.org/10.1016/j.jiph.2016.08.007

Gaietta, G., Deerinck, T. J., Adams, S. R., Bouwer, J., Tour, O., Laird, D. W., Sosinsky, G. E., Tsien, R. Y., & Ellisman, M. H. (2002). Multicolor and electron microscopic imaging of connexin trafficking. *Science*, *296*(5567). https://doi.org/10.1126/science.1068793

Garg, A. D., Nowis, D., Golab, J., Vandenabeele, P., Krysko, D. V., & Agostinis, P. (2010). Immunogenic cell death, DAMPs and anticancer therapeutics: An emerging amalgamation. *Biochimica et Biophysica Acta—Reviews on Cancer*, *1805*(1). https://doi.org/10.1016/j.bbcan.2009.08.003

Ghoodarzi, R., Changizi, V., Montazerabadi, A. R., & Eyvazzadaeh, N. (2016). Assessing of integration of ionizing radiation with Radachlorin-PDT on MCF-7 breast cancer cell treatment. *Lasers in Medical Science*, *31*(2). https://doi.org/10.1007/s10103-015-1844-0

Gollmer, A., Felgenträger, A., Bäumler, W., Maisch, T., & Späth, A. (2015). A novel set of symmetric methylene blue derivatives exhibits effective bacteria photokilling—A structure-response study. *Photochemical and Photobiological Sciences*, *14*(2). https://doi.org/10.1039/c4pp00309h

Gouterman, M. (1961). Spectra of porphyrins. *Journal of Molecular Spectroscopy*, *6*(C). https://doi.org/10.1016/0022-2852(61)90236-3

Hamblin, M. R., & Luke Newman, E. (1994). New trends in photobiology. On the mechanism of the tumour-localising effect in photodynamic therapy. *Journal of Photochemistry and Photobiology, B: Biology*, *23*(1). https://doi.org/10.1016/S1011-1344(94)80018-9

Hu, X., Huang, Y. Y., Wang, Y., Wang, X., & Hamblin, M. R. (2018). Antimicrobial photodynamic therapy to control clinically relevant biofilm infections. *Frontiers in Microbiology*, *9*(Jun). https://doi.org/10.3389/fmicb.2018.01299

Iluz, N., Maor, Y., Keller, N., & Malik, Z. (2018). The synergistic effect of PDT and oxacillin on clinical isolates of Staphylococcus aureus. *Lasers in Surgery and Medicine*, *50*(5). https://doi.org/10.1002/lsm.22785

Imlay, J. A. (2013). The molecular mechanisms and physiological consequences of oxidative stress: Lessons from a model bacterium. *Nature Reviews Microbiology*, *11*(7). https://doi.org/10.1038/nrmicro3032

Kästle, M., Grimm, S., Nagel, R., Breusing, N., & Grune, T. (2011). Combination of PDT and inhibitor treatment affects melanoma cells and spares keratinocytes. *Free Radical Biology and Medicine*, *50*(2). https://doi.org/10.1016/j.freeradbiomed.2010.11.012

Khdair, A., Chen, D., Patil, Y., Ma, L., Dou, Q. P., Shekhar, M. P. V., & Panyam, J. (2010). Nanoparticle-mediated combination chemotherapy and photodynamic therapy overcomes tumor drug resistance. *Journal of Controlled Release*, *141*(2). https://doi.org/10.1016/j.jconrel.2009.09.004

Kuppusamy, P., & Zweier, J. L. (1989). Characterization of free radical generation by xanthine oxidase. Evidence for hydroxyl radical generation. *Journal of Biological Chemistry*, *264*(17).

Lam, M., Jou, P. C., Lattif, A. A., Lee, Y., Malbasa, C. L., Mukherjee, P. K., Oleinick, N. L., Ghannoum, M. A., Cooper, K. D., & Baron, E. D. (2011). Photodynamic therapy with Pc 4 induces apoptosis of Candida albicans. *Photochemistry and Photobiology*, *87*(4). https://doi.org/10.1111/j.1751-1097.2011.00938.x

Laxminarayan, R., van Boeckel, T., Frost, I., Kariuki, S., Khan, E. A., Limmathurotsakul, D., Larsson, D. G. J., Levy-Hara, G., Mendelson, M., Outterson, K., Peacock, S. J., & Zhu, Y. G. (2020). The lancet infectious diseases commission on antimicrobial resistance: 6 years later. *The Lancet Infectious Diseases*, *20*(4). https://doi.org/10.1016/S1473-3099(20)30003-7

Lin, S. L., Hu, J. M., Tang, S. S., Wu, X. Y., Chen, Z. Q., & Tang, S. Z. (2012). Photodynamic inactivation of methylene blue and tungsten-halogen lamp light against food pathogen listeria monocytogenes. *Photochemistry and Photobiology*, *88*(4). https://doi.org/10.1111/j.1751-1097.2012.01154.x

Ma, L. W., Berg, K., Danielsen, H. E., Kaalhus, O., Iani, V., & Moan, J. (1996). Enhanced antitumour effect of photodynamic therapy by microtubule inhibitors. *Cancer Letters*, *109*(1–2). https://doi.org/10.1016/S0304-3835(96)04437-0

Maeda, H., Nakamura, H., & Fang, J. (2013). The EPR effect for macromolecular drug delivery to solid tumors: Improvement of tumor uptake, lowering of systemic toxicity, and distinct tumor imaging *in vivo*. *Advanced Drug Delivery Reviews*, *65*(1). https://doi.org/10.1016/j.addr.2012.10.002

Mai, B., Gao, Y., Li, M., Wang, X., Zhang, K., Liu, Q., Xu, C., & Wang, P. (2017). Photodynamic antimicrobial chemotherapy for staphylococcus aureus and multidrug-resistant bacterial burn infection *in vitro* and *in vivo*. *International Journal of Nanomedicine*, *12*. https://doi.org/10.2147/IJN.S138185

Mai, B., Jia, M., Liu, S., Sheng, Z., Li, M., Gao, Y., Wang, X., Liu, Q., & Wang, P. (2020). Smart hydrogel-based DVDMS/bFGF nanohybrids for antibacterial phototherapy with multiple damaging sites and accelerated wound healing. *ACS Applied Materials and Interfaces*, *12*(9). https://doi.org/10.1021/acsami.0c00298

Maisch, T (2015). Resistance in antimicrobial photodynamic inactivation of bacteria. *Photochemical and Photobiological Sciences*, *14*(8). https://doi.org/10.1039/c5pp00037h

Maisch, T., Baier, J., Franz, B., Maier, M., Landthaler, M., Szeimies, R. M., & Bäumler, W. (2007). The role of singlet oxygen and oxygen concentration in photodynamic inactivation of bacteria. *Proceedings of the National Academy of Sciences of the United States of America*, *104*(17). https://doi.org/10.1073/pnas.0611328104

Maisch, T., Eichner, A., Späth, A., Gollmer, A., König, B., Regensburger, J., & Bäumler, W. (2014). Fast and effective photodynamic inactivation of multiresistant bacteria by cationic riboflavin derivatives. *PLoS ONE*, *9*(12). https://doi.org/10.1371/journal.pone.0111792

Maisch, T., Szeimies, R. M., Jori, G., & Abels, C. (2004). Antibacterial photodynamic therapy in dermatology. *Photochemical and Photobiological Sciences*, *3*(10). https://doi.org/10.1039/b407622b

Makdoumi, K., & Bäckman, A. (2016). Photodynamic UVA-riboflavin bacterial elimination in antibiotic-resistant bacteria. *Clinical and Experimental Ophthalmology*, *44*(7). https://doi.org/10.1111/ceo.12723

McArthur, D. B. (2019). Emerging infectious diseases. *Nursing Clinics of North America*, *54*(2). https://doi.org/10.1016/j.cnur.2019.02.006

Mcdonnell, G., & Russell, A. D. (1999). Antiseptics and disinfectants: Activity, action, and resistance. *Clinical Microbiology Reviews*, *12*(1). https://doi.org/10.1128/cmr.12.1.147

Meimandi, M., Ardakani, M. R. T., Nejad, A. E., Yousefnejad, P., Saebi, K., & Tayeed, M. H. (2017). The effect of photodynamic therapy in the treatment of chronic periodontitis: A review of literature. *Journal of Lasers in Medical Sciences*, *8*. https://doi.org/10.15171/jlms.2017.s2

Mizuno, K., Zhiyentayev, T., Huang, L., Khalil, S., Nasim, F., Tegos, G. P., Gali, H., Jahnke, A., Wharton, T., & Hamblin, M. R. (2011). Antimicrobial photodynamic therapy with functionalized fullerenes: Quantitative structure-activity relationships. *Journal of Nanomedicine and Nanotechnology*, *2*(2). https://doi.org/10.4172/2157-7439.1000109

Morley, S., Griffiths, J., Philips, G., Moseley, H., O'Grady, C., Mellish, K., Lankester, C. L., Faris, B., Young, R. J., Brown, S. B., & Rhodes, L. E. (2013). Phase IIa randomized, placebo-controlled study of antimicrobial photodynamic therapy in bacterially colonized, chronic leg ulcers and diabetic foot ulcers: A new approach to antimicrobial therapy. *British Journal of Dermatology*, *168*(3). https://doi.org/10.1111/bjd.12098

Murphy, M. P. (2009). How mitochondria produce reactive oxygen species. *Biochemical Journal*, *417*(1). https://doi.org/10.1042/BJ20081386

Nagata, J. Y., Hioka, N., Kimura, E., Batistela, V. R., Terada, R. S. S., Graciano, A. X., Baesso, M. L., & Hayacibara, M. F. (2012). Antibacterial photodynamic therapy for dental caries: Evaluation of the photosensitizers used and light source properties. *Photodiagnosis and Photodynamic Therapy*, *9*(2). https://doi.org/10.1016/j.pdpdt.2011.11.006

Nam, W., Lim, M. H., Oh, S. Y., Lee, J. H., Lee, H. J., Woo, S. K., Kim, C., & Shin, W. (2000). Remarkable anionic axial ligand effects of iron(III) porphyrin complexes on the catalytic oxygenations of hydrocarbons by H_2O_2 and the formation of oxoiron(IV) porphyrin intermediates by m-chloroperoxybenzoic acid. *AngewandteChemie—International Edition*, *39*(20). https://doi.org/10.1002/1521-3773(20001016)39:20<3646::AID-ANIE3646>3.0.CO;2-Q

Nammour, S., Zeinoun, T., Bogaerts, I., Lamy, M., Geerts, S. O., Saba, S. B., Lamard, L., Peremans, A., & Limme, M. (2010). Evaluation of dental pulp temperature rise during photo-activated decontamination (PAD) of caries: An *in vitro* study. *Lasers in Medical Science*, *25*(5). https://doi.org/10.1007/s10103-009-0683-2

Nezhadi, J., Eslami, H., Fakhrzadeh, V., Moaddab, S. R., Zeinalzadeh, E., & Kafil, H. S. (2019). Photodynamic therapy of infection in burn patients. *Reviews in Medical Microbiology*. https://doi.org/10.1097/MRM.0000000000000188

Nieves, I., Hally, C., Viappiani, C., Agut, M., & Nonell, S. (2020). A porphycene-gentamicin conjugate for enhanced photodynamic inactivation of bacteria. *Bioorganic Chemistry*, *97*. https://doi.org/10.1016/j.bioorg.2020.103661

Omolo, C. A., Kalhapure, R. S., Agrawal, N., Jadhav, M., Rambharose, S., Mocktar, C., & Govender, T. (2018). A hybrid of mPEG-b-PCL and G1-PEA dendrimer for enhancing delivery of antibiotics. *Journal of Controlled Release*, *290*. https://doi.org/10.1016/j.jconrel.2018.10.005

Oyim, J., Omolo, C. A., & Amuhaya, E. K. (2021). Photodynamic Antimicrobial Chemotherapy: Advancements in Porphyrin-Based Photosensitize Development. In *Frontiers in Chemistry* (Vol. 9). Frontiers Media S.A. https://doi.org/10.3389/fchem.2021.635344

Penha, C. B., Bonin, E., da Silva, A. F., Hioka, N., Zanqueta, É. B., Nakamura, T. U., de Abreu Filho, B. A., Campanerut-Sá, P. A. Z., & Mikcha, J. M. G. (2017). Photodynamic inactivation of foodborne and food spoilage bacteria by curcumin. *LWT - Food Science and Technology*, *76*. https://doi.org/10.1016/j.lwt.2016.07.037

Queiroz, A. C., Suaid, F. A., de Andrade, P. F., Novaes, A. B., Taba, M., Palioto, D. B., Grisi, M. F. M., & Souza, S. L. S. (2014). Antimicrobial photodynamic therapy associated to nonsurgical periodontal treatment in smokers: Microbiological results. *Journal of Photochemistry and Photobiology B: Biology*, *141*. https://doi.org/10.1016/j.jphotobiol.2014.10.017

Ren, W., & Ai, H. W. (2013). Genetically encoded fluorescent redox probes. In *Sensors (Switzerland)*, *13*(11). https://doi.org/10.3390/s131115422

Rice, L. B. (2012). Mechanisms of resistance and clinical relevance of resistance to β-lactams, glycopeptides, and fluoroquinolones. *Mayo Clinic Proceedings*, *87*(2). https://doi.org/10.1016/j.mayocp.2011.12.003

Ryumina, A. P., Serebrovskaya, E. O., Shirmanova, M. V., Snopova, L. B., Kuznetsova, M. M., Turchin, I. V., Ignatova, N. I., Klementieva, N. V., Fradkov, A. F., Shakhov, B. E., Zagaynova, E. V., Lukyanov, K. A., & Lukyanov, S. A. (2013). Flavoprotein miniSOG as a genetically encoded photosensitizer for cancer cells. *Biochimica et Biophysica Acta—General Subjects*, *1830*(11). https://doi.org/10.1016/j.bbagen.2013.07.015

Samreen, Ahmad I., Malak, H. A., & Abulreesh, H. H. (2021). Environmental antimicrobial resistance and its drivers: a potential threat to public health. *Journal of Global Antimicrobial Resistance*, *27*. https://doi.org/10.1016/j.jgar.2021.08.001

Sathyendranath, S., Abdulaziz, A., Menon, N., George, G., Evers-King, H., Kulk, G., Colwell, R., Jutla, A., & Platt, T. (2020). Building capacity and resilience against diseases transmitted via water under climate perturbations and extreme weather stress. *Studies in Space Policy*, *22*. https://doi.org/10.1007/978-3-030-21938-3_24

Sederholm, C., Hillerdal, G., Lamberg, K., Kölbeck, K., Dufmats, M., Westberg, R., & Gawande, S. R. (2005). Phase III trial of gemcitabine plus carboplatin versus single-agent gemcitabine in the treatment of locally advanced or metastatic non-small-cell lung cancer: The Swedish Lung Cancer Study Group. *Journal of Clinical Oncology*, *23*(33). https://doi.org/10.1200/JCO.2005.01.2781

Serebrovskaya, E. O., Edelweiss, E. F., Stremovskiy, O. A., Lukyanov, K. A., Chudakov, D. M., & Deyev, S. M. (2009). Targeting cancer cells by using an antireceptor antibody-photosensitizer fusion protein. *Proceedings of the National Academy of Sciences of the United States of America*, *106*(23). https://doi.org/10.1073/pnas.0904140106

Sharma, S. K., Chiang, L. Y., & Hamblin, M. R. (2011). Photodynamic therapy with fullerenes *in vivo*: Reality or a dream? *Nanomedicine*, *6*(10). https://doi.org/10.2217/nnm.11.144

Shrestha, A., & Kishen, A. (2012). The effect of tissue inhibitors on the antibacterial activity of chitosan nanoparticles and photodynamic therapy. *Journal of Endodontics*, *38*(9). https://doi.org/10.1016/j.joen.2012.05.006

Silva, A. F., Borges, A., Giaouris, E., Graton Mikcha, J. M., & Simões, M. (2018). Photodynamic inactivation as an emergent strategy against foodborne pathogenic bacteria in planktonic and sessile states. *Critical Reviews in Microbiology*, *44*(6). https://doi.org/10.1080/1040841X.2018.1491528

Skwor, T. A., Klemm, S., Zhang, H., Schardt, B., Blaszczyk, S., & Bork, M. A. (2016). Photodynamic inactivation of methicillin-resistant Staphylococcus aureus and Escherichia coli: A metalloporphyrin comparison. *Journal of Photochemistry and Photobiology B: Biology*, *165*. https://doi.org/10.1016/j.jphotobiol.2016.10.016

Snell, S. B., Foster, T. H., & Haidaris, C. G. (2012). Miconazole induces fungistasis and increases killing of Candida albicans Subjected to photodynamic therapy. *Photochemistry and Photobiology*, *88*(3). https://doi.org/10.1111/j.1751-1097.2011.01039.x

Sobotta, L., Skupin-Mrugalska, P., Piskorz, J., & Mielcarek, J. (2019). Porphyrinoid photosensitizers mediated photodynamic inactivation against bacteria. *European Journal of Medicinal Chemistry*, *175*. https://doi.org/10.1016/j.ejmech.2019.04.057

Spaeth, A., Graeler, A., Maisch, T., & Plaetzer, K. (2018). Cure Cuma–cationic curcuminoids with improved properties and enhanced antimicrobial photodynamic activity. *European Journal of Medicinal Chemistry*, *159*. https://doi.org/10.1016/j.ejmech.2017.09.072

Spranley, T. J., Winkler, M., Dagate, J., Oncale, D., & Strother, E. (2012). Curing light burns. *General Dentistry*, *60*(4).

Takahashi, J., Misawa, M., & Iwahashi, H. (2015). Transcriptome analysis of porphyrin-accumulated and X-ray-irradiated cell cultures under limited proliferation and non-lethal conditions. *Microarrays*, *4*(1). https://doi.org/10.3390/microarrays4010025

Takemoto, K., Matsuda, T., McDougall, M., Klaubert, D. H., Hasegawa, A., Los, G. V., Wood, K. V., Miyawaki, A., & Nagai, T. (2011). Chromophore-assisted light inactivation of HaloTag fusion proteins labeled with eosin in living cells. *ACS Chemical Biology*, *6*(5). https://doi.org/10.1021/cb100431e

Taraszkiewicz, A., Fila, G., Grinholc, M., & Nakonieczna, J. (2013). Innovative strategies to overcome biofilm resistance. *BioMed Research International*, *2013*. https://doi.org/10.1155/2013/150653

Tardivo, J. P., Adami, F., Correa, J. A., Pinhal, M. A., Parecida, S., & Baptista, M. S. (2014). A clinical trial testing the efficacy of PDT in preventing amputation in diabetic patients. *Photodiagnosis and Photodynamic Therapy*, *11*(3). https://doi.org/10.1016/j.pdpdt.2014.04.007

Theuretzbacher, U., Outterson, K., Engel, A., & Karlén, A. (2020). The global preclinical antibacterial pipeline. *Nature Reviews Microbiology*, *18*(5). https://doi.org/10.1038/s41579-019-0288-0

Thornber, K., Verner-Jeffreys, D., Hinchliffe, S., Rahman, M. M., Bass, D., & Tyler, C. R. (2020). Evaluating antimicrobial resistance in the global shrimp industry. *Reviews in Aquaculture*, *12*(2). https://doi.org/10.1111/raq.12367

Tortik, N., Spaeth, A., & Plaetzer, K. (2014). Photodynamic decontamination of foodstuff from Staphylococcus aureus based on novel formulations of curcumin. *Photochemical and Photobiological Sciences*, *13*(10). https://doi.org/10.1039/c4pp00123k

Toyokuni, S. (1999). Reactive oxygen species-induced molecular damage and its application in pathology. *Pathology International*, *49*(2). https://doi.org/10.1046/j.1440-1827.1999.00829.x

Tripathy, B. C., & Oelmüller, R. (2012). Reactive oxygen species generation and signaling in plants. *Plant Signaling and Behavior*, *7*(12). https://doi.org/10.4161/psb.22455

Vatansever, F., de Melo, W. C. M. A., Avci, P., Vecchio, D., Sadasivam, M., Gupta, A., Chandran, R., Karimi, M., Parizotto, N. A., Yin, R., Tegos, G. P., & Hamblin, M. R. (2013). Antimicrobial strategies centered around reactive oxygen species - bactericidal antibiotics, photodynamic therapy, and beyond. *FEMS Microbiology Reviews*, *37*(6). https://doi.org/10.1111/1574-6976.12026

Vohra, F., Al-Rifaiy, M. Q., Lillywhite, G., Abu Hassan, M. I., & Javed, F. (2014). Efficacy of mechanical debridement with adjunct antimicrobial photodynamic therapy for the management of peri-implant diseases: A systematic review. *Photochemical and Photobiological Sciences*, *13*(8). https://doi.org/10.1039/c4pp00083h

Vzorov, A. N., Dixon, D. W., Trommel, J. S., Marzilli, L. G., & Compans, R. W. (2002). Inactivation of human immunodeficiency virus type 1 by porphyrins. *Antimicrobial Agents and Chemotherapy*, *46*(12). https://doi.org/10.1128/AAC.46.12.3917-3925.2002

Wainwright, M. (1998). Photodynamic antimicrobial chemotherapy (PACT). *Journal of Antimicrobial Chemotherapy*, *42*(1). https://doi.org/10.1093/jac/42.1.13

Wainwright, M. (2003). Local treatment of viral disease using photodynamic therapy. *International Journal of Antimicrobial Agents*, *21*(6). https://doi.org/10.1016/S0924-8579(03)00035-9

Wainwright, M., Maisch, T., Nonell, S., Plaetzer, K., Almeida, A., Tegos, G. P., & Hamblin, M. R. (2017). Photoantimicrobials—are we afraid of the light? *The Lancet Infectious Diseases*, *17*(2). https://doi.org/10.1016/S1473-3099(16)30268-7

Wang, X., Meng, G., Zhang, S., & Liu, X. (2016). A reactive 1O2-responsive combined treatment system of photodynamic and chemotherapy for cancer. *Scientific Reports*, *6*. https://doi.org/10.1038/srep29911

Wiehe, A., O'Brien, J. M., & Senge, M. O. (2019). Trends and targets in antiviral phototherapy. *Photochemical and Photobiological Sciences*, *18*(11). https://doi.org/10.1039/c9pp00211a

Wilkinson, F., Helman, W. P., & Ross, A. B. (1993). Quantum yields for the photosensitized formation of the lowest electronically excited singlet state of molecular oxygen in solution. *Journal of Physical and Chemical Reference Data*, *22*(1). https://doi.org/10.1063/1.555934

Wilson, B. C., & Patterson, M. S. (2008). The physics, biophysics and technology of photodynamic therapy. *Physics in Medicine and Biology*, *53*(9). https://doi.org/10.1088/0031-9155/53/9/R01

Wozniak, A., & Grinholc, M. (2018). Combined antimicrobial activity of photodynamic inactivation and antimicrobials-state of the art. *Frontiers in Microbiology*, *9*(May). https://doi.org/10.3389/fmicb.2018.00930

Xiong, Y., Tian, X., & Ai, H. W. (2019). Molecular tools to generate reactive oxygen species in biological systems. *Bioconjugate Chemistry*, *30*(5), 1297–1303. https://doi.org/10.1021/acs.bioconjchem.9b00191

Zeina, B., Greenman, J., Purcell, W. M., & Das, B. (2001). Killing of cutaneous microbial species by photodynamic therapy. *British Journal ofDermatology*, *144*(2). https://doi.org/10.1046/j.1365-2133.2001.04013.x

Zhang, J., Liang, H., Zheng, Y., Wang, D., Xia, J., Peng, W., Cheng, K., Wang, L., Liu, Y., Peng, W., & Li, Q. (2019). Photodynamic therapy versus systemic antibiotic for the treatment of periodontitis in a rat model. *Journal of Periodontology*, *90*(7). https://doi.org/10.1002/JPER.18-0305

Zhang, Q., & Li, L. (2018). Photodynamic combinational therapy in cancer treatment. *JBUON*, *23*(3), 561–567.

Zhou, Z., Song, J., Nie, L., & Chen, X. (2016). Reactive oxygen species generating systems meeting challenges of photodynamic cancer therapy. *Chemical Society Reviews*, *45*(23). https://doi.org/10.1039/c6cs00271d

12 Antimicrobial Photodynamic Inactivation of Oral Bacteria

Pitchaipillai Sankar Ganesh, Monal Yuwanati, and Sathish Sankar
Saveetha Institute of Medical and Technical Sciences (SIMATS), Chennai, India

Esaki Muthu Shankar
Central University of Tamil Nadu, Thiruvarur, India

CONTENTS

12.1 INTRODUCTION

The oral cavity contains several microbes having the potential to cause diseases as well as can act as foci of infection. It includes common *Streptococcus mutans* to rare *Streptococcus sobrinus*. These microbes can cause dental caries, periodontal, and gingival infection which may lead to infection of

DOI: 10.1201/9781003345299-12

bone and tooth loss. Further, there is evidence of chronic periodontitis predisposing the person to diabetes mellitus. Sometimes, it can cause bacteraemia, septicaemia, and sub-acute bacterial endocarditis. Prolonged oral infection affects the person's overall immunity making the person susceptible to bacterial infections.

Oral microbes reside mainly on tooth surfaces and gingival crevices. Usually, these bacteria do not cause infection. However, they create a survival environment to form a biofilm. Once adhered to the surface, the bacteria releases myriad virulence traits and cause infection in the oral cavity. Due to the production of virulence factors (exopolysaccharide, glucosyltransferases toxins etc.) and biofilm formation, microbes have developed resistance to several conventional antibiotics. Further, these microbes have changed their structure to avoid the action of standard antimicrobials resulting in drug resistance and evolving into more virulent strains. This has created a challenge for oral healthcare settings for the management of these oral infections.

Antimicrobial photodynamic therapy (aPDT) has recently emerged as a vital tool as an alternative to standard antimicrobial therapy. aPDT involves the use the ultraviolet, infrared, and visible light along with photosensitiser for killing the microbes. aPDT can kill microbes but avoid any side effects on host tissue. Further, it has an advantage over conventional antimicrobial therapy by not developing any resistance as well as shorter duration.

12.2 ORAL MICROBIAL DISEASES

12.2.1 Caries and Pulpitis

Caries/pulpitis is the oldest and most common oral disease affecting the teeth. The aetiology is multifactorial, although it results mainly from microbes that often lead to localised destruction of teeth structure. Furthermore, caries may extend into the pulp causing pulp infection/inflammation. Numerous bacteria can cause caries, however, the most prevalent strains found in carious lesions are *Streptococcus mutans* and *Lactobacilli*. Similarly, *Dialister invisus*, *Porphyromonas gingivalis*, *Streptococcus salivarius*, and *Treponema denticola* are reported in pulpitis. Oral microbes work in symbiosis to survive and cause disease. Considering the polymicrobial nature of dental caries and pulpitis, clinicians have difficulty controlling the progress of the disease. Further, the majority of the microbes are part of the oral microbial flora. Caries and pulpitis are usually treated with dental fillings and endodontic treatment, respectively.

12.2.2 Apthous Ulcer

Aphthous ulcers are the most common ulcerative condition of the oral mucosa. These usually present as a painful punched-out sore on oral mucous membranes, and heal within 7–10 days. They are usually small, round to oval ulcers, with a central halo area. Its aetiology is unknown although factors have been suggested. One of the main factors is bacteria such as *Streptococcus sanguinis* 2A (Barile et al., 1963). Bacteria can be pathogens or a source of antigens that trigger antibody production and can cross-react with the oral mucosa. Other bacteria which can cause the aphthous ulcer includes *Streptococcus mitis* etc. (Bankvall et al., 2014). Apthous ulcer treatment is mostly symptomatic which includes anti-inflammatory ointment and gels. Although these treatments can give relief to a person, in case of bacterial infection, antibiotics are used in treatment. Furthermore, bacterial infection extends the duration of treatment as well as healing period.

12.2.3 Acute Necrotising Ulcerative Gingivitis

Acute necrotising ulcerative gingivitis (ANUG) is another bacterial infection of oral mucosa especially affecting the gingiva. ANUG is characterised by sudden onset of inflammation, pain, rapid destruction of the gingiva, and the presence of "punched-out" crater-like lesions. The precise aetiology of ANUG is not known, however, it is believed to be a polymicrobial infection with the implicated organisms such as *Treponema* sp., *B. melaninogenicus* ssp., *S. intermedius* and *Fusobacterium* sp. which are commensals in the oral cavity. There is no specific treatment other than evasion of the underlying risk factors for infection along with empirical use of antimicrobials such as metronidazole.

12.2.4 Thrush (Candidiasis)

Candidiasis, or oral thrush, is an opportunistic infection usually seen in several situations including decreased immunity, poor oral hygiene, denture wearing etc. Candidiasis is caused by *Candida albicans*, a commensal fungus. It often causes change or loss in taste, burning, and a whitish appearance of the oral mucosa. It can be treated with oral antifungal drugs such as fluconazole, clotrimazole, nystatin, itraconazole, and amphotericin B (in severe cases).

12.2.5 Chronic Periodontitis and Gingivitis

Periodontal inflammation frequently occurs in the oral region. It usually involves the inflammation of periodontal ligaments and adjacent structures resulting in loss of tooth structure. The disease process involves multiple organisms which are mostly commensal oral bacteria. These microorganisms associated with periodontitis include bacteroidetes, firmicutes, proteobacteria, spirochaetes and synergistetes (Rosier et al., 2018).

The periodontitis treatment varies from person to person and depends upon the underlying causative factor. However, the presence of bacteria can complicate or delay recovery or healing. In that situation, topical or oral antibiotics can help control bacterial infection.

Gingivitis is inflammation of the gingiva due to irritation resulting in redness and swelling. Bacterial plaque is the most common cause of gingivitis. Similar to periodontitis, it is polymicrobial, hence, difficult to treat with any specific single drug.

12.2.6 Dental Abscess

A dental abscess is a presence of pus in dental pulp, or in the gingiva, or within an alveolar socket. Similar to gingivitis, it is polymicrobial, involving facultative anaerobes such as streptococci, *Prevotella*, and *Fusobacteria* (Shweta and Prakash, 2013). These abscesses are routinely treated with surgery while antimicrobials are used to limit the spread of infections.

12.3 MECHANISM OF BACTERIAL INFECTION IN THE ORAL CAVITY

The oral cavity harbours diverse microbial flora including bacteria, viruses, and fungi that can infect the oral mucosa. Bacterial phyla that inhabit the oral cavity include bacteroidetes, firmicutes, tenericutes, Actinobacteria, and others. These highly complex microbial communities play a central role in the maintenance of good health and the immune system. However, this oral flora can also cause dental and periodontal infections especially when the immune system is compromised. Several other infections such as oral lesions manifest in the oral cavity in the presence of other microbial agents. These bacteria occupy the gingival crevices and form biofilms on the outer surface of the tooth causing dental plaques. The acid produced by the bacteria destroys the teeth leading to cavities and infections. Bacterial infections arise primarily following the damage to the oral epithelium as a cause of trauma, poor oral hygiene, and smoking. Other major factors that predispose oral cavity infections are increasing age, and a diet rich in sugar and hormones, especially during puberty, menstruation, and pregnancy. The frequency of bacteraemia following invasive dental treatments such as extraction is the other major risk factor for bacterial endocarditis. Many dental procedures such as tooth extraction, endodontic, and periodontal surgical procedures and scaling make the predominant pathogens invade, resulting in bacteraemia (Li et al., 2000) due to damage of mucocutaneous barriers. A temporal association of invasive dental procedures has been associated with infective endocarditis in individuals with people at higher risk (Thornhill et al., 2022). Studies indicate about 14–20% of bacterial endocarditis is related to poor oral hygiene. The oral infection especially periodontitis causes bacteraemia which in turn causes hyperactive mononuclear phagocytes, the release of bacterial products such as endotoxin and heat shock proteins, inflammatory products such as cytokines and C-reactive proteins, hypercoagulation state which includes increased fibrinogen and WBC count, *S. sanguis* and *P. gingivalis* mediated platelet aggregation and the loss of teeth. All these conditions eventually affect the vascular pathology leading to cardiovascular diseases. The association between oral infection and atherosclerosis has been attributed to total dental index, bone levels, plaque index and probing depth (Li et al., 2000).

Other infections such as HIV, cancer, and immunosuppressed patients may predispose to an increased risk of bacterial oral infections. Most bacterial infections with involvement in oral mucosa cause sexually transmitted infections, cardiac involvement, and granulomatous diseases such as tuberculosis.

Common oral infections arise from dental caries that extend to the alveolar bone involving the face and oral cavity in different planes. The mandible and the maxilla are more vulnerable to infections on the buccal side involving incisors and canine teeth.

Dental caries and periodontal diseases appear most commonly in adults compared to children and adolescents. These are reported to be important causes of tooth loss in geriatric patients. Dental caries was the predominant cause of tooth loss in individuals less than 35 years of age. Periodontal diseases are highly prevalent in older patients. The prevalence ranges from moderate to severe periodontal disease with 20% of patients being more than 75 years of age.

The bacterial cause is more prevalent in endodontic infections and other pathogens including fungi, archaea, and viruses are associated with apical periodontitis. However, endodontic infections are most commonly associated with bacterial aetiologies. Pulp necrosis due to irreversible pulpitis occurs in the root canal when has access to the periradicular tissues (Siqueira and Rôças, 2022).

Bacteria, especially Gram-negative rods, are found in dental plaque and cause periodontal disease. The most commonly found bacteria include *Actinobacillus actinomycetemcomitans*, *Porphyromonas gingivalis*, and *Bacteroides forsythus* that causes periodontitis.

Understanding the landscape of the distinct oral microbial species is important in understanding the pathobiology of several diseases and therapeutic options. Changes in the oral cavity or any oral manifestations often precede gastrointestinal disorders (Vasovic et al., 2016). Recent studies suggest the role of reservoirs of *Helicobacter pylori* in causing dental plaques and subsequent infection of the gastric mucosa. A significant association of *H. pylori* is also observed with periodontal disease, periodontitis and gingivitis with a prevalence of up to 83% (López-Valverde et al., 2022). Pathogens such as *H. pylori* are associated with pancreatic cancer and the exact mechanism is still elusive. The combined effect of multiple pathogens and their virulence factors now add to the tumorigenesis process of pancreatic cancer. Further extensive studies on the potential role of these pathogens in tumorigenesis process would be valuable for the design of newer preventive and therapeutic strategies (Sun et al., 2019). Desulfovibrio species, sulphate-reducing, anaerobic, Gram-negative bacilli, and an inhabitant of the human oral cavity and intestinal tract cause bacteraemia and abdominal infections.

The inflammatory response of oral diseases leads to systemic inflammation involving the vascular system and in the neuropathogenesis of many different neurodegenerative disorders. The inflammation in the vascular system is associated with cognitive impairment (Ranjan et al., 2018).

The oral infection has also been shown to have a role in the process of respiratory infection. Pathogens causing periodontitis such as *Porphyromonas gingivalis*, *Actinobacillus actinomycetemcomitans* act as a reservoir and causes pneumonia while entering the lungs. Enzymes present in saliva promote adhesion and colonisation aggravates the inflammatory response along with cytokines released from periodontal tissues. This helps the respiratory pathogens to cause diseases (Scannapieco, 1999).

Dental caries present with lesions on the outer layer of the dental pulp. Complications with the involvement of deep tissues might arise leading to bacteraemia, if untreated. The cariogenic microbes such as *Streptococcus mutans*, Streptococcus, and Actinomyces species form biofilms and produce acids. This further leads to demineralisation of the enamel and dental calculus.

In addition to the teeth, the bacteria also affect the surrounding structures resulting in infections, called periodontal diseases. The periodontal diseases is characterised by inflammation and damage of the bone and the soft tissues surrounding the teeth. Inflammation of the gums leads to irritation and bleeding called gingivitis. Gingivitis is caused by different bacteria that lead to gum disease. The bacteria that colonise the gingival space, especially the anaerobes, including Porphyromonas, Streptococcus, and Actinomyces predominate. Several virulence factors such as lipopolysaccharides, proteases, lipoteichoic acids, and many other exotoxins induce inflammation and gum damage. Gingivitis presents as a chronic and serious condition of periodontitis.

Various environmental factors including oral microbiota help to maintain homeostasis in the oral cavity. These organisms act on the oral mucosa directly and have innate immunity in the oral cavity. The oral microbiome and different immune molecules such as human cathelicidin possess active antimicrobial properties and immunomodulation ability with cell surface receptors. These proteins are also now found to be involved in oral dysbiosis and many bacterial and viral infections and tumours (Tokajuk et al., 2022).

Biofilms formed on the subgingival layer have a high bacterial load. The lipopolysaccharides produced by Gram-negative bacteria with more contact with the periodontal tissues cause bacteraemia. The upregulation of endothelial cell adhesion molecules, IL-1 and TNF-alpha by the lipopolysaccharides leads to increased inflammation.

12.4 PHOTODYNAMIC THERAPY (PDT)

Photodynamic therapy (PDT) was discovered hundred years ago, in 1903, in ancient Greece, Egypt, and India, where the reaction between a photon (light) and photosensitiser (PS) (dyes), connected with oxygen (O_2) molecules, was called PDT (Cieplik et al., 2014). The first report on the use of PDT by a voltaic arc or arc light to treat *Lupus vulgaris* (*L. vulgaris*) was released by Danish physician Niels Finsen (Finsen, 1896). Antimicrobial photodynamic therapy (aPDT) has been investigated in several *in vitro* studies over the past few decades (Ganesh and Rai, 2018), particularly, in the oral domain. Further, PDT has evolved in recent years to target bacterial (Kang et al., 2019), viral (Nunes et al., 2021), and fungal infections of the oral cavity (Roomaney et al., 2021).

PDT act by cleaving the bacterial cell membrane and preventing uncontrolled cell proliferation in the host cells. In addition, it can oxidise bacterial cellular components such as the plasma membrane and DNA, and photo-activate PS, by generating reactive oxygen species and causing cell death (Castano et al., 2004; Konopka and Goslinski, 2007). This reaction is termed an aPDT. Contrary to conventional and traditional antimicrobial therapies, PDT was found to be effective against Gram-positive and Gram-negative bacteria resistant to antibiotics in pulpal disorders, periodontitis, dental caries, dental plaque, and mucositis. It also had no deleterious effects on the host tissue (Diogo et al., 2015).

The PDT process involves the reaction of photosensitisers (PS) (e.g., non-toxic dyes) caused by a specific wavelength of visible light that can directly absorb and activate the PS. The PS can be activated and form the lowest energy excited singlet state, followed by the long-lived excited triplet state (Abrahamse and Hamblin, 2016). Triplet PS can undergo photochemical reaction in the occurrence of ambient O_2 to form highly reactive singlet O_2 species (ROS) and cytotoxic ROS (such as hydroxyl radicals), that can penetrate through bacterial cell wall or membrane and induce oxidation of membrane, amino acids, lipid peroxidation, inhibition of enzymatic reaction, oxidative damage of nucleic acid and agglutination of proteins (Hamblin and Hasan, 2004; Li et al., 2021). Similarly, it can harm bacteria by oxidising cellular components and thereby destroying oral bacteria (Džunková et al., 2018).

12.4.1 Photosensitiser (PS)

Photosensitisers (PS) are chemical agents (natural and synthetic dye) that absorb photons released by visible light at a specific wavelength and convert them into useful energy. When this energy reacts with molecular oxygen, it creates cytotoxic chemicals that kill the bacteria (Rocha, 2015). Synthetic (Table 12.1) and natural dyes (Table 12.2) exhibit a high binding affinity for specific oral pathogens, as well as a broad spectrum of antibiotic action. PSs damage the cell membrane and cellular organelles, destroying bacterial pathogens (Hu et al., 2018). Additionally, PSs accumulate very quickly in targeted oral pathogens and exhibit a very low interaction with host cells, reducing the possibility of collateral damage or mutagenesis (Sobczyński, 2017).

PS are mostly unsaturated conjugated organic molecules with a high absorption coefficient in the visible light and the red near-infrared (IR) region for efficient penetration into the host tissue (Mallidi et al., 2016). Visible spectrum (photons) induced inactivation of bacterial pathogens has been established more effective on oral bacterial pathogens. Gram-positive pathogens have a thick peptidoglycan layer and do not have an outer lipid membrane, allowing a faster and deeper penetration of the PS which was achieved by mono-cationic PS (Dahl et al., 1989). This mono-cationic PS more predominantly activates and kills the bacteria (both resistant and biofilm-forming bacteria). However, Gram-negative bacteria require poly-cationic derivatives (PS) to kill bacteria because these bacteria have negatively charged lipopolysaccharide and highly packed asymmetric lipid bilayers, which limit the penetration of PS through the cell membrane (Sperandio et al., 2013). Unlike Gram-positive bacteria, Gram-negative bacterial pathogens produce various virulence factors and biofilms which are mainly coordinated and controlled by the QS system (Rutherford and Bassler, 2012). By preventing the entry of antibiotics, these mechanisms hinder the ability of bacteria to resist conventional antibiotics and also survive even in harsh environments. Antimicrobial PDT has a reliable method used to kill oral pathogens without having any side effects on host cells (Huang et al., 2012).

12.4.2 Mechanisms of Light Sources, Oxygen (O_2), and ROS Generation for Antimicrobial Photodynamic Therapy

12.4.2.1 Light Sources

Light is an important source of PDT to treat various bacterial diseases. A vast variety of light sources are used for aPDT such as visible light (Xenon Lamps), Laser, diode laser, light-emitting diodes (LEDs), fibre optic devices etc.

A xenon lamp spectrum covers a wide range of light sources that is very closest to that of the sunlight spectrum, and typically it illuminates a different spectrum such as near-infrared (IR) wavelength is 700–810, green (550–600), blue (400–490 nm), yellow (550–600 nm), red (600–700 nm), and ultraviolet A (330–400 nm) and these wavelengths are widely applied for aPDT (Tables 12.1 and 12.2). Xenon short-arc lamp provides power levels of 150–1600 W. Blue xenon lamp (wavelength: 400–500 nm) 60 s (68 J/cm^2) penetrated H_2O_2 which reduced the viable cells of *S. mutans* biofilm growth within 30 min of irradiation (Santin et al., 2014).

In the dental field, laser radiation is used to kill oral pathogens that cause infections in the oral cavity. These bacteria are mainly responsible for periodontal disease, endodontic disease, and tooth loss. There are different types of lasers employed in dentistry. High-power laser diodes are mainly used for scaling and root canal treatment (RCT) and are also involved in some invasive surgery. Similarly, the low-power output laser diodes are predominantly involved in cellular stimulation, and activation of antimicrobial drugs following the treatment of root planning, and scaling (Aoki et al., 2004; Cobb, 2017).

Diode laser irradiation (1.5W) releases incident energy that is converted into heat which is absorbed by the pigmented cell membrane of endodontic and periopathogenic bacteria. A diode laser is highly monochromatic, and it emits radiation at wavelengths for dental applications ranging

TABLE 12.1
Synthetic Photosensitisers Used for Antimicrobial Photodynamic Therapy against Dental Pathogens

S. No	Study Population	Site of Lesions	Mean Age / (Male/Female)	Organisms	Light	Photosensitiser	Concentration of Dye	Wavelength	Power	Duration of PDT	Energy	References
1.	—	Cervical, occlusal, and proximal caries	—	*Streptococcus mutans*	Gallium-aluminium-arsenide Laser	Fotoenticine	0.6 mg/mL	660 nm	50 mW	300 s	15 J	Terra-Garcia et al. (2021)
2.	—	—	—	*Streptococcus mutans*	helium/neon (HeNe) laser or a light-emitting diode (LED)	Toluidine blue O	100 mg/L	632.8 nm	32 mW	5, 15 or 30 min	49 J/cm^2 (5 min), 147 J/cm^2 (15 min) or 294 J/cm^2 (30 min)	Zanin (2005)
3.	01	Carious lesions & Dentin tissue	4-year Female	*Streptococcus mutans* Block pigmented bacilli	Laser	Methylene blue dye	50 µg/mL	660 nm	—	3–5 min	40 J/cm^2	da Silva Barbosa et al. (2014)
4.	—	—	—	*Streptococcus mutans* ATCC25175	Halogen	Erythrosine	20 µmol·L^{-1}	400–520 nm	600 mW·cm^{-2}	—	J·cm^{-2}	Lee et al. (2012)
5.	21 healthy volunteers	Caries lesions	19–38 years (11 female and 10 male)	*Streptococcus mutans*	Light-emitting diode	Toluidine blue ortho	100 µg ml^{-1}	638.8 nm	40 mW	15 min	55 J cm^{-2}	Teixeira et al. (2012)
6.	21 patients	Oral cavity (Gingival region (around brackets))	Adults	Gram-positive and Gram-negative bacteria	Red Light-emitting diode	Methylene Blue + Toluidine Blue	12.5 µg/mL	640 ± 5 ηm	110 mW	294 sec	30 J/cm^2	Soares et al. (2019)
7.	212 patients	Dental caries	18–35 years	*Streptococcus mutans*	Diode laser	Toluidine blue and indocyanine blue	1290 µM/mL and 370 µM/mL	635 nm	220 mW/cm^2	1 min	31.2 J/cm^2	Pourhajibagher et al. (2018)
8.	21 healthy volunteers	Oral cavity (sub-gingival)	Age ≥ 20 years	Species not listed	GaAlAs diode laser	phenothiazine chloride	1%	670 nm	0.445 W/cm^2	2 min (60 s buccally, 60 s palatally/lingually)	—	Katsikanis et al. (2020)
9.	90 patients (chronic periodon)	Mesiobuccal, distobuccal, mesiolingual, or distolingual sites	39.6 ± 8.7 years Female 51 and Male 39	Species not listed	Diode laser	Methylene blue	10 mg/ml	655 nm,	60 mW/cm^2	60 sec	—	Betsy et al. (2016)
10.	45 patients	Caries lesion	>18 years	*Streptococcus spp. Lactobacillus spp*	Red light-emitting diode	Toluidine blue ortho	100 µg.mL^{-1}	630 nm	150 mW	5 min	94 J.cm^{-2}	Melo et al. (2015)

TABLE 12.2
Natural Photosensitisers Used for Photodynamic Therapy against Dental Pathogens

S. No	Organisms	Light	Photosensitiser	Concentration of Dye	Wavelength	Power	Duration of PDT	Energy	References
1.	*Streptococcus mutans* and Lactobacillus	Visible blue light	Curcumin	0.8%	455	40 mW.cm^{-2}	935 s or 1,870 s)	3.75 J and 7.5 J	Cusicanqui Méndez et al. (2018)
2.	*Streptococcus mutans*	Visible light	Chlorella	0.5 mg/ml	405	59 mW	5 min	17.7 J	Hwang et al. (2021)
3.	*Streptococcus mutans* ATCC 35668	Diode laser	Chlorophyllin-phycocyanin mixture	100 µg/mL	635±10 nm	220 mW	3 min	104 J/cm^2	Afrasiabi et al. (2020)
4.	*Streptococcus mutans* UA159	Lite emitting diode	chlorin e6	25 µM	660 nm	19 mW/cm^2	13.25 min	15 J/cm^2	Nie et al. (2020)

from 810 to 1064 nm. This wavelength easily penetrates (2 to 3 mm) the soft tissue and cleaves the bacterial cell membrane of oral pathogens. The wavelength of 810 and 940 nm diode laser has reduced *E. faecalis* and *Escherichia coli* counts by 98% when including access cavities in irradiation. Laser irradiation has remarkably reduced the intracanal bacterial count to 810 nm compared to 980 nm laser group (Asnaashari et al., 2016). Diode laser (665 nm wavelength) exposed 60s (energy density 21.2 J/cm^2) along with photosensitiser (Methylene blue), and ≈ 95% of *Aggregatibacter actimonycetemcomitans*, and *Fusobacterium nucleatum* were eliminated. Similarly, more than 99% of the black–pigmented pathogens such as *Prevotella intermedia*, *Porphyromonas gingivalis*, and *Streptococcus sanguis* were also destroyed. Further, an appropriate wavelength diode laser has prevented the establishment of bacterial colonisation of subgingival lesions (Chan and Lai, 2003).

In 1962, a light-emitting diode (LED) was invented for its beneficial effects on human health which was confirmed by the National Aeronautics and Space Administration (NASA) (Whelan et al., 2001). LED has produced a narrow spectrum of visible light in a non-coherent manner and is able to transmit the suitable wavelength, and specific intensity for the treatment of bacterial infection (Calderhead, 2007). LEDs are used in dentistry as PDT and are one of the safest methods of treating many oral diseases. LED technology has evolved steadily over the last few years, and today different light sources such as red (630–700 nm), blue (400–495 nm), yellow (570–590 nm), and near-infrared (800–1200 nm) monochromatic energy are used in many dental applications (Oh and Jeong, 2019). Blue light irradiation at 60 and 90 J/cm^2 destroyed the bacterial growth of *P. gingivalis* (Chui et al., 2013). Morio et al. (2021) have demonstrated that ultraviolet-C (UV-C) LED at 265 nm (22.5 mW) potentially destroyed both bacterial and fungal colonies (*E. faecalis*, *Staphylococcus aureus*, methicillin-resistant *S. aureus* (MRSA), and *Candida albicans*) in the 30s of exposure in the root canals of extracted teeth.

12.4.2.2 Oxygen (O_2) and ROS Generation

Oxygen (O_2) is one of the main components of PDT. During PDT, light penetrates with a photosensitiser and temporarily depletes oxygen in the surrounding environment. As a result of O_2 depletion, ROS production can lead to the damage of bacterial cell membrane and finally, it leads to cell death. ROS are more effective against oral pathogens.

12.5 MECHANISM OF NATURAL PHOTOSENSITISERS AGAINST ORAL PATHOGENS

Plants produce a vast variety of bioactive compounds which help them to prevent the entry of pathogens as well as protect against harsh environments. Plant species are usually enriched with chlorophyll pigment, which exerts promising PDT effects when it releases high quantum yields of singlet oxygen ($^1O_2$) and shows greater absorption under visible spectrum (sunlight). Natural PSs have been derived from various plant species, bacteria, fungi, and algae. The natural extracts have potentially used as PSs for aPDT (Table 12.2). Naturally derived PSs are less expensive, non-toxic, more effective, and relatively rapid clearance from normal tissues, thereby reducing side effects of phototoxicity than synthetic PS compounds (Figure 12.1). Currently, a very limited number of naturally occurring photosensitisers have been identified, but naturally, plants contain a wide variety of PSs which can identify and use aPDT (Abrahamse and Hamblin, 2016). Photosensitisers are being used for treating various oral cavity infections. Some of the photosensitisers that have been used to treat various pathogens in the oral cavity include *S. mutans* (Lee et al., 2012), *Enterococcus faecalis* (Tennert et al., 2014), *F. nucleatum*, *P. gingivalis*, and *A. actinomycetemcomitans* (Park et al., 2020), *Capnocytophaga gingivalis*, *Fusobacterium nucleatum*, *P. intermedia*, *P. gingivalis*, *S. sanguis*, and *Actinomyces israelli*, (Pfitzner et al., 2004; Chan and Lai, 2003). Most of the PSs have strongly inhibited bacterial growth greater than 90%. Gerola et al., 2011 reported that chlorophyll derivative 4f exhibited the highest dark cytotoxicity and showed promising antimicrobial activity against oral pathogens.

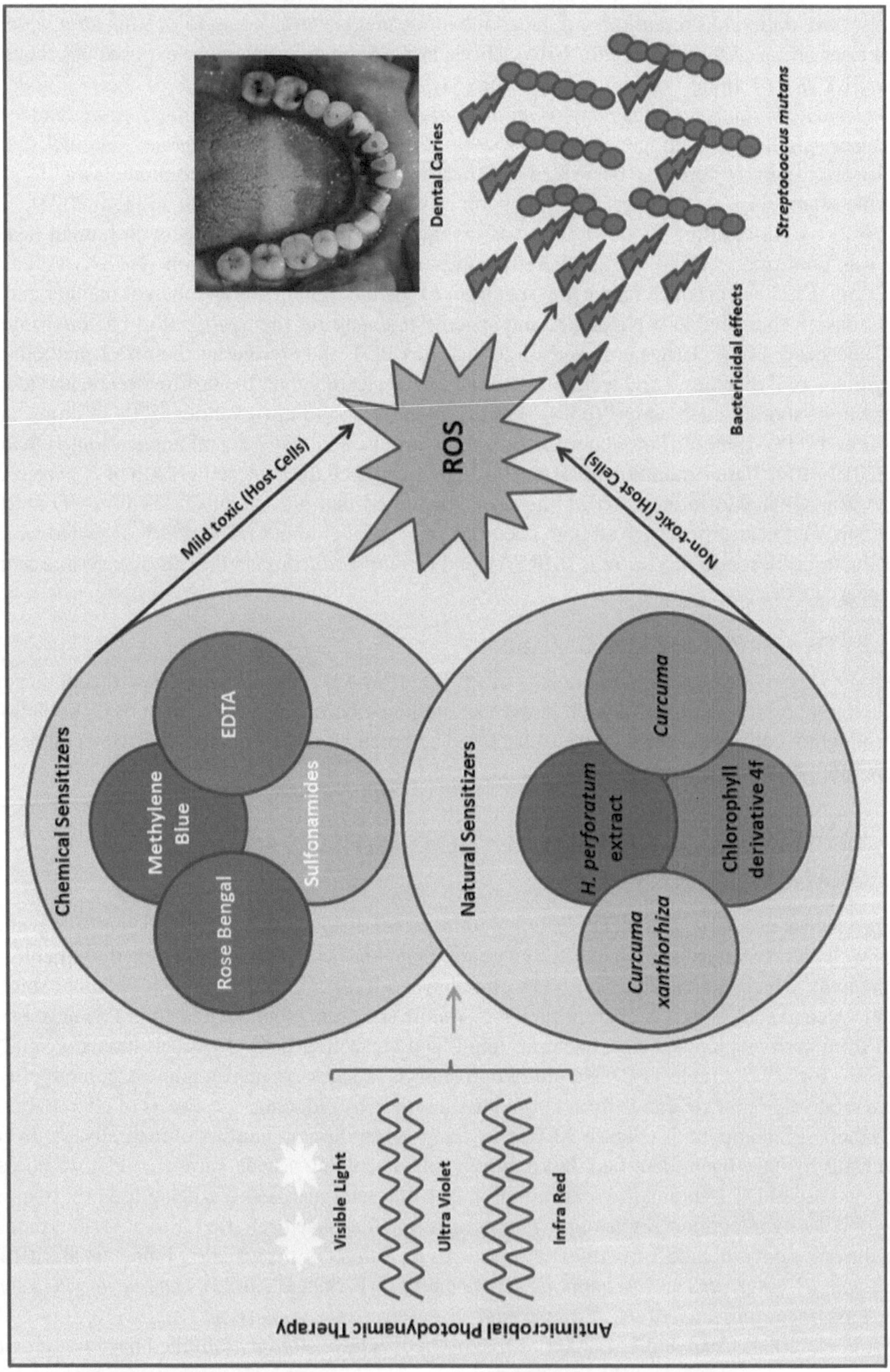

FIGURE 12.1 List of synthetic and natural photosensitising agents for aPDT against oral pathogens.

A natural compound of curcumin is derived from perennial herbaceous plant of *Curcuma longa* (*C. longa*) which mediates and is activated by blue light. It increases the quantum yield ranges from 405 to 435. This PS can strongly inactivate the growth of *S. mutans* bacterial colonies in the oral cavity (Azizi et al., 2019). Similarly, *C. longa* extract has a potential photosensitiser, which has the potential to increase the percentage of *Listeria innocua* death in the oral cavity (Bonifácio et al., 2018). A combination of *Curcuma xanthorrhiza* extract and curcumin has remarkably reduced *S. mutans* growth during irradiation with 405 nm light. The results show that LED light passes a mixture of compounds (curcumin and CXE) which increases the phototoxicity of the photosensitiser and leads to damage to the bacterial cell membrane of *S. mutans* (Lee et al., 2017) (Figure 12.1). Oregano essential oil is a natural PS and effective for PDT, destroying the oral bacterial infection and reducing inflammation in the periodontal region (Dascalu Rusu et al., 2022). The lowest concentrations of *H. perforatum* extract (32 mg/ml) exposed visible light to the oral cavity and released PS agents that destroyed bacterial biofilms nearly 92% completely. Based on that *H. perforatum*-mediated aPDT for the adjunctive treatment of QS controlled virulence factors and biofilm-associated oral bacterial diseases (Vollmer et al., 2019).

12.6 STUDIES ON THE APPLICATION OF APDT IN ORAL BACTERIAL INFECTIONS

In past decades, there has been an increased interest in treatment options with minimal toxicity for several oral pathogens involved in oral disease albeit with limited evidence. Lactobacillus or *Streptococcus mutans* are the most common oral microbes involved in dental caries. Traditional methods fail to remove the all-infected dentin or pulp along with microbes due to practical limitations in identifying the infected part of the tooth even with the help of disclosing dyes. As a result, there is a continuous progression of caries leading to loss of teeth. A study on the effect of aPDT on *Lactobacillus acidophilus* using different lasers indicate a reduction in colonies within 24 hr as compared to the control (Yoshii et al., 2021). Similarly, Aziz et al. (2019) observed that aPDT had equivalent effects on *L. acidophilus* and that aPDT with methylene blue has greater antibacterial action against *L acidophilus*. Giusti et al. (2008) demonstrated antibacterial activity not only against *L. acidophilus* but also against *S. mutants*. LED energy in conjunction with Photogen or toluidine blue O (TBO) can decrease bacteria in carious dentin especially *S. mutans* and *L. acidophilus*. In contrast, *in vitro* study by Gabaran et al. (2022) indicated that the growth rate of *S. mutans* cannot be greatly reduced by diode laser. Rolım et al. (2012) assess the antibacterial activity of PDT against *S. mutans* with various photosensitisers at the same dose. According to this study, TBO is efficient in decreasing (99.9%) bacteria count.

Another study found similar results when cariogenic bacteria were exposed to aPDT. There were significant reductions in *S. mutans*, *L. casei*, and *A. viscosus* (to 13%, 30%, and 55%, respectively). Bargrizan et al. (2019) investigated the efficacy of aPDT with TBO and diode laser on salivary mutans streptococci obtained from caries in 5–6-year-old children. However, they discovered that antimicrobial TBO+ laser was more effective than diode laser or TBO alone. Similarly, fluoride varnish + diode laser antibacterial effectiveness was observed to be greater than fluoride varnish or diode laser alone (Darmiani et al. 2022).

Paracoccidioidomycosis (PCM) is a fungal infection with oral ulcers caused by *Paracocdioides brasiliensis* and *Paracoccidioides lutzii* and is clinical manifest as ulcer in the oral mucosa (Santos et al., 2022). aPDT has been found effective against these microorganisms (Santos et al., 2022). Similar findings have been reported on treated oral paracoccidioidomycosis (Dos Santos et al., 2017). In another case of oral paracoccidioidomycosis therapy, the patient had complete regression of the oral lesion and pain.

aPDT was also evaluated against *Porphyromonas gingivalis* and *Tannerella forsythia*, which is usually involved in pericoronitis, periodontitis and gingivitis (Elsadek et al., 2020). aPDT can reduce

TNFF-α and microbial load in acute severe pericoronitis. Similarly, it can decrease the *P. gingivalis* and *T. forsythia* as compared to periodontal debridement (Alshahrani et al., 2020).

12.7 ADVANTAGES AND LIMITATIONS

12.7.1 No Microbial Resistance

The microbe over the period develops resistance to currently administered antimicrobials. This results in effective drug therapy and severe disease presentation or progression. PDT can overcome this limitation by specifically targeting the microbe with higher sensitivity and specificity. Furthermore, this will reduce the economic burden.

12.7.2 Short Treatment Duration

Most microbial drug treatments require prolonged intake to cure the infection. Not only is this time-consuming, but it is also responsible for higher levels of morbidity, and on rare occasions, mortality. PDT therapy can reduce the treatment duration. It can be done on an outpatient single-visit basis.

12.7.3 Non-Toxicity

The standard local microbial drugs can have more toxicity and can exert unwanted toxic effects on host cells. It may range from excessive oral mucosal inflammation to tissue degeneration such as an ulcer. The PSs usually will not show toxicity even if synthetic PSs are used in PDT. Considering the non-toxicity of PSs, it can be more acceptable to dentists as well as patients for treatment of gingivitis, periodontitis, and endodontic treatment.

12.7.4 Enhancement of Host Immune Response

PDTs have a stimulatory effect on the host immune system, where they can activate the immune cells and help in controlling the infection. PDTs are not exempted from disadvantages. It includes photosensitivity and chemical reactions. The patient can develop photosensitivity to natural or synthetic PSs. There may be redness and a tingling or burning sensation at the site of application. Sometimes, it can cause generalised hypersensitivity reactions. Hence, before treatment, it is necessary to check any history of hypersensitivity reaction or allergy to any photosensitiser substance.

PDT over the decade evolved to its present status; however, it has not completely developed as a good treatment option due to several reasons discussed above.

12.8 FUTURE SCOPE

PDT is a good alternative treatment option for the treatment of several oral diseases. It can minimise the emergence of resistant, more virulent microbes. Future research in PS having more sensitivity and specificity is required. Also, there is a need for portable handheld devices that can deliver PDT at outpatient clinics.

REFERENCES

Abrahamse, H., and Hamblin, M.R. 2016. New photosensitizers for photodynamic therapy. *The Biochemical Journal* 473: 347–364.

Afrasiabi, S., Pourhajibagher, M., Chiniforush, N., Aminian, M., Rasi Varaei, S.S., and Bahador, A. 2020. Effects of sub-lethal dose of antimicrobial photodynamic therapy on major virulence traits of *Streptococcus mutans*. *Photodiagnosis Photodynamic Therapy* 32: 102044.

Alshahrani, A., Togoo, R.A., Kamran, M.A., and Alshahrani, I. 2020. Clinical periodontal, bacterial, and immunological outcomes of antimicrobial photodynamic therapy in orthodontic treatment-induced gingival enlargement. Photodiagnosis Photodynamic Therapy 31: 101934. doi: 10.1016/j.pdpdt.2020.101934.

Aoki, A., Sasaki, K.M., Watanabe, H., and Ishikawa, I. 2004. Lasers in nonsurgical periodontal therapy. *Periodontology 2000* 36: 59–97.

Asnaashari, M., Ebad, L.T., and Shojaeian, S. 2016. Comparison of antibacterial effects of 810- and 980-nanometer diode lasers on *Enterococcus Faecalis* in the root canal system—An *in vitro* study. *Laser Therapy* 25: 209–214.

Azizi, A., Shohrati, P., Goudarzi, M., Lawaf, S., and Rahimi, A. 2019. Comparison of the effect of photodynamic therapy with curcumin and methylene blue on *Streptococcus mutans* bacterial colonies. *Photodiagnosis Photodynamic Therapy* 27: 203–209.

Bankvall, M., Sjöberg, F., Gale, G., Wold, A., Jontell, M., and Östman, S. (2014). The oral microbiota of patients with recurrent aphthous stomatitis. *Journal of Oral Microbiology* 6: 25739.

Bargrizan, M., Fekrazad, R., Goudarzi, N., and Goudarzi, N. 2019. Effects of antibacterial photodynamic therapy on salivary mutans streptococci in 5- to 6-year-olds with severe early childhood caries. *Lasers in Medical Sciences* 34: 433–440.

Barile, M.F., Graykowski, E.A., Driscoll E.J., and Riggs D.B. 1963. L form of bacteria isolated from recurrent aphthous stomatitis lesions. *Oral Surgery Oral Medical Oral Pathology* 16: 1395–1402.

Betsy, J., Prasanth, C.S., Baiju, K.V., Presanthila, J., and Subhash, N. 2016. Patients' perceptions of antimicrobial photodynamic therapy in the management of chronic periodontitis. *Photodiagnosis Photodynamic Therapy*. 14: 84–90.

Bonifácio, D., Martins, C., David, B., Lemos, C., Neves, M.G.P.M.S., Almeida, A., Pinto, D.C.G.A., Faustino, M.A.F., and Cunha, Â. 2018. Photodynamic inactivation of *Listeria innocua* biofilms with food-grade photosensitizers: A curcumin-rich extract of *Curcuma longa* vs commercial curcumin. *Journal of Applied Microbiology* 125: 282–294.

Calderhead, R.G. 2007. The photobiological basics behind light-emitting diode (LED) phototherapy. *Laser Therapy* 16: 97–108.

Castano, A.P., Demidova, T.N., and Hamblin, M.R. 2004. Mechanisms in photodynamic therapy: Part one-photosensitizers, photochemistry and cellular localization. *Photodiagnosis Photodynamic Therapy* 1: 279–293.

Chan, Y., and Lai, C.H. 2003. Bactericidal effects of different laser wavelengths on periodontopathic germs in photodynamic therapy. *Lasers Medical Science* 18: 51–55.

Chui, C., Aoki, A., Takeuchi, Y., Sasaki, Y., Hiratsuka, K., Abiko, Y., and Izumi, Y. 2013. Antimicrobial effect of photodynamic therapy using high-power blue light-emitting diode and red-dye agent on *Porphyromonas gingivalis*. *Journal of Periodontal Research* 48: 696–705.

Cieplik, F., Tabenski, L., Buchalla, W., and Maisch, T. 2014. Antimicrobial photodynamic therapy for inactivation of biofilms formed by oral key pathogens. *Frontiers in Microbiology* 5: 405.

Cobb, C.M. 2017. Lasers and the treatment of periodontitis: The essence and the noise. *Periodontology 2000* 75: 205–295.

Cusicanqui Méndez, D.A., Gutierres, E., José Dionisio, E., Afonso Rabelo Buzalaf, M., Cardoso Oliveira, R., Andrade Moreira Machado, M.A., and Cruvinel, T. 2018. Curcumin-mediated antimicrobial photodynamic therapy reduces the viability and vitality of infected dentin caries microcosms. *Photodiagnosis Photodynamic Therapy* 24: 102–108.

da Silva Barbosa, P., Duarte, D.A., Leite, M.F., and de Sant'Anna, G.R. 2014. Photodynamic therapy in pediatric dentistry. *Case Reports in Dentistry* 2014: 217172.

Dahl, T.A., McGowan, W.M., Shand, M.A., and Srinivasan, V.S. 1989. Photokilling of bacteria by the natural dye curcumin. *Archives of Microbiology* 151: 183–185.

Darmiani, S., Yousefi, M., and Rad, M.S. 2022. Comparison of the effect of diode laser irradiation and fluoride varnish on salivary *Streptococcus mutans* bacterial colonies counts: A randomized controlled clinical trial. *International Journal of Clinical Pediatric Dentistry* 15(Suppl 2): S239–S241.

Dascalu Rusu, L.M., Moldovan, M., Sarosi, C., Sava, S., Dreanca, A., Repciuc, C., Purdoiu, R., Nagy, A., Badea, M.E., Paun, A.G., Badea, I.C., and Chifor, R. 2022. Photodynamic therapy with natural photosensitizers in the management of periodontal disease induced in rats. *Gels (Basel, Switzerland)* 8: 134.

Diogo, P., Gonçalves, T., Palma, P., and Santos, J.M. 2015. Photodynamic antimicrobial chemotherapy for root canal system asepsis: A narrative literature review. *International Journal of Dentistry* 2015: 1–26.

Dos Santos, L.F.M., Melo, N.B., de Carli, M.L., Mendes, A.C.S.C., Bani, G.M.A.C., Verinaud, L.M., Burger, E., de Oliveira I Moraes, G., Pereira, A.A.C., Brigagão, M.R.L., Hanemann, J.A.C., and Sperandio, F.F. 2017. Photodynamic inactivation of *Paracoccidioides* brasiliensis helps the outcome of oral paracoccidiodomycosis. *Lasers in Medical Sciences* 32: 921–930.

Džunková, M., Martinez-Martinez, D., Gardlík, R., Behuliak, M., Janšáková, K., Jiménez, N., Vázquez-Castellanos, J.F., Martí, J.M., D'Auria, G., Bandara, H.M.H.N., Latorre, A., Celec, P., and Moya, A. 2018. Oxidative stress in the oral cavity is driven by individual-specific bacterial communities. *NPJ Biofilms Microbiomes* 27(4): 29.

Elsadek, M.F., Ahmed, B.M., and Eskandrani, R.M. 2020. Level of pain intensity, cytokine profiling and microbial load after photodynamic therapy in acute severe pericoronitis. *Photodiagnosis Photodynamic Therapy* 31: 101830.

Finsen, N.R. 1896. *Om Anvendelse I Medicinen AF koncentrerede kemiske Lysstraaler*. Copenhagen: Gyldendalske Boghandels Forlag.

Gabaran, Z.M., Taram, S., Karimi, J.R., and Khashab, F. 2022. 980 nm diode laser irritation on *Streptococcus mutans* growth inhibition: *In vitro* study. *Laser Dental Science* 6: 27–30.

Ganesh, P.S., and Rai, R.V. 2018. Alternative strategies to regulate quorum sensing and biofilm formation of pathogenic Pseudomonas by quorum sensing inhibitors of diverse origins. In: Kalia, V. (eds) *Biotechnological Applications of Quorum Sensing Inhibitors*. Singapore: Springer.

Gerola, A.P., Santana, A., França, P.B., Tsubone, T.M., De Oliveira, H.P., Caetano, W., Kimura, E., and Hioka, N. 2011. Effects of metal and the phytyl chain on chlorophyll derivatives: Physicochemical evaluation for photodynamic inactivation of microorganisms. *Photochemistry and Photobiology* 87: 884–894. doi: 10.1111/j.1751-1097.2011.00935.x.

Giusti, J.S., Santos-Pinto, L., Pizzolito, A.C., Helmerson, K., Carvalho-Filho, E., Kurachi, C., and Bagnato, V.S. 2008. Antimicrobial photodynamic action on dentin using a light-emitting diode light source. *Photomededicine and Laser Surgery* 26(4): 281–287.

Hamblin, M.R., and Hasan T. 2004. Photodynamic therapy: A new antimicrobial approach to infectious disease? *Photochemical and Photobiological Sciences* 3: 436–450.

Hu, X., Huang, Y.Y., Wang, Y., Wang, X., and Hamblin, M.R. 2018. Antimicrobial photodynamic therapy to control clinically relevant biofilm infections. *Frontiers in Microbiology* 9: 1299.

Huang, L., Xuan, Y., Koide, Y., Zhiyentayev, T., Tanaka, M., and Hamblin, M.R. 2012. Type I and type II mechanisms of antimicrobial photodynamic therapy: An *in vitro* study on Gram-negative and gram-positive bacteria. *Lasers in Surgery and Medicine* 44: 490–499.

Hwang, H.R., Lee, E.S., Kang, S.M., Chung, K.H., and Kim, B.I. 2021. Effect of antimicrobial photodynamic therapy with chlorella and curcuma extract on *Streptococcus mutans* biofilms. *Photodiagnosis Photodynamic Therapy*. 35: 102411.

Kang, S.M., Jung, H.I., and Kim, B.I. 2019. Susceptibility of oral bacteria to antibacterial photodynamic therapy. *Journal of Oral Microbiology* 11: 1644111.

Katsikanis, F., Strakas, D., and Vouros, I. 2020. The application of antimicrobial photodynamic therapy (aPDT, 670 nm) and diode laser (940 nm) as adjunctive approach in the conventional cause-related treatment of chronic periodontal disease: A randomized controlled split-mouth clinical trial. *Clinical Oral Investigation* 24: 1821–1827.

Konopka, K., and Goslinski, T. 2007. Photodynamic therapy in dentistry. *Journal of Dental Research* 86(8): 694–707.

Lee, H.J., Kang, S.M., Jeong, S.H., Chung, K.H., and Kim, B.I. 2017. Antibacterial photodynamic therapy with curcumin and *Curcuma xanthorrhiza* extract against *Streptococcus mutans*. *Photodiagnosis Photodynamic Therapy* 20: 116–119.

Lee, Y.H., Park, H.W., Lee, J.H., Seo, H.W., and Lee, S.Y. 2012. The photodynamic therapy on *Streptococcus mutans* biofilms using erythrosine and dental halogen curing unit. *International Journal of Oral Sciences* 4: 196–201.

Li, H., Zhou, X., Huang, Y., Liao, B., Cheng, L., and Ren, B. 2021. Reactive oxygen species in pathogen clearance: The killing mechanisms, the adaption response, and the side effects. *Frontiers in Microbiology* 11: 622534.

Li, X., Kolltveit, K.M., Tronstad, L., and Olsen, I. 2000. Systemic diseases caused by oral infection. *Clinical Microbiology Reviews* 13: 547–558.

López-Valverde, N., Macedo de Sousa, B., López-Valverde, A., Suárez, A., Rodríguez, C., and Aragoneses, J.M. 2022. Possible association of periodontal diseases with helicobacter pylori gastric infection: A systematic review and meta-analysis. *Frontiers in Medicine* 9: 822194.

Mallidi, S., Anbil, S., Bulin, A.L., Obaid, G., Ichikawa, M., and Hasan, T. 2016. Beyond the barriers of light penetration: Strategies. *Perspectives and Possibilities for Photodynamic Therapy Theranostics* 6: 2458–2487.

Melo, M.A., Rolim, J.P., Passos, V.F., Lima, R.A., Zanin, I.C., Codes, B.M., Rocha, S.S., and Rodrigues, L.K. 2015. Photodynamic antimicrobial chemotherapy and ultraconservative caries removal linked for management of deep caries lesions. *Photodiagnosis Photodynamic Therapy* 12: 581–586.

Morio, K.A., Sternowski, R.H., and Brogden, K.A. 2021. Dataset of endodontic microorganisms killed at 265 nm wavelength by an ultraviolet C light emitting diode in root canals of extracted, instrumented teeth. *Data in Brief* 40: 107750.

Nie, M., Deng, D.M., Wu, Y., de Oliveira, K.T., Bagnato, V.S., Crielaard, W., and Rastelli, A.N.S. 2020. Photodynamic inactivation mediated by methylene blue or chlorin e6 against *Streptococcus mutans* biofilm. *Photodiagnosis Photodynamic Therapy* 31: 101817.

Nunes, L.P., Chalub, L.O., Strazzi-Sahyon, H.B., Dos Santos, P.H., Cintra, L., and Sivieri-Araujo, G. 2021. Photodynamic therapy as a potential oral disinfection protocol during COVID-19 outbreak. *Photodiagnosis and Photodynamic Therapy* 33: 102187.

Oh, P.S., and Jeong, H.J. 2019. Therapeutic application of light emitting diode: Photo-oncomic approach. *Journal of Photochemistry and Photobiology B* 192: 1–7.

Park, D., Kim, M., Choi, J.W., Baek, J.H., Lee, S.H., and Baek, K. 2020. Antimicrobial photodynamic therapy efficacy against specific pathogenic periodontitis bacterial species. *Photodiagnosis Photodynamic Therapy* 30: 101688.

Pfitzner, A., Sigusch, B.W., Albrecht, V., and Glockmann, E. 2004. Killing of periodontopathogenic bacteria by photodynamic therapy. *Journal of Periodontology* 75: 1343–1349.

Pourhajibagher, M., Beytollahi, L., Ghorbanzadeh, R., and Bahador, A.2018. Analysis of glucosyltransferase gene expression of clinical isolates of *Streptococcus mutans* obtained from dental plaques in response to sub-lethal doses of photoactivated disinfection. *Photodiagnosis Photodynamic Therapy* 24: 75–81.

Ranjan, R., Abhinay, A., and Mishra, M. 2018. Can oral microbial infections be a risk factor for neurodegeneration? A review of the literature. *Neurology India* 66: 344.

Rocha, L.G.B. 2015. Development of a Novel Photosensitizer for Photodynamic Therapy of Cancer. Ph.D. Thesis, University of Coimbra, Coimbra, Portugal.

Rolim, J.P., de-Melo, M.A., Guedes, S.F., Albuquerque-Filho, F.B., de Souza, J.R., Nogueira, N.A., Zanin, I.C., and Rodrigues, L.K. 2012. The antimicrobial activity of photodynamic therapy against *Streptococcus mutans* using different photosensitizers. *Journal of Photochemistry Photobiology B* 5(106): 40–46.

Roomaney, I.A., Holmes, H.K., and Engel, M.M. 2021. Treatment of oral fungal infections using photodynamic therapy: Systematic review and meta-analysis. *Clinical Experimental Dental Research* 7: 354–364.

Rosier, B.T., Marsh, P.D., and Mira, A. 2018. Resilience of the oral microbiota in health: Mechanisms that prevent dysbiosis. *Journal of Dental Research* 97: 371–380.

Rutherford, S.T., and Bassler, B.L. 2012. Bacterial quorum sensing: Its role in virulence and possibilities for its control. *Cold Spring Harbor Perspectives in Medicine* 2: a012427.

Santin, G.C., Oliveira, D.S., Galo, R., Borsatto, M.C., and Corona, S.A. 2014. Antimicrobial photodynamic therapy and dental plaque: A systematic review of the literature. *The Scientific World Journal* 2014: 824538.

Santos, D.S.A., Galvão, G.S., Ribas, P.F., Peres, M.P.S.M., and Franco, J.B. 2022. Photodynamic therapy in the treatment of oral lesions caused by paracoccidiomycosis. *Photodiagnosis Photodynamic Therapy* 37: 102648.

Scannapieco, F.A. (1999). Role of oral bacteria in respiratory infection. *Journal of Periodontology* 70: 793–802.

Shweta, and Prakash, S.K. 2013. Dental abscess: A microbiological review. *Dental Research Journal* 10: 585–591.

Siqueira, J.F., and Rôças, I.N. 2022. Present status and future directions: Microbiology of endodontic infections. *International Endodontic Journal* 55: 512–530.

Soares, L.G.P., Crugeira, P.J.L., Nunes, I.P.F., Santos, A.S., Cangussú, M.C.T., de Almeida, P.F., Pinheiro, A.L.B., and Habib, F.A.L. 2019. Oral microbiological control by photodynamic action in orthodontic patients. *Photodiagnosis Photodynamic Therapy* 28: 221–225.

Sobczyński, Jan. 2017. Nanostructures for Antimicrobial Therapy. In: *Nanocarriers for Photosensitizers for Use in Antimicrobial Photodynamic Therapy*, 481–502.

Sperandio, F.F., Huang, Y.Y., and Hamblin, M.R. 2013. Antimicrobial photodynamic therapy to kill Gram-negative bacteria. *Recent Patents on Anti-Infective Drug Discovery* 8: 108–120.

Sun, Z., Xiong, C., Teh, S.W., Lim, J.C.W., Kumar, S., and Thilakavathy, K. 2019. Mechanisms of oral bacterial virulence factors in pancreatic cancer. *Frontiers in Cellular and Infection Microbiology* 9: 412.

Teixeira, A.H., Pereira, E.S., Rodrigues, L.K., Saxena, D., Duarte, S., and Zanin, I.C. 2012. Effect of photodynamic antimicrobial chemotherapy on *in vitro* and in situ biofilms. *Caries Research* 46: 549–554.

Tennert, C., Feldmann, K., Haamann, E., Al-Ahmad, A., Follo, M., Wrbas, K.T., Hellwig, E., and Altenburger, M.J. 2014. Effect of photodynamic therapy (PDT) on Enterococcus faecalis biofilm in experimental primary and secondary endodontic infections. *BMC Oral Health* 14: 132.

Terra-Garcia, M., de Souza, C.M., Ferreira Gonçalves, N.M., Pereira, A.H.C., de Barros, P.P., Borges, A.B., Miyakawa, W., Strixino, J.F., and Junqueira, J.C. 2021. Antimicrobial effects of photodynamic therapy with fotoenticine on *Streptococcus mutans* isolated from dental caries. *Photodiagnosis Photodynamic Therapy* 34: 102303.

Thornhill, M.H., Crum, A., Rex, S., Campbell, R., Stone, T., Bradburn, M., Fibisan, V., Dayer, M.J., Prendergast, B.D., Lockhart, P.B., Baddour, L.M., and Nicholl, J. 2022. Infective endocarditis following invasive dental procedures: IDEA case-crossover study. *Health Technol Assess* 26: 1–86.

Tokajuk, J., Deptuła, P., Piktel, E., Daniluk, T., Chmielewska, S., Wollny, T., Wolak, P., Fiedoruk, K., and Bucki, R. 2022. Cathelicidin LL-37 in Health and Diseases of the Oral Cavity. *Biomedicines* 10: 1086.

Vasovic, M., Gajovic, N., Brajkovic, D., Jovanovic, M., Zdravkovaic, N., and Kanjevac, T. 2016. The relationship between the immune system and oral manifestations of inflammatory bowel disease: A review. *Central European Journal of Immunology* 3: 302–310.

Vollmer, A., Al-Ahmad, A., Argyropoulou, A., Thurnheer, T., Hellwig, E., Attin, T., Vach, K., Wittmer, A., Ferguson, K., Skaltsounis, A.L., and Karygianni, L. 2019. Antimicrobial photoinactivation using visible light plus water-filtered infrared-A (VIS + wIRA) and hypericum perforatum modifies in situ oral biofilms. *Scientific Report* 9: 20325.

Whelan, H.T., Smits, R.L. Jr, Buchman, E.V., Whelan, N.T., Turner, S.G., Margolis, D.A., Cevenini, V., Stinson, H., Ignatius, R., Martin, T., Cwiklinski, J., Philippi, A.F., Graf, W.R., Hodgson, B., Gould, L., Kane, M., Chen, G., and Caviness, J. 2001. Effect of NASA light-emitting diode irradiation on wound healing. *Journal of Clinical Laser Medical Surgery* 19: 305–314.

Yoshii, D., Katsuragi, H., and Shinkai, K. 2021. Bactericidal effect of antimicrobial photodynamic therapy (aPDT) on dentin plate infected with *Lactobacillus acidophilus*. *Odontology* 109: 67–75.

Zanin, I.C.J. 2005. Susceptibility of *Streptococcus mutans* biofilms to photodynamic therapy: An *in vitro* study. *Journal of Antimicrobial Chemotherapy* 56: 324–330.

Index

Pages in *italics* refer to figures, pages in **bold** refer to tables, and pages followed by "n" refer to notes.

A

B